COOK–CHILL CATERING:
TECHNOLOGY AND MANAGEMENT

COOK–CHILL CATERING: TECHNOLOGY AND MANAGEMENT

NICHOLAS LIGHT
Devro Ltd, Moodiesburn, Chryston, Glasgow, UK
and
ANNE WALKER
Hotel and Catering Training Board, Ealing, London, UK

ELSEVIER APPLIED SCIENCE
LONDON and NEW YORK

ELSEVIER SCIENCE PUBLISHERS LTD
Crown House, Linton Road, Barking, Essex IG11 8JU, England

Sole Distributor in the USA and Canada
ELSEVIER SCIENCE PUBLISHING CO., INC.
655 Avenue of the Americas, New York, NY 10010, USA

WITH 58 TABLES AND 73 ILLUSTRATIONS

British Library Cataloguing in Publication Data

Light, Nicholas D.
Cook–chill catering.
1. Cook–chill catering
I. Title II. Walker, Anne
641.7'9

ISBN 1-85166-437-8

Library of Congress Cataloging-in-Publication Data

Light, N. D. (Nicholas D.)
Cook–chill catering: technology and management/Nicholas Light and Anne Walker.
p. cm.
Includes bibliographical references.
ISBN 1-85166-437-8
1. Quantity cookery. 2. Caterers and catering. I. Walker, Anne.
II. Title.
TX820.L53 1990
641.5'7--dc20
89-26073
CIP

Photoset by Enset (Photosetting), Midsomer Norton, Bath, Avon
Printed in Great Britain by Page Bros (Norwich) Ltd

Preface

Cook–chill is the name coined for a method used by both the catering and food retail industries for producing food for large numbers of people. This technique, which rationalises food production by means of batch (or, in some cases, continuous) production of large ranges of menu items, utilises the principles of pasteurisation, rapid chilling and chilled storage to allow the short term preservation of food-stuffs prior to re-heating and consumption. Although the method attracted considerable specialist and general press attention from its inception in the early 1960s, in the late 1980s it received unprecedented scrutiny from a number of quarters, e.g. caterers, equipment manufacturers, consumer groups, the media, government bodies and the general public.

The reasons for this upsurge in the public profile of this far-from-new food production method are diverse. In the UK, factors such as the general focussing of consumer attention on food quality and food ingredients, food poisoning scares, the apparent reticence of the food industry and government bodies to make definitive statements on food production and quality have all helped to fuel the food debate. In addition, the publicity afforded the food poisoning bacterium *Listeria monocytogenes* in the late 1980s and the realisation that it can grow at chilled temperatures has led to some sweeping and often inaccurate statements regarding cook–chill technology. Thus, cook–chill catering has received what may be considered undue and unbalanced public attention. In fact, as we shall see later, such attention, though often leading to misapprehensions, has brought the method into the public eye and forced a proper discussion of its potential problems.

In fact, there are real and pressing reasons why cook–chill catering should be publicly discussed and more thoroughly researched. First, the method has often been introduced to cut costs with little thought of final product quality. In the catering industry this has led both to loss of jobs and discontent amongst customers who, in poorly run operations, may find themselves having to eat low quality foods. Second, as with all food preservation methods, if food products are improperly processed, stored or handled there is a danger that they may cause food poisoning, although, to date, there has not been a single recorded incident in the UK of food poisoning related to the consumption of cook–chill foods.

In practice, cook–chill catering offers many advantages to the consumer as well as to the hospitality and leisure industries, businesses with their own catering facilities, schools, hospitals, social service operations and the forces. It is a systematic technique which can provide very large numbers of high quality meals to customers at the site of food production or even at sites distantly removed from the production kitchen. It provides the caterer with the potential to rationalise food production and, because it adopts a systematic strategy for production, to install a fully capable quality assurance system to ensure total safety of products as well as high standards of organoleptic food quality.

As with any complex technological system, there is room for abuse and, in the past, the provision of poor, unsuited or inadequate equipment has caused problems. In addition, the growth of the cook–chill catering industry world-wide virtually unsupported by a research culture has meant that many of the potential and real problems of the method are only just now coming to light. However, the body of research on this method is growing and our understanding and knowledge is expanding concomitantly.

This book, then, is an attempt to bring together the disparate elements common to cook–chill catering systems. It does not address in detail the fundamental aspects of chilling technology and science which have been covered in depth in several previous texts. Rather, it deals with the factors which are now known to influence the functioning of cook–chill systems and the quality of the meals produced by them; for example microbiological contamination, sensory properties, equipment, staff training, communication and so on. We have made no attempt to enter directly into the debate about the rights and wrongs of the cook–chill catering system—instead, we have tried to present the facts as they are seen to exist today. In addition, the book draws heavily on the only in-depth analysis of working UK cook–chill catering sites carried out to date. It is hoped that this will provide a comprehensive guide to how a modern cook–chill catering operation runs and the pit-falls and dangers involved in setting up such systems. Finally, we have drawn on a wide body of research evidence in piecing together the chapters of this book and have, therefore, included a comprehensive reference list in the appendices as well as guiding the reader to key works in the bibliographies to each chapter.

We hope that this book will provide the knowledge required by students of food science, catering and related disciplines as well as forming the basis for the establishment of successful new cook–chill catering operations.

NICHOLAS LIGHT
ANNE WALKER

Acknowledgements

The authors would like to gratefully acknowledge the help of the following people which was kindly given in the course of researching and writing this book. The catering managers, staff and customers at the eighty UK cook–chill units which so freely and openly participated in the UK Cook–Chill Survey whose data contributed so greatly to Part 2 of this book; Ivor Church, Philippa Hudson and Philip Creed who all contributed to the compiling of the reference list; Andrew Lewis who also helped with the reference list but also gave unpublished data on menu items used in UK cook–chill units (Appendix 1); Joachim Schafheitle who helped compile the reference list but also was a constant support on the issue of *sous-vide* cuisine; and last but certainly not least, Sally Light who typed, compiled, prepared drawings and data and generally supported the project from start to finish. To all these people, many thanks.

Contents

PART 1.

THE SCIENCE AND TECHNOLOGY OF COOK–CHILL CATERING SYSTEMS

CHAPTER 1

An Overview of Cook–Chill Catering

INTRODUCTION

Cook–Chill catering is a technologically-based method of preparing food in advance of need which utilises pasteurisation of food during prime cooking, rapid chilling after prime cooking and carefully controlled chilled storage of the food below 3°C before controlled re-heating and service. By using chilling for food preservation, the technique allows the separation of food preparation and service and the distribution of pre-cooked, chilled foods to sites remote from the site of production thereby enabling the rationalisation of a catering process (Fig. 1.1).

The use of cook–chill catering began in countries such as France, Sweden and the USA as early as the beginning of the 1960s. It was nearly ten years later that it was first used in a UK catering establishment. In this country, throughout the 1970s and early 1980s, it remained a little known system of mass catering, which was very often categorised alongside its fore-runner, cook–freeze catering. By the beginning of 1986, the use of the cook–chill technique as a method of mass catering had increased in popularity in the UK and there were approximately 240 working cook–chill central production units (CPU) in operation. However, even though interest in cook–chill catering was growing rapidly, it only represented approximately 1–2% of the UK catering industry as a whole. The method had been more readily adopted as a total system in countries other than in the UK, particularly in its application in very large scale operations, i.e. 10–30,000 meals/day. In France, for example, there were over 300 systems operating as early as 1976. These were predominantly hospital catering outlets but there were a substantial number in schools and industrial catering facilities also.

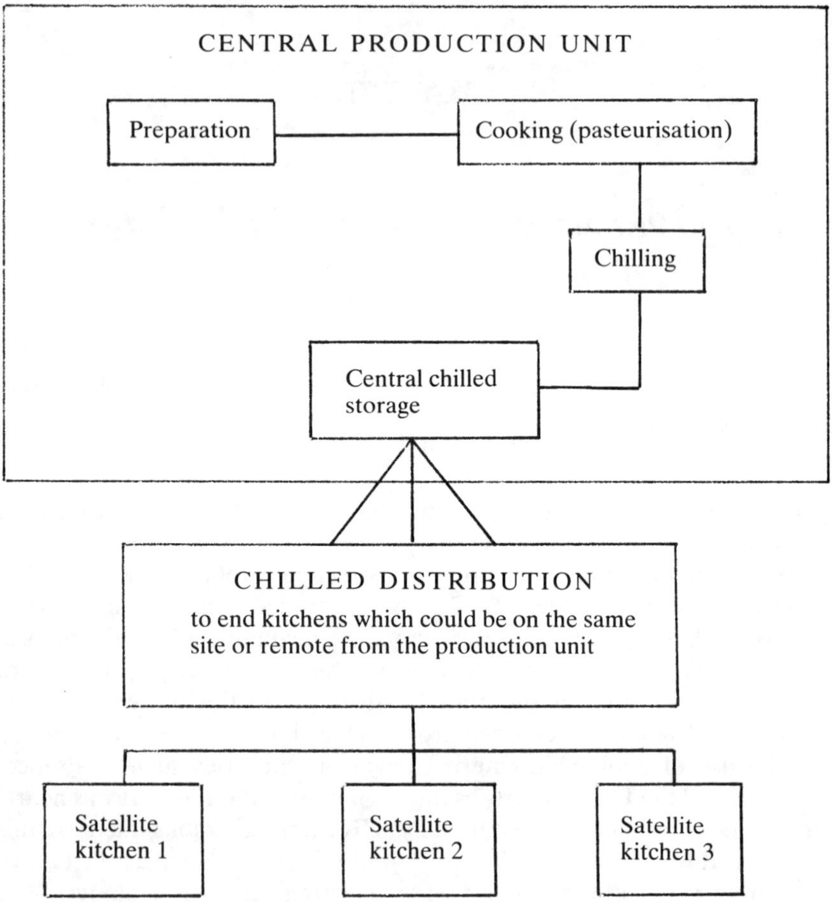

Fig. 1.1. Diagrammatic representation of the overall layout and structure of a generalised cook–chill catering operation.

Similarly, cook–chill catering was being widely used in The Netherlands, Switzerland and Australia to cope with the problems of supplying a 'meals-on-wheels' service to elderly people in rural communities. Sweden was the first country to develop large scale vacuum pack methods (see Chapters 7 and 8) of cook–chill catering and these were widely adopted in Germany. In addition, large scale systems were also developed in the USA during the 1970s, including one ambitious system in a hospital which used a mono-rail transportation service to deliver chilled food throughout the hospital. The development of cook–chill catering using cryogenics, i.e. low temperature gases to chill the food (see Chapter 5), was seen at an early stage in Australia, although this system has only slowly been taken up in the UK.

Throughout the last thirty years many adaptations and variations of the method have evolved, particularly in Europe and the USA. These have included the use of specialised packaging systems (Chapter 6), large scale food manufacturing-style production systems (Chapter 5) and in-pack pasteurisation (Chapters 7 and 8).

In early 1987 cook–chill catering became popular news in the UK; it made headlines in the national and local press, on radio and television. The main reason for this was the heightened public awareness of food matters in general and the pre-occupation of the media with such topics. Specifically, food poisoning had become a matter of public concern and cook–chill catering (and the production of cooked, chilled foods for retail sale) was targeted as a 'high-risk' method. As we shall see later in this book, cook–chill food production, when used within stringent guidelines set down for its correct practice, is no more risky than any other method of food production utilising a 'time buffer' between production and service, i.e. a method of food preservation.

The aim of this book, therefore, is to allow a dispassionate view of the method and by so doing, provide a basic understanding of the principles which underlie the method and the managerial, financial and social implications of its use. This introductory chapter will deal with the historical background of the method but before such a review it would be useful to survey the different methods of catering and the nature of the catering industry itself.

DIFFERENT METHODS OF CATERING

Catering is the commercial or non-commercial production of food for consumption by groups of people. Figure 1.2 summarises the major methods of catering used today. These can be described as:

1. classical cook–serve,
2. cook–warm-hold,
3. convenience/fast food catering,
4. cook–freeze catering,
5. cook–chill catering.

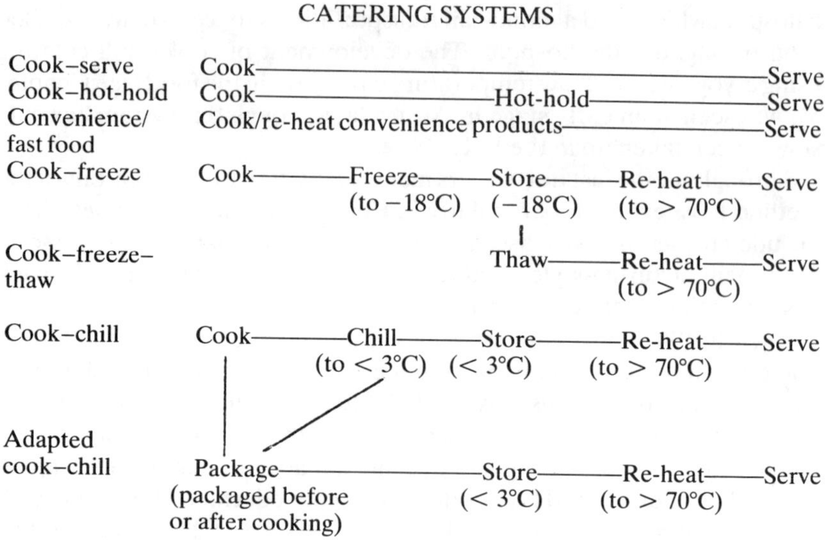

Fig. 1.2. Summary of the major types of catering systems. Cook–freeze–thaw involves a step in which frozen food is thawed before re-heating; adapted cook–chill could be a more complex form of cook–chill catering but is shown as two systems, one utilising packaging after cooking and chilling, the other utilising packaging before cooking and chilling.

The classical and preferred method of catering is to obtain fresh raw food materials and prepare them and cook them directly before service. In other words, with no temporal gap between cooking, assembly of the meal and service to the consumer. This method, though preferable, is not possible in most catering operations simply because the space needed for storage, preparation, cooking and assembly of freshly cooked food and the number of staff needed to carry out the task for large numbers of people is prohibitive in terms of cost and organisation.

Because of this problem and the needs of the commercial caterer to make profits (and the subsidised caterer in hospitals, schools, social services, etc. to break even—i.e. the cost-orientated caterer) alternative methods have evolved over the last century. Perhaps the two most common today are the use of so-called convenience products (i.e. pre-cooked frozen or chilled foods, dried foods, etc.) and the practice of warm- or hot-holding. The former relies on having a stock of partly prepared foods that can be quickly converted to a plate-ready form. The latter consists of preparing foods and then holding them at a high temperature (to prevent the growth of micro-organisms) until they are served.

Both methods have their drawbacks. Bought-in convenience foods can be expensive and their preparation and cooking is outside of the caterer's control. The method of warm-holding is open to abuse, especially with respect to temperature control and re-heating unused meals, and the practice leads to a poor quality product both in terms of flavour, texture and appearance (i.e. organoleptic quality), and nutritional status. Finally, the practice of allowing cooked food simply to cool on a display and then be re-heated later for service is not unknown and is uncontrolled and risky.

Because of this, many caterers have sought to improve on the structure of their operations and the optimisation of the use of staff and equipment by introducing new methods. Thus, fast food catering has grown out of the use of convenience products and the systems methods (Chapter 4) of cook–freeze catering and cook–chill catering have emerged to rationalise large scale catering operations.

Cook–freeze catering can be sub-divided into two approaches; one in which food is re-heated from frozen and another in which food is re-heated after thawing (cook–freeze–thaw catering). There are many variations on the cook–chill theme. These will be briefly outlined later in this chapter and discussed in depth throughout this book.

NATURE OF THE CATERING INDUSTRY

The catering industry comprises a diverse and proliferative number of operations which may be sub-divided into two main groups, i.e. commercial or profit-orientated and non-commercial or cost-orientated. In the profit-orientated group are the hotels, restaurants, fast food outlets, cafes, bars, public houses/taverns, leisure centres, country clubs, health spas and travel caterers (in-flight catering, catering on trains and ships, etc.). Many public service caterers can be categorised in the cost-orientated group, e.g. hospitals (patients), schools, social services, forces (army, navy, air-force), old peoples homes, fire stations, police, prisons. Each of the sub-divisions within the two main groups may be referred to as a sector.

A major sector not mentioned above and which is hard to categorise under the headings of profit- or cost-orientated is that of industrial catering, i.e. providing meals for people at work. For example, the provision of meals for hospital workers, hotel workers, school workers falls into this group. Also, forces workers (i.e. soldiers, sailors and airmen) could equally be included. Thus, this sector falls both into the commercial and non-commercial groups in that some businesses require that their catering section is profit-making whilst others subsidise their catering unit heavily and regard it as a part of the commitment of the company to its staff.

The kind of catering service supplied in these two groups has been, traditionally, quite different. In the commercial group food quality has always been a major criterion as customers who have to pay for a service expect standards which reflect their expenditure. In the non-commercial group food quality has not always been of paramount importance as reducing costs has often been the main criterion. In a world in which consumers are increasingly aware and in which our understanding of nutritional requirements has increased substantially, quality has started to become the main criterion across the board in catering. Also, as food quality has become increasingly important so has job quality and pay, as we shall see below.

The catering industry has long been established as craft-based and labour-intensive with little need for the benefits of high technology. It plays an important role in providing employment for a large proportion of the working population in most developed countries. For example, in the UK the hotel and catering industry employs in excess of 2 400 000 people, i.e. approximately 10% of the working population, and creates around 45 000 new jobs per year. Current forecasts estimate that by 1993, the number of people employed in the commercial group alone will have increased by 230 000. Similarly, in other European countries and the USA the hotel and catering industry is a major provider of work. In France approximately 3·4 million people (15% of the total working population) work in catering, in Germany the figure is nearer 4·2 million (15%) and in the USA over 22·8 million people are employed by the industry, a massive 19% of the total working population. It is fair to say that, in developed countries, the catering industry is one of the largest employers.

Contrary to these figures, the industry also remains one of the lowest paid, often offering very poor conditions of work, resulting in high staff turnover rates and the spectre of long-term unfilled vacancies. For example, in the UK, staff turnover is estimated at 30% per year; that is about 790 000 job vacancies which need refilling each year. Also, in February 1989, 20 000 catering job vacancies remained unfilled. Not surprisingly, given these data, in the UK there has been an increase in trade-unionisation in the industry and, alongside the limited work of Wages

TABLE 1.1
Pay Levels and Hours Worked by Employees in the Hotel and Catering Industry for 1980 and 1986

	Average weekly pay (£)		*Average hours worked over year*		*% Inflation (average)*	
	1980	*1986*	*1980*	*1986*	*1980*	*1986*
F/T manual male	77·5	125·1	42·7	42·8	18·0	4·0
F/T non-manual male	98·0	188·8	43·2	41·6	18·0	4·0
F/T manual female	56·4	88·3	39·6	38·2	18·0	4·0
F/T non-manual female	68·7	116·9	39·4	39·3	18·0	4·0

Source: *Employment Gazette* (1986); *New Earnings Survey* (1980, 1987). F/T = full time.

Councils, this change has led to some improvements in working conditions and pay. However, Table 1.1 shows that workers in the hotel and catering industry remain some of the lowest paid in the UK.

Despite the fact that catering workers, relative to workers in other industries, are still low paid, their pay has improved substantially over the last ten years; at times over and above the level of inflation of the day. However, alongside the increasingly competitive nature of the industry, this has led to the introduction of 'new' systems catering methods by many organisations. These new technologies have been sought out in an effort to reduce costs often at the expense of labour. As in all cases of technological change (technology transfer) the introduction of new technology in the catering industry has resulted in the claim that it has (a) tended to lower the skill requirement of the work-force and (b) caused job loss. This is comparable to experiences of the introduction of new technology in other industries e.g. the coal and steel industries.

The traditional skills, methods and labour usage of the catering industry is, of course, still apparent in most sectors, especially small hotels and restaurants. However, many sectors have invested in new technologies so that there are now examples of cook–freeze and cook–chill technology in every sector of the industry. In some of these cases investment has been on such a large scale as to result in operations which more closely resemble those in the food manufacturing industry. On the other hand, some establishments have taken up the technology in only a small way so that just a few meals per day are produced by a systems method whilst the majority of meals are produced traditionally.

A major difference between the food manufacturing and catering industries should be highlighted here—namely, that the former, in general, produces individual food items (e.g. sausages, burgers, frozen peas, canned sweet corn) whilst the latter has to produce full meals (starter, main course and sweet). The task of the catering industry, especially in terms of producing food for preservation, is often more complex than that of the food manufacturing industry. Inevitably, the individual components of each meal will need to be treated differently to optimise their preservation, thus making the task complex and difficult.

The motivation for technological change is often a result of financial pressure. In many developed countries both public and private sector caterers have experienced such pressures in recent years. In the UK, public sector caterers have also experienced the threat of privatisation through contracts put out to tender and this has acted as a catalyst for technological change. Similarly, in industrial catering units there has been a shift from subsidised canteens to commercially viable catering units. Simultaneously, as mentioned above, in the wake of increased consumer awareness there has been a general move to improve the image of both public and private sector catering. In labour-intensive situations an obvious choice for managements trying to become more cost effective has been to reduce their burden of constant staff costs by turning to methods such as cook–chill catering which are technologically-based and require less staff.

In the following sections we will define the cook–chill catering method, describe some of the reasons behind the gradual introduction of the method into the UK between 1968 and 1980 and review the potential advantages of the method. However, as cook–chill catering relies on the use of a food preservation method, i.e. chilling, it is important to gain an overview of the concepts of preservation and, specifically, chilling.

HISTORICAL BACKGROUND OF FOOD PRESERVATION

Because since pre-civilisation it was necessary and useful for mankind to be able to preserve food, we learned, very early in our evolution, the advantages of cooking foods in terms of their preservation (see Chapter 2) and also of drying, pickling and salting foods. Much later it became possible to store foods after high temperature treatment (pasteurisation or sterilisation) in sealed containers (bottling and canning) or at reduced temperatures (chilling and freezing).

All of these methods are connected with slowing unwanted chemical reactions which lead to food decay (see Chapter 2) and inhibiting the growth of micro-organisms in the food which will lead to its spoilage or the risk of food poisoning (see Chapter 3). However, many of these methods lead to inevitable changes in the nature of the food. For example, drying, salting and pickling change grapes, pork and red cabbage to sultanas or raisins, ham and pickled red cabbage respectively. Similarly, bottled fruit is a very different product from the fresh fruit and canned vegetables have a totally different quality to their freshly cooked counter-parts.

The French inventor Nicolas-Francois Appert won a prize of 12 000 Francs in 1809 for his innovative process of bottling. In the 180 years since this invention, technological advances have made it possible to preserve foods on a massive scale, as in modern bottling and canning, and to preserve the natural features of the food in the case of storage at reduced temperatures. Foods can be preserved in their fresh state simply by chilling them. However, whereas bottling and canning can assure storage for months if not years without any notable change in the quality of the stored food, chilling merely slows the process of change which leads inevitably to the decay of food. Freezing provides a longer storage life but brings about changes in the texture of food due to the formation of ice crystals in the products during the freezing process. It is also often necessary to pre-cook many foods before freezing.

Chilling also provides an important route for the short-term preservation of *cooked* foods. In this case, once again, the nature of the food is not changed by the chilling process as it is by freezing, canning, salting, dehydrating and pickling. Thus, it is not surprising that, in the search for methods which allow the preservation of foods in their natural state, chilling has become increasingly popular.

PRINCIPLES AND PROBLEMS OF CHILLING

Chilling food is an effective method of preserving cooked food over a short period of time (several days at $< 3°C$) because it results in the retention of the main quality attributes (i.e. flavour, texture, appearance) of the food whilst retarding growth of harmful micro-organisms (Chapter 3) and deteriorative chemical changes (Chapter 2) in the food. In this section we will give a brief summary of the principles and pitfalls of chilling as an introduction to the method of cook–chill catering. However, as other texts have successfully covered the subject of the principles of chilling in some detail this section serves only as a cursory overview.

It is important to begin with an understanding of the differences between chilled storage used in cook–chill catering applications and simple refrigeration. The latter employs equipment capable of maintaining temperatures below 10°C and often below 5°C but which operates to relatively low tolerance levels. Thus, commercial and domestic refrigerators rarely operate below 4°C and often fluctuate as high as 8–10°C. Chillers used for the storage of food in cook–chill catering must operate between 0 and 3°C, preferably with a median temperature around 1–2°C. The reasons for these strict specifications revolve around the inhibition of unwanted chemical changes in the food (Chapter 2) but mainly, in slowing the growth of harmful micro-organisms (Chapter 3). In addition, freezing should be avoided as this brings about changes in the textural properties of many foods as mentioned above and discussed in Chapter 2. Chilling is used primarily to reduce the overall use of energy (and thereby cost) in preservation and because food quality is much better retained.

When heat is removed from cooked food during the chilling process it does not change its state, i.e. from liquid to solid, as in the freezing process. However, the food does experience a rapid, controlled decrease in temperature. This is referred to as the removal of 'sensible' heat. The latent heat of fusion (i.e. the heat required to change the state of a substance from, say, a solid to a liquid without a temperature change in the substance but with an increase in the temperature of the surroundings) only becomes important in the process of freezing. Thus, in order to meet the needs of rapid chilling without freezing, the operating temperature range of the chilling cabinet/room needs to reach temperatures well below 0°C but in a highly controlled manner such that removal of latent heat, and thus freezing, is avoided.

The chilling rate depends on:

(a) Factors external to the food being chilled such as:
1. the nature of the chilling medium (air, other gases such as nitrogen, iced water),
2. the temperature of the chilling medium,
3. the agitation of the chilling medium, i.e. hot spots can quickly develop around the food and slow overall chilling if the medium is not continuously circulated,
4. the geometrical shape of the food package or container,
5. the filled height of the food in its package or container,
6. the temperature differential between the hot food and the chilling medium,

7. the head space in the package or container,
8. whether or not the package or container is lidded.

(b) Factors internal to the food such as its:
1. heat conductivity and heat capacity,
2. density,
3. initial temperature,
4. bulk and total volume,
5. moisture content.

Many studies have been carried out on these parameters but it is only possible to cite a few examples here. In terms of the importance of covering the food product, it was shown that chill time is increased by about 15% if the food container has a cover. However, uncovered foods, although chilled more easily, have a propensity to dry out in systems which circulate gases (e.g. air, nitrogen).

In another set of experiments which compared foods of different densities it was shown that each kilogram of goulash chilled at a rate of 1·8 min/mm depth compared to each kilogram of Brussels sprouts which chilled at a rate of 1·2 min/mm depth. Chilling time was also found to be proportional to the filled height of the product in its container. Thus, chilling rate is highly product dependent.

In terms of the importance of the head space volume in the food container (i.e. the space between the upper surface of the food and the top of the container), this was also shown to be critical as it has an insulating effect. For example a head space of 10 mm gives a 15% saving in chilling time over a head space of 20 mm.

As we have said, there are many organoleptic (taste, odour, texture and appearance) qualities to be gained by chilling foods rather than canning, dehydrating or even freezing them. However, chilling does have some disadvantages. These include excessive dehydration of food products due to drying in the chilling and re-heating process, loss of crispness in many pastry and battered products due to water migration from wet to dry parts of a product during storage, short shelf-life expectations for most foods compared with those gained by drying, canning or freezing (i.e. a general deterioration in quality in most foods over a five day period). Other, less relevant problems arise in the chilled storage of many fresh fruits (chill injury) and bread products (staling) but these will not be dealt with here. In general, not all foods may be suitable for processing through a cook–chill catering system.

WHAT IS COOK–CHILL CATERING?

As we have already seen, cook–chill catering is a system of catering which divorces production from service by the introduction of a preservation step or 'time buffer', i.e. chilling.

Food is, therefore, prepared for cooking by a range of methods which may include classical (i.e. preparation of fresh vegetables, meats, fish and complex products) and convenience methods (i.e. the use of pre-packed convenience foods such as frozen peas, frozen puff pastry). The product which may be simple (e.g. peas, beans) or complex (e.g. meat in sauce, quiche, flan) is then cooked (prime cooked) to temperatures which will lead to pasteurisation of the food (see Chapter 3). The product is then rapidly chilled, in a pre-determined time, using specialised equipment (see Chapters 4 and 5) and stored at a pre-determined and carefully controlled temperature in a specialised chill store (see Chapters 4 and 5). Storage time can vary considerably from product to product and recommended maximum storage times tend to vary from country to country (see Chapters 4 and 6). The chilled product can be transported, in a constant chilled state, to a kitchen for re-heating (regeneration) prior to service and consumption. The finishing or end (satellite) kitchen can be on the same site as the production unit or may be geographically distant.

Early systems in France and the USA were based on the assumption that temperatures below 5°C were adequate for safe storage of cooked, chilled menu items. Research has shown, however, that to maximise quality and minimise the risk of microbiological food poisoning (a risk with all foods), storage should be below 3°C. Indeed, some workers have suggested temperatures between 0 and 2°C as a proper working range.

Almost all cook–chill catering systems differ from each other as each is built to the specific requirements of the individual catering operation. However, there are several broad categories of the main variants. Thus, within the category of classical cook–chill operations, systems may vary with respect to the type of chillers used (electro-mechanical or cryogenic, see Chapter 5), or the form of regeneraton unit used for re-heating the food after storage. Another broad category encompasses those units which produce primarily for service through food vending machines (see Chapter 4), a rather specialised method. Other categories include those systems which utilise some form of advanced packaging system to enhance food quality and, perhaps, to extend the shelf-life of foods (e.g. vacuum packaging, see Chapter 6), those which utilise large scale in-pack pasteurisation (see Chapter 7) and those which use the new high quality method of *sous-vide* production (see Chapter 8). It is interesting to note at this point that these latter more sophisticated systems have been allocated longer shelf-life maxima in certain countries than the classical method of cook–chill catering (i.e. 21–45 days compared with only 5–6 days). These differences will be explored in detail later in this book.

HISTORICAL REASONS FOR THE GROWTH OF COOK–CHILL CATERING IN THE UK

Cook–chill catering is practiced world-wide but little research has been carried out on its use in most countries. Thus, we will investigate the growth of the method using the UK as a specific example as one or two major studies have been carried out in this country.

In the UK the provision of school meals (and hence mass catering in the education sector) started in the early years of this century. However, the origins of much mass catering in the industrial sector stems from the Second World War. Following the war, institutional feeding was seen as a vital element of a thrust to maintain a healthy population. This was supported by the introduction, in 1948, of a National Health Service (NHS) and, thus, a need to provide mass catering for staff and patients. However, despite the good intentions of mass catering operators, there were many problems associated with the provision of meals to a large number of people in one place over a limited meal time period. Food was often prepared long before it was needed, resulting in products being warm-held for excessive amounts of time. This led to wholly unacceptable meals with little remaining in terms of nutritional quality (Chapter 2).

A report in 1963 drew attention to the poor fare served to patients in hospitals. One result of this was the initiation of a cook–freeze catering trial in the Hospital for Women in Leeds. This trial instigated the first move towards more systematic methods of food production within NHS catering.

In 1968 the London Royal Homeopathic Hospital explored the possibility of an alternative system to meet their catering needs. What resulted was one of the earliest cook–chill catering units in the UK. The decision to innovate in this way was motivated by two main factors; first a pressing need to relieve nurses of non-nursing ward duties (such as meal preparation) and second to improve the organoleptic properties of the food provided. At this time there were no guidelines on which to base the introduction of a systemised catering method using cooked, chilled foods.

Similarly, even though, by now, cook–freeze catering was becoming popular, there were no suitable guidelines available on which to base the safe use of this other system. This is, perhaps, not surprising in an industry historically associated with reactive, traditionalist attitudes and a poor record of not widely supporting research. Indeed, apart from isolated projects carried out by some pioneering workers, there seemed no urgency in the industry to mount a proper research effort to underpin these methods.

There were few known (or appreciated) hazards of any great concern amongst the early catering adopters of the cook–chill catering method and the few reports from equivalent systems in Europe and the USA were favourable (if a little uncritical). Thus, the original system established at the London Royal Homeopathic Hospital differed from cook–chill systems of the late 1980s and would be wholly unacceptable (in terms, mainly, of temperature specifications and control) in the light of current knowledge of food poisoning organisms.

In the early 1970s the OPEC oil crisis had a dramatic effect on energy costs. In addition, research had shown that there were considerable energy losses in conventional catering systems. Cook–chill and cook–freeze catering offered the ability to exercise greater control over energy usage by their systemised method of food production. However, it became apparent that the energy usage in cook–freeze systems could be far greater even than a poorly controlled conventional system and, coupled with its need for high initial capital outlay and its relatively limited menu repetoire, its popularity declined. Cook–chill catering, on the other hand, seemed to offer the best of both worlds. It was less dramatic a change from conventional catering, it was more economical on fuel costs and it was claimed that it could process all food products; a claim which, as we shall later see, has since been often refuted.

The poor reputation of cook–freeze catering in the UK influenced the early take up of cook–chill catering systems in the late 1970s. Confusion over differences between the two systems and stories of bad experiences in the catering press (coupled with only a slowly emerging research literature) combined to slow the up-take of the method. As we shall see in the second part of this book, financial influences have proved a major factor in determining the up-take of cook–chill catering. There is little doubt that costs were an important force in driving the up-take of the method.

A number of factors combined to lead to the upsurge in popularity of the system in the early 1980s. These included government policies putting public sector contracts out to private tender, the withdrawal of school meals subsidies, the effect of the economic recession of the early 1980s and the emergence of complete 'systems' of equipment from manufacturers who had recognised a new market niche.

At present caterers in all sectors are faced increasingly with the need to be competitive and the demands of consumers for ever cheaper high quality products. The growth of cook–chill catering in the UK (and worldwide) goes on apace and it seems likely that this is a method set to stay for a long time.

CAUSES FOR CONCERN IN COOK–CHILL CATERING

In its infancy, cook–chill catering was often seen as a total system—a method of catering which had in-built control of food quality, control of staff and equipment utilisation and control of raw materials and waste. People envisage a central production unit supplying food to a number of smaller satellite units with a minimal alteration of food quality and only a small input of technology and skill at the satellite kitchens. Overdemand could be provided for through the four day buffer stock and peaks and troughs of production and service could be largely removed. In theory, it was seen as a total system. As we shall see later in this book, although cook–chill catering is an important and useful method and can be successfully applied in a wide range of catering situations, it is not a panacea.

In 1987 an independent report was published in the UK on the method. It is important to recognise that this was a largely journalistic report based almost exclusively on anecdotal evidence from a limited number of sites and on a limited number of visits by the author to working cook–chill operations. However, it did provide one of the first realistic views of the method and voiced concern over some of its shortcomings. What followed was an ill-thought out attack of the method by the popular press based on its disadvantages outlined in the report. In response caterers, consultants and equipment manufacturers strongly criticised the report. In effect, two polarised groups emerged. One apparently opposed to cook–chill on principle and the other whole heartedly in its support.

Needless to say, the truth probably lies along the middle path and a dispassionate reading of this report shows it to be a reasonably balanced view of the method which highlights, for perhaps the first time, some of its shortcomings. It points out that it is not a cure-all; it cannot produce high quality foods of every sort because some menu items are not suited to the method (see later chapters); it has the potential (as has every large scale systematic method) for mis-management which could result in very poor quality food; if introduced wrongly it can lead to de-motivation of staff and thereby precipitate industrial problems.

Other causes for concern have centred around microbiological contamination and the lack of adequate temperature control. In particular, the bacterium *Listeria monocytogenes*, which can grow at low temperatures, has been implicated as a potential food poisoning risk in poorly run cook–chill systems (Chapter 3). Self-evidently, if a system relies on preventing the growth of such organisms by the use of strictly controlled storage temperatures and times then, should these conditions be abused, there might be a risk of food poisoning resulting. This issue, one which faces all large scale caterers (indeed, all caterers in general) using any system, not merely cook–chill catering, is dealt with in Chapter 3.

ADVANTAGES AND DISADVANTAGES OF COOK–CHILL CATERING

Having broadly reviewed the main characteristics of the cook–chill catering system and pointed to some of the areas of concern about the method it would be useful to summarise the perceived advantages and disadvantages of the method. However, before doing so, it is important to emphasise that many of the advantages can only be gained through good management and proper preparation for installation of a new cook–chill catering system. Similarly, many of the disadvantages can be obviated by proper management and by use of more than one system (i.e. cook–chill, cook–freeze and conventional catering in tandem).

The main advantages of cook–chill catering are:

1. It divorces production from service and consumption and allows elimination of peaks and troughs of the traditional catering working day. This results in more efficient staff and equipment usage and more time can be made for the optimal use of specialist staff skills (recipe development and attention to details, e.g. of service).
2. The introduction of a systemised process allows its accurate documentation and thereby the introduction of a systematic quality assurance programme such as Hazard Analysis Critical Control Points (HACCP—see Chapter 9).
3. Production can be incorporated in an eight hour shift day on a five

day a week basis allowing staff to work 'normal' hours, reducing labour costs by eliminating or cutting overtime payments and contributing to higher job satisfaction through better hours.

4. Short-term staff shortages in the production area do not cause an immediate crisis as they can in a conventional catering situation.
5. The method allows for greater management control over the whole production process so that problems can be solved (potentially) in advance of them affecting consumers.
6. Less skilled staff may be required at satellite (end) kitchens, although this point has been challenged (see Chapter 13).
7. A smaller range of equipment and less space are needed at end kitchens than in conventional units.
8. Higher productivity can be achieved by concentrating activities and skills in a central production unit rather than spreading them over many smaller satellite kitchens each producing its own quota of meals on a much smaller scale.
9. Bulk buying of food raw materials can contribute to reduced costs.
10. Labour costs can be significantly reduced through optimising staff usage by centralisation and the requirement for less skilled staff.
11. The smaller floor area required at satellite units allows food to be regenerated and served at units which otherwise could not support a full catering function.
12. Energy savings are possible through efficient use of equipment by day and a reduction of the need for equipment at the satellite units.
13. Improved palatability of the food products due to the reduced need for extensive warm-holding.
14. A more extensive range of menu items is available with cook–chill catering than its sister system, cook–freeze catering.
15. Chilled food, unlike frozen food is pliable and therefore can easily be sub-portioned at chill temperatures whereas frozen food needs to be thawed first.
16. Chilled foods often look more attractive than frozen foods and therefore can be successfully used in cook–chill vending situations.

The disadvantages of the method, apart from its inflexibility in not being able to process a small range of specific food products, are largely based on the need for capital investment and the danger of improper management. These include:

1. Large capital investment *may* be needed for some systems, including the need to build new units or even the central production unit.

Such costs can be in the million of pounds sterling.
2. Improper preparation can lead to the installation of inadequate or unsuited equipment.
3. Poor staff relations can lead to industrial problems on the installation of a new and foreign system.
4. Need for maintenance of strict standards of hygiene, training and management in central production unit, distribution, and at the end kitchens.
5. On removal from the time buffer, food has to be sometimes transported and always 'regenerated' before service.
6. Risk to public health if not managed as a system.
7. Poor management can potentially lead to bigger 'disasters' simply due to the possible size of cook–chill catering units.

It should, perhaps, be emphasised here that most, if not all, of these criticisms apply equally to all other forms of catering when they are set up as new systems. However, establishing a successful catering unit of any sort is a complex and difficult process and so we will end this chapter with a brief discussion of the factors which can affect success in implementing a new cook–chill catering system.

FACTORS AFFECTING THE SUCCESS OF COOK–CHILL CATERING

The majority of research carried out to date on cook–chill catering centres on the technical, organoleptic and microbiological aspects of the method. There has been little in-depth research carried out on the use of the method in practice. However, one study investigated the setting up and running of 80 UK cook–chill units in considerable detail. It was the first systematic study of its kind and analysed the differences between those organisations which were relatively successful and those which were relatively unsuccessful throughout the whole process of feasibility study, decision-making, implementation and development of a cook–chill system. We will discuss the results of this survey in some depth in Part 2 of this book. However, it is worth noting here that it provided objective empirical evidence of activities in working cook–chill catering situations which, to date, anecdotal reports failed to do. The survey showed that many cook–chill operations were striving for success by building into their systems strict controls and high standards. At the other end of the spectrum there were operations in which there was ignorance of many of the basic requirements for successful use of cook–chill catering, even the need for basic controls and several of these operations were abandoning cook–chill catering as a method.

The results of the survey showed that ten key points could be targeted in terms of setting up and running a successful cook–chill catering unit. These were:

1. Awareness and compliance with approved guidelines for cook–chill catering.
2. Awareness of the need for and operation of a proper quality control system.
3. Awareness and enforcement of strict temperature and time controls throughout the process.
4. Relevant training for all the staff prior to and after the introduction of the method.
5. Employment stability amongst managers and decision-makers.
6. The perception of a real need for the introduction of the method.
7. Use of many sources of information during the feasibility study.
8. Thorough development work and an ongoing organisation-specific research and development programme.
9. Adequate capital expenditure.
10. Effective communication throughout feasibility study, installation and running with staff and where appropriate, consumers.

In brief, there can be no short cuts to success in cook–chill catering as in the introduction of any new technology.

FURTHER READING

American Society of Heating, Refrigeration and Air Conditioning Engineers (ASHRAE) (1977) *Handbook of Fundamentals*, New York.

Anon. (1976) Wide acceptance of cook–chill except in the UK. *Staff & Welfare Caterer*, May, 29–32.

Ciobanu, A., Lascu, G., Bercescu, V. & Niculescu, L. (1976) *Cooling Technology in the Food Industry*, Tunbridge Wells, UK, Abacus Press.

Fennema, O. R. (1975) *Principles of Food Science: Part II Physical Principles of Food Preservation*, New York, Marcel Dekker.

Fusco, L. (1987) Is cook–chill the saviour? *Hospital Caterer*, Nov./Dec., 9–10.

Glew, G. (ed.) (1980) *Advances in Catering Technology*, London, Applied Science Publishers.

Hotel and Catering Training Board (1987) *New Employment Forecasts for the Hotel and Catering Industry 1988–1993*, London, HCTB Research Unit/ MMD Publications.
James, S. J. & Bailey, C. (1985) Principles and problems in cooling foodstuffs. In: *Proc. Chill Foods*, 7–9th May, Stratford-upon-Avon. Chipping Campden, UK, Campden Food Preservation Research Association.
Parker, L. (1988) *Ready Meals – an International Perspective*, Leatherhead, UK, Leatherhead Food Research Association.
Platt, B. S., Eddy, T. P. & Pellet, P. L. (1963) *Food in Hospitals*, Oxford, Oxford University Press.
Sheppard, J. (1987) The big chill: a report on the implications of cook–chill catering for the public services. London Food Commission. Report no. 15.
Walker, A. E. (1988) The transfer of technology in the catering industry: A study of UK cook–chill catering operations. PhD Thesis, Dorset Institute, Poole, Dorset UK.
Yorkshire Regional Health Authority (1988) Cook–chill catering systems: The report to the YRHA by the expert group on the safety and implementation of cook–chill systems.

CHAPTER 2

Food Preservation: the Sensory and Nutritional Quality of Cook–Chill Foods

INTRODUCTION

The definition of cook–chill methods as laid out in Chapter 1 makes it clear that the most important aspects of these systems are (a) the principle of the 'time buffer' which allows the use of prepared, cooked foods on a flexible temporal and locational basis, thereby allowing the rationalisation of food production and service and (b) the preservation of cooked food in a safe and nutritionally viable state but so that it is also still good to eat. In the following chapters we will explore the parameters, principles and details of the actual process of cook–chill food production but first it is necessary to gain an understanding of the background in terms of food quality and food preservation that underlies these methods.

To discuss the details of chilled food preservation, the basic principles of food microbiology and sensory and nutritional quality would require a whole book. Therefore, as there are many excellent text books which can be used as basic reference sources on these essential topics (and several of these will be mentioned in the bibliography to this chapter and that of Chapter 3), the examination of these subjects in this book will be restricted to the salient features of these general subjects and those areas of most interest with respect to cook–chill catering.

The present chapter will deal with the essential principles of those aspects of food chemistry, sensory quality and nutrition that are implicated in the study and understanding of food preservation. Thus, we will examine why foods deteriorate if not preserved in some way, the basis of measuring food sensory quality and the principles of nutrition. We will also examine the research work to date which has sought to understand the changes brought about in foods during chilled storage which affect its sensory and nutritional quality.

In terms of the fundamentals of food chemistry, sensory quality and nutrition, the reader is strongly advised to consult other texts for more detailed information.

FOOD PRESERVATION

All food decays if it is not preserved in some way, i.e. prevented from decaying. This process of decay originates both inside and outside of the food through:

(a) natural enzymic and non-enzymic reactions within the food,
(b) oxidation of food chemicals by oxygen dissolved in the food or oxygen in the air,
(c) reactions accelerated by the influence of external factors such as light, and
(d) the attack of micro-organisms.

The result of this process is that the food becomes inedible due to the production of off-flavours and colours (food that looks unpleasant is hard to eat even if it does not taste bad and is not toxic) and, ultimately, poisons originating from micro-organisms living on the food. Fresh animal products like meat and fish, as well as fruit and vegetables are all particularly susceptible to decay but cooked foods too are not immune.

The two basic processes used in the method of cook–chill catering are cooking and chilling. Both have an effect on the storage life of the food as we will see in detail below. The process of heating food (or using thermal processing), i.e. blanching, cooking, pasteurisation or sterilisation (see below) brings about changes in its consistency or texture, its colour, its taste and its smell. Man, therefore, has used the method to:

(a) make food more palatable by changing its texture, e.g. cooking vegetables,
(b) produce flavours (and aromas) in foods which would otherwise be rather bland, e.g. meat and fish,
(c) make food easier to eat, i.e. more digestible,
(d) to destroy natural poisons in food, e.g. red kidney beans contain a protein toxin which is inactivated by heat,
(e) to inactivate enzymes which might cause unacceptable changes in the food during storage,
(f) to preserve food from spoilage which might be caused by the growth of micro-organisms found normally on or in the food, and

(g) to destroy living micro-organisms in or on the food which might cause diseases, i.e. pathogenic micro-organisms.

Only the last three of these lead to preservation, at least in part, of the food.

The second major function carried out in cook–chill catering is the process of chilling. As virtually all chemical reactions speed up as the temperature is raised above ambient (i.e. normal room temperature, say about 20°C) and slow down when the temperature is lowered, and most reactions in food are unwanted in terms of storage and lead to unacceptable changes in quality, if follows that cooling will lengthen the shelf-life of foods. Shelf-life is a common term in food preservation and simply means the time between production and consumption of the food. Some observers define it as the time between production and sale of the food but the first definition is accepted by most as being the most accurate and flexible of the two. The concept of shelf-life and its importance in terms of cooked, chilled foods will be discussed in detail in Chapter 6.

Chilling, then, leads to further preservation effects in the foods which have been cooked by the cook–chill caterer as it slows down the chemical reactions of food components and those which take place as part of the life processes of micro-organisms, thereby slowing their growth. The latter will be considered in the next chapter but we will now discuss the effect of heat and cooling on the enzymic and non-enzymic reactions that occur in foods.

There are many enzymic reactions which will occur naturally in foods if not checked. Many of these are of little consequence to food quality but some are unwanted because they cause unfavourable changes in the food. Such changes include colour and flavour and generally involve enzymes which cause browning (i.e. enzymic browning). Other enzymes catalyse the production of specific 'off' food flavours by the breakdown of natural substances in the food like fats and oils, proteins and carbohydrates. All of these deleterious reactions can be stopped by irreversibly destroying the ability of the enzymes to act on food chemicals.

All enzymes are proteins and all proteins are denatured at high temperatures (i.e. their shape is so physically changed that their biological activity is destroyed). This means that the ability of enzymes in food to catalyse unwanted reactions is destroyed by heating to temperatures of 70–90°C and above for relatively short times. However, there can be problems if the heating process is not carried out properly. If the food is slowly heated to 90°C the enzymes may become more active before they are destroyed (denatured). Figure 2.1 shows the activity of a typical food enzyme at various temperatures. The enzyme is far more active between 30 and 70°C than it is at ambient temperature and when it is denatured at 72°C. Therefore, this enzyme would remain highly active and could cause damage in the food by slow heating. Thus, the process of blanching (placing foods in hot water i.e. 90–100°C, for 2–3 min) is used commonly prior to other preservation methods and cooking itself leads to irreversible denaturation of food enzymes. The prevention or inhibition of enzyme activity in food, therefore, is easily achieved in the cook–chill process as the foods are, by definition, cooked in the process prior to preservation, i.e. chilling, and may be blanched prior to cooking or storage if necessary.

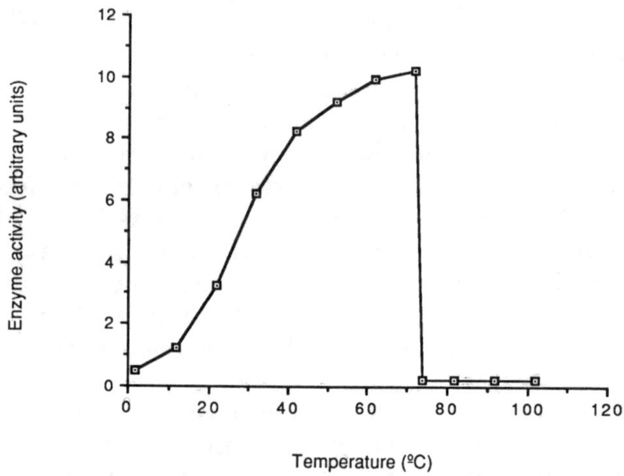

Fig. 2.1. Idealised graph showing the activity of a typical food enzyme at various temperatures.

The main non-enzymic chemical reactions in foods which lead to deterioration are the destruction of vitamins and some other essential compounds (these will be dealt with later in this chapter), and the oxidation of fats and oils leading to rancidity. The browning of sugars (caramelisation and Maillard reactions) normally occurs only at high temperatures and is usually desirable as it leads to acceptable colour and flavour changes. The rancid-flavoured products of the oxidation of fats and oils, however, are one of the major reasons for stored foods becoming unacceptable from a sensory point of view, and is difficult to prevent.

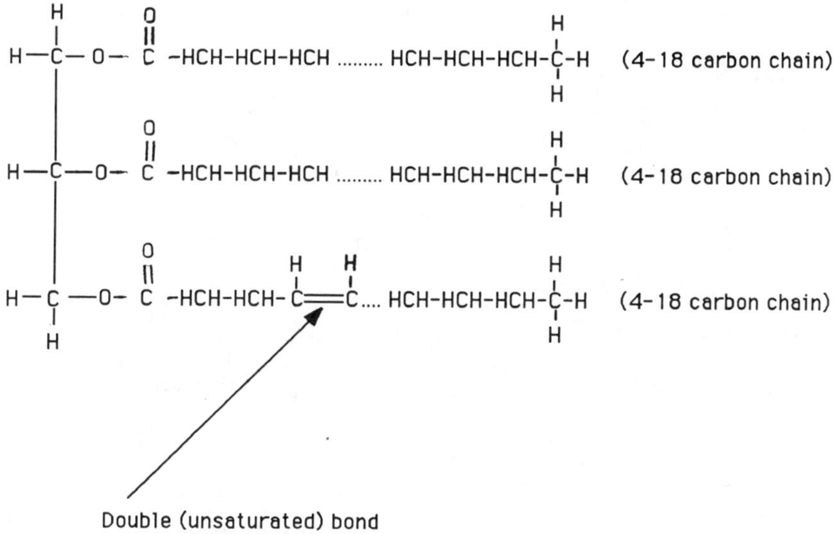

Fig. 2.2. Diagrammatic representation of a triglyceride showing both saturated (no double bonds between carbon atoms) and unsaturated (some double bonds between carbon atoms) fatty acid components.

The oxidation of fats and oils (generally in the form of triglycerides) is a complex process in which the unsaturated fatty acid parts of a fat or oil (Fig. 2.2) are oxidised by oxygen in the atmosphere or oxygen dissolved in the food to give very reactive chemicals called radicals. These then react with other fatty acids to give peroxides and, eventually hydroperoxides, which are also very reactive and can go on to form a multitude of chemicals, in reaction with other food components, many of which have 'off' flavours, odours and colours. In the process of the production of peroxides the first reaction, i.e. the formation of another radical, is re-initiated and so the cycle starts again. As the process is self-perpetuating through this cyclic mechanism (Fig. 2.3) it is known as auto-oxidation. The reactions only need oxygen in order to proceed and even removal of water, which prevents many unwanted food reactions, or cooking cannot entirely stop them. Thus, rancidity in stored cooked, chilled foods can be a major problem.

Because auto-oxidation is accelerated by high temperatures, by light and some other factors such as heavy metals (iron, copper, etc.) which act as catalysts to the process, it can also be slowed. Thus, by keeping the

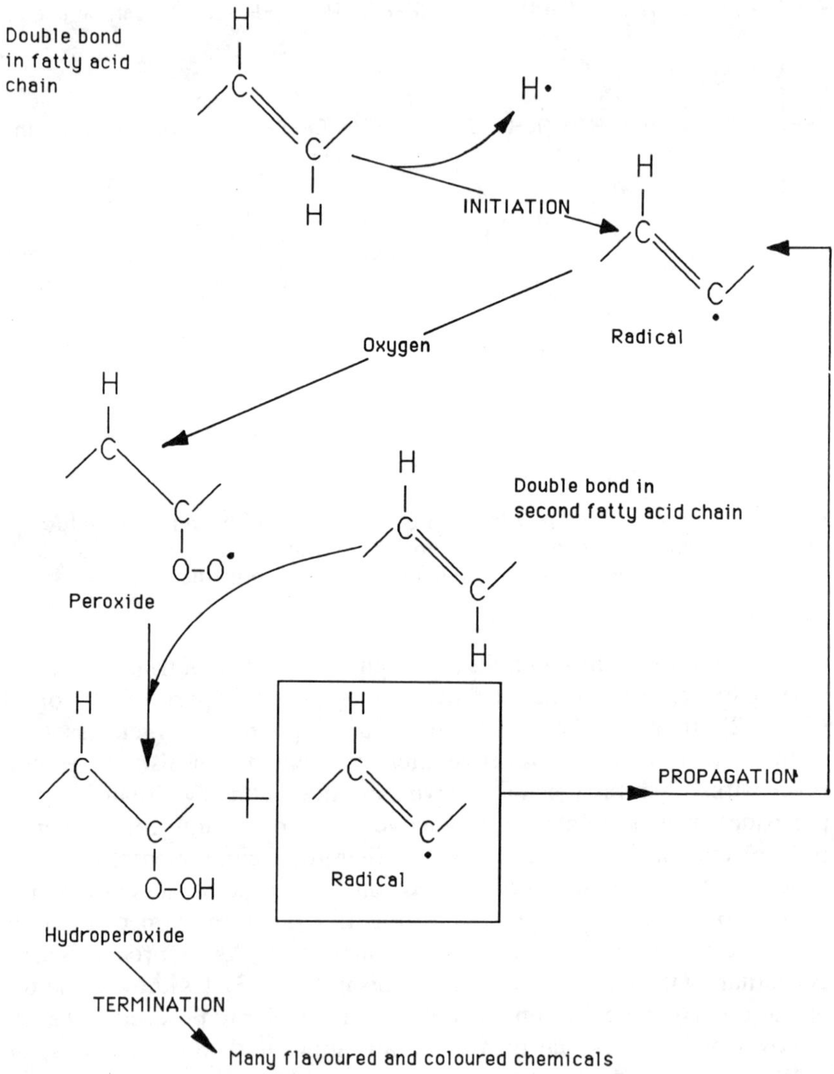

Fig. 2.3. Diagram illustrating lipid (fat or oil) oxidation (auto-oxidation).

foods chilled (0–3°C) and stored in the dark the process can be controlled such that, within a few days of chilled storage (3–5 days depending on the food product) the food will remain edible. These reactions do not produce toxins but do eventually cause unacceptable deterioration of the foods. Experiments with modified atmosphere packing and vacuum packing, in which the oxygen required for auto-oxidation to proceed is largely removed from the food, have shown these methods to be a great advantage in slowing the process of rancidity (see Chapter 6).

Most food reactions require water, as mentioned above and so a classical way to preserve foods has been to dehydrate them. This can be seen in the case of meats and fruits dried slowly and naturally in the sun such as 'biltong' (dried meat in Africa), raisins and sultanas. The modern equivalent of this process is freeze, spray or roller drying. These methods, though common, are not relevant to cook–chill production and so will not be discussed further here.

As we have mentioned above, another very efficient way to slow down chemical reactions in foods is to lower their temperature. This, of course, is the method used in cook–chill catering but in this case the food has always been cooked first. The cooking process, as we have seen above, actually stops all enzymic reactions, but as other reactions proceed more quickly at high temperatures (e.g. non-enzymic browning reactions), it encourages many reactions to occur which otherwise would proceed very slowly. Thus, new flavours, aromas and colours are produced by the cooking process. The chilling process used after cooking in cook–chill catering serves to inhibit or slow down any further reactions which might occur in the food to destroy or change the chemicals responsible for these flavours, aromas and colours.

The main changes in food which lead to loss of quality other than those caused by enzymes, oxidation or other chemical reactions are those brought about by micro-organisms growing on or in the food. These will be discussed in the next chapter.

SENSORY QUALITY

The sensory properties of a food are those which are perceived by the senses, i.e. taste, odour, appearance and texture. The sensory quality of a food, therefore, is made up of all the characteristics and properties of the parts of that food concerned with stimulation of the senses. In other words, those properties which influence its taste, odour, appearance, mouth feel and texture, i.e. its organoleptic properties.

Sensory characteristics in food are determined by the chemical content of the food material which itself is affected by various physical and chemical effects such as cooking (heating), oxidation, ultra-violet light and so on. The texture and flavour of a food are sensed in the mouth respectively by pressure sensors (specialised nerve endings) and taste buds (special kinds of cell found in the tongue with receptors on their surface). The receptors are proteins or other macromolecules on the surface of the taste bud cells which interact with and recognise certain substances or molecules. Receptors are also present on the surface of cells in the oropharyngeal tract (the upper and rear surfaces of the nasal tract and upper reaches of the throat) and are involved in sensing aroma. The four basic tastes in foods (i.e. sweet, sour, bitter and salt), are recognised by taste bud receptors and signals corresponding to these flavours are sent to the brain by nerves. The mouth also senses pain and hot and cold, all of which contribute to the overall sensation of 'taste'.

Another, extremely important feature or attribute of foods is their eye appeal. This refers simply to the appearance of a food, i.e. those qualities in a food (e.g. colour, shape, visible texture), which stimulate the sense of sight. Many decisions made by consumers about whether a food is good or not are heavily influenced by the eye appeal of the food.

In nutrition or microbiology it is possible to measure objectively the amount of a nutrient or the number of bacteria per gram of food by using well-defined chemical or biological tests. Measuring organoleptic quality is not so easy. Over the years, measurements of sensory quality have developed along two separate lines. First, the taste panel or consumer test, in which humans are used as the instrument with which quality is measured. Secondly, chemical and physical tests, particularly of texture, have evolved to enable objective and absolute measurements of certain (though not all) characteristics of sensory quality.

Methods in which humans are used as the measuring 'device' for assessment of quality are generally referred to as sensory analysis techniques. As stated above, the human food tasters (indeed, they may only be asked to view the food or just smell it; tasting is not a necessary pre-requisite of this type of test) may be consumers or a taste panel. A taste panel is a small (normally 6–20 individuals) group of people who test a food with their own senses and may be either trained to test the food or not trained. Taste panel training usually takes place in various steps. These include checking that panellists are able to discriminate flavour, smell and texture, are consistent in the way they do this (and in the way they give scores for the attributes of the food that are being measured) and an introduction to the product to be tested and the methods to be used. A fully trained panel may have worked together for years.

A consumer panel is normally much larger (e.g. 50–several hundreds in size), is not trained and is normally used only to check the acceptibility of or preference for certain foods. Thus, using quantitative scoring methods, trained taste panels can be used to measure sensory characteristics in a food whilst the consumer panel gives information on whether or not it will be acceptable (i.e. if it will sell).

There are many types of sensory tests that are in common use. For example, the discrimination (or difference) test in which the difference between food samples is judged. Examples of this type of test are the paired comparison test (in which a person is asked if there is a difference between two samples), the Triangle Test (in which two samples are the same and one is different and the judge is asked to find the odd one out) and the duo–trio test (in which the judge is asked to say which of two samples is different from a control sample). These tests are non-quantitative.

In quantitative descriptive (sensory) analysis (QDA), a trained taste panel technique, the characteristics of a food are described by the taste panel and descriptors (or words used to describe a quality in a food sample) are proposed and agreed by the panel. Each panellist then tastes the food and gives scores to each of the attributes, as defined by the agreed descriptors, on a proforma (see Fig. 2.4). Various types of QDA exist, for example, the ranking test (in which judges are asked to show their order of preference of a certain attribute in a set of samples), the scoring test (in which judges are asked to show the differences between a set of samples by giving a score to each sample) and the hedonic scale (in which a scale of words or phrases or even pictures are used to show preference for a food sample in an acceptance–preference test).

Using QDA, flavour profile sensory analysis and texture profile sensory analysis techniques have been developed in which the major flavour and texture attributes of a food sample can be defined and quantified to build up a picture of the quantitative sensory value of a food. As we will see below, many research workers have applied both taste panel and consumer analysis methods to the examination of stored, cook–chill foods.

Other techniques for measuring food quality are less widely used in connection with catering methods. The most important of these is physical analysis. The types of method used for physical analysis include micros-

Place a vertical line across each horizontal line at the point representing your evaluation of each particular attribute. Product – deep fried chicken drumsticks

Sample Code No.	**Attribute**	
	Appearance of skin	
	No shrivelling	Very shrivelled
....	....	
....	....	
....	....	
	Natural fried chicken odour	
	Strong	Weak
....	....	
....	....	
....	....	
	Juiciness of flesh	
	Not juicy	Very juicy
....	....	
....	....	
....	....	
	Degree of chewing	
	Low	High
....	....	
....	....	
....	....	
	Resistance to first bite	
	Less resistant	More resistant
....	....	
....	....	
....	....	
	Fresh fried chicken flavour	
	Strong	Weak
....	....	
....	....	
....	....	
	Rancid flavour	
	Weak	Strong
....	....	
....	....	
....	....	

Fig. 2.4. Example of a sensory analysis score sheet used to obtain a quantitative profile of the major characteristics (attributes) of a food product.

copy and textural tests using apparatus like the Warner–Bratzler shear apparatus (for measuring the tenderness of meat), the penetrometer (used to measure the textural quality of meat or plant foods), the Magness–Taylor pressure tester (for measuring vegetable texture) and the farinograph (used to measure the stability and consistency of dough).

The onset of rancidity in cooked meats within a few hours of cooking is well documented as a result of both sensory and chemical analyses. Cooked chicken, for example, is especially vulnerable to unwanted flavour changes associated with rancidity during refrigerated storage. In cooked meats rancid flavours are frequently described as 'warmed over flavour' (WOF) and are organoleptically different from rancidity in uncooked meats, probably because of the differences in the side reactions after hydroperoxide decomposition.

Studies of the chemical and sensory characteristics of cooked, chilled foods started in the 1960s and 70s. The overall conclusion of many of these studies was that the odour, texture, juiciness, flavour and appearance of a freshly prepared product was always preferable (and scored more highly) than these characteristics of the same product after it had been stored at chill temperatures for any length of time. Also, the extent of chemically measurable rancidity, normally by the thiobarbituric acid (TBA) test, was always higher in the stored product than in fresh. Given what is known about chemical changes in food and especially the effects of oxygen during food preservation, these results were hardly surprising.

In an early study, the effect of chilled storage on cooked peas showed that very little change occurred in sensory quality, as judged by a large trained taste panel, during the first three days of storage but that after the third day deterioration was rapid. In contrast, cooked beef was shown to exhibit a significant deterioration in flavour after only one day's storage at chill temperatures although the changes in quality were not great (Table

TABLE 2.1
Mean Scores for Various Sensory Attributes of Beef Stew Prepared by the Cook–Chill Method and Stored for up to 3 Days at Chill Temperatures (adapted from Nicholanco & Matthews, 1978)

Days storage	*Appearance*	*Flavour*	*Gravy texture*	*Gravy consistency*	*Meat texture*	*Vegetable texture*	*Overall acceptability*
1	3·90	3·45	3·80	3·95	3·55	3·80	3·85
2	3·60	3·15	3·60	3·50	3·30	3·65	3·45
3	3·95	3·80	3·75	3·80	3·55	3·30	3·60

2.1). Another study agreed with this latter finding and showed profound decreases in sensory quality in beef loaves over nine day's chilled storage. On the other hand, another research group were unable to show any real differences in the overall acceptability scores for beef/soy loaves over three day's storage. These data seem to be conflicting and confusing but most researchers agree that one day's chilled storage (at 1–3°C) does not change the sensory characteristics of the cooked food very much but that by day three of storage most characteristics are either rapidly changing or have already changed significantly.

The types of comment made by taste panellists and consumers with reference to the sensory quality of cook–chill foods stored at 1–3°C for up to 10 days were as follows. Flavours became increasingly flat and masked by those of spices and condiments with time of storage. At the same time perceivable increases in acid, bitter and rancid flavours were noted in many products. Vegetable colour became successively more grey/brown or, in the case of carrots and Brussels sprouts, paler. The odours of most foods also became flat and bland with time or were replaced with rancid odours. Potatoes had a slightly stale/bitter flavour which increased with storage time. A typical experiment is illustrated in Fig. 2.5 in which the deterioration of cooked beef, beans and potatoes on chilled storage is shown. These changes were also noted in mashed potato, potatoes in sauce and potato salad.

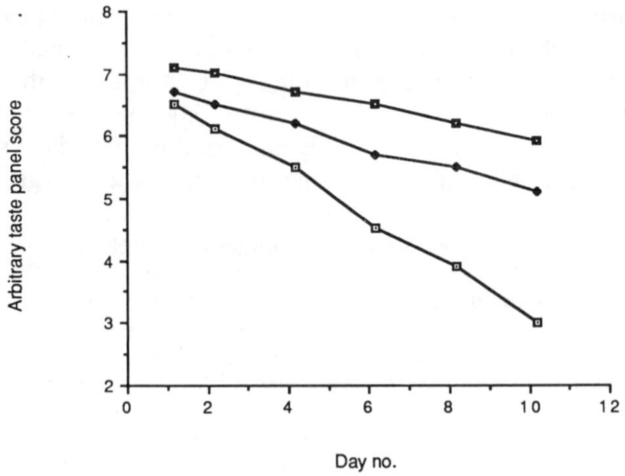

Fig. 2.5. Sensory changes in cooked beef (■), green beans (◆) and starchy foods (□) stored at chill temperatures.

In general then, the changes in the sensory quality of foods with time at chilled storage temperatures are not pleasant and lead, eventually, to the complete ruin of the food. What should be emphasised is that cook–chill is a balance of preserving food under safe conditions (see Chapters 3 and 9) for a reasonable time such that rationalisation of the process and, thereby, maximal benefit to the caterer is obtained whilst maintaining the eating quality of the food. This may mean that some products can only be stored for 1–2 days whereas others may be kept for up to 5 days without serious damage to their sensory quality. Several new techniques, such as vacuum and modified atmosphere packaging, allow the optimisation of the retention of sensory quality during storage (see Chapter 6). One or two recent studies have set out to examine the effects of packaging on the sensory qualities of cooked, chilled foods after extended chilled storage. These studies are discussed in depth in Chapter 6.

NUTRITIONAL CONCEPTS IN RELATION TO COOK–CHILL CATERING

The nutritional value of a food is determined by its content of certain vital chemical factors which cannot be obtained by human beings from any source other than food. These include protein (which provides essential amino acids), essential fats, vitamins, minerals and energy. All of these except energy are chemicals and even energy is obtained from the chemicals in food. Therefore, the kind of chemical changes brought about in food during storage which have been discussed earlier are likely to affect the nutritional quality of the food if these reactions involve any of these above named nutritionally vital chemicals. Protein quality can be easily diminished if a protein in radically changed chemically during thermal treatment and storage through chemical interactions with other active compounds in food like sugars. Minerals are less likely to be affected but essential fatty acids (which are unsaturated and, therefore, prone to oxidative degradation) and several vitamins are all very susceptible to such changes. Indeed, heat alone or other factors such as ultra-violet light (UV light) can cause the breakdown and, therefore, loss of some nutrients. In this section we will discuss the kind of changes that can be brought about in terms of the nutritional value of food using some of the sensitive (or chemically labile) vitamins and essential amino acids as examples. However, before embarking on such a discussion it is important to begin with a definition of these nutrients.

A vitamin is an organic compound needed in very small amounts by living cells. Many vitamins are used in cells as co-enzymes in metabolic reactions and, as such, are indispensable. Vitamins are found in many different foods, especially green vegetables, fruit, meat, eggs, fish and milk. Although they are not related to one another, the vitamins can be divided into two classes; the fat soluble vitamins and the water soluble vitamins. The former are vitamins which are soluble in lipids (oils and fats) or organic solvents (like ether and chloroform), e.g. vitamins A, D, E and K. Water soluble vitamins are, as the name suggests, soluble in water e.g. vitamin C (ascorbic acid), niacin and thiamin, and are prone to leaching, a process in which they are lost from a food during preparation, cooking or blanching by literally being dissolved out into cooking fluids or being lost in 'drip' from foods. If the intake of vitamins in a diet is not sufficient to match the requirements of the person concerned then this can result in vitamin deficiency diseases, some of which can be fatal (see Table 2.2). Because vitamins are needed for so many reactions in all cells in the body, vitamin deficiency nearly always leads to illness. Indeed, certain populations are more prone to illness if they are deprived of vitamins, e.g. old people and growing children.

Table 2.2 presents a list of some of the vitamins, with their chemical names and their sensitivities to heat processing as well as their sources and their associated deficiency diseases.

Proteins are large organic molecules which are made from long chains of much smaller molecules called amino acids. There are about 20 such amino acids commonly found in food proteins and the body liberates these from food proteins by digestion in the stomach and small intestine. These free amino acids are then taken up into the blood and transported all over the body to cells where new proteins are made from them. Some amino acids can be made in the body from other chemicals but eight (nine in young children) cannot. These are known as the essential amino acids. These must be eaten in the diet or new proteins, which are used in the body to build new structures and carry out the reactions of metabolism, cannot be made and illness or even death can result.

Vitamins can be lost from foods primarily in four ways. Firstly, they can be chemically degraded during the cooking process or combined with other chemicals in the food in such a way as to become unavailable for use by the person who has eaten the food in which they are found (this latter case is known as being biologically unavailable). Secondly, if they are water soluble they can be leached out of the food during the preparation, cooking or storage processes as described above. Thirdly, they can be

TABLE 2.2
A List of Some Fat and Water Soluble Vitamins, Their Sources, Chemical Names and Properties

Vitamin	*Chemical name*	*Sources*	*Solubility*	*Heat stability*	*Deficiency disease*
A	Retinol	Fish liver oils, animal liver, butter, margarine, cheese and eggs	Fat	Fairly stable	Night blindness, slow rate of growth, skin disease, xero-phthalmia
D_1	Cholecalciferol	Fish liver oil, margarine, butter, eggs	Fat	Stable	Rickets, osteomalcia
E	Tocopherol	Vegetable oils, wheat germ, eggs and milk	Fat	Stable	None known
B_1	Thiamin	Germ and bran of cereal grains, meat, many vegetables, nuts and milk	Water	Labile	Beri beri
B_2	Riboflavin	Liver, kidney, cheese and eggs	Water	Labile	Soreness of eyes and mouth
B_6	Pyridoxine, pyridoxal, pyridoxamine	Meat, cereals and pulses	Water	Labile	Rare—can cause fits in infants
	Niacin	Yeast, meat, liver, fish, cheese, cereals and pulses	Water	Labile	Pellagra
C	Ascorbic acid	Green plants	Water	Labile	Scurvy

changed in some way or degraded by the effect of thermal processing. Finally, they can be reacted upon or chemically affected by other agents (such as UV light) or chemicals in the food or its atmosphere during storage.

In cook–chill catering the processes of preparation, prime cooking, chilled storage and re-heating can all lead, in theory, to a lowering of the amounts of some, if not all, chemically labile (easily destroyed) vitamins in foods treated in this way. Much has been said on this subject, particularly in respect of cook–chill foods used in schools and hospitals (i.e. with groups at nutritional risk) and most guidelines on the use of cook–chill specify that minimal treatments should be used and certain precautions should be taken to prevent inflicting unacceptable nutritional damage on foods prepared by the cook–chill method (see Chapter 4). However, very little fundamental scientific work has been done to investigate the effects of the method on these important parameters.

Early studies indicated that 55–95% of vitamin C (ascorbic acid) is lost during only 1 day's chilled storage of certain vegetables and potato. However, many of these studies were poorly controlled and later work holding food at 2–4°C showed that ascorbic acid losses ranged from 3–20% per day of chilled storage whilst losses of retinol and thiamin remained lower than

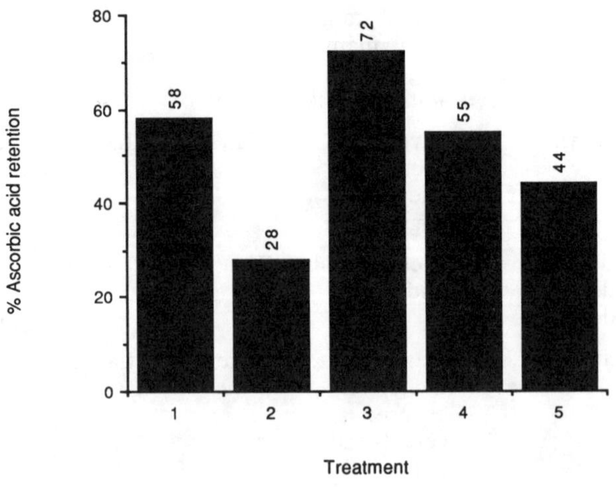

Fig. 2.6. Comparison of ascorbic acid levels in frozen, warm-held and chilled potatoes stored for various times. 1, Chilled potatoes stored for 1 day; 2, chilled potatoes stored for 3 days; 3, frozen potatoes stored for 2 months; 4, warm-held potatoes (heated container) stored for 5 h; 5, warm-held potatoes (insulated container) stored for 5 h.

10% even after storage at this temperature for 3 days. Further work showed that, when foods which had been cooked, chilled, stored for 1 day at 0–3°C and re-heated were compared with freshly cooked foods, thiamin fell by 20–50%, riboflavin fell by about 50% and β-carotene (a precursor of vitamin A, retinol) was largely unaffected. Vitamin C was reduced by 30–90% after chilled storage for 1 day when compared with freshly prepared foods, the fluctuations observed being ascribed to the variations in behaviour of different green vegetable sources and the different methods of different producers.

In a study of potatoes used for meals-on-wheels services it was shown that ascorbic acid levels were best maintained by frozen storage (72% after two months). However, use of warm-holding was shown to be quite harmful to vitamin C levels; 55 and 44% only of original levels were left after 5 hour's holding in heated or hot insulated containers respectively. After 1 day's storage under cook–chill conditions the potatoes retained 58% of the original ascorbic acid levels and after 3 days only 28%. These data are illustrated in Fig. 2.6.

In another study of potato, beef loafs and peas it was shown that vitamin C levels in potatoes were reduced to 57% of original levels after 1 day's storage, corroborating the above mentioned study (Fig. 2.7). However, this work showed quite clearly that it was the re-heating stage that

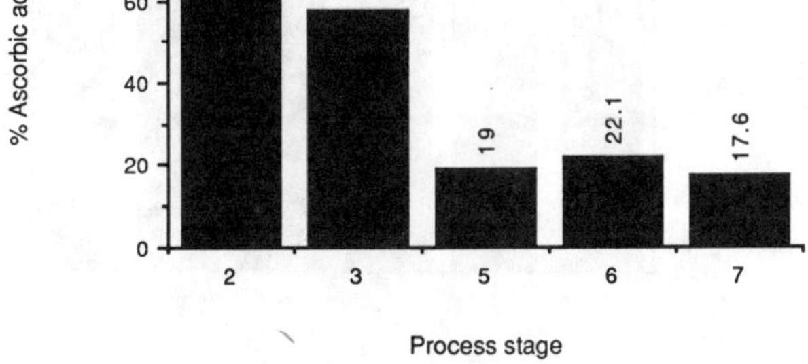

Fig. 2.7. Ascorbic acid levels in potato after cooking (2), 1 day's chilled storage (3), conduction re-heating (5), convection re-heating (6) and microwave re-heating (7).

led to the highest loss of vitamin C, reducing it to about 20% of original levels no matter what method of re-heating was used. In beef loaf and peas the loss of thiamin was much less. After 1 day's chilled storage 75 and 91% of the vitamin were left in beef loaf and peas respectively and even re-heating did little to reduce the levels in either product to less than 67 and 85% respectively (see Figs 2.8 and 2.9).

Similar results have been obtained from other limited studies leading to the general conclusion that chilled storage results in further losses of the labile vitamins, like ascorbic acid, after initial losses inflicted by cooking (pasteurisation). This view is supported by the evidence obtained to date but does not allow for the newer methods of storage which include the use of vacuum and modified atmosphere packaging (MAP) methods as mentioned above (see also Chapter 5). Ascorbic acid is particularly prone to oxidation so the removal of air from the environment of the chilled food will cut the loss of this vitamin dramatically. Other precautionary methods can be taken to avoid undue vitamin loss, for example, using liquors (gravies, etc.) with the meals, thereby retaining leached vitamins, pasteurising for optimal times but avoiding over-cooking (thereby reducing thermal damage), re-heating for minimal (safe) times to avoid further heat-induced losses. Also, if these losses are recognised, compensation can be made by making the diet complete, i.e. serving a green salad with a cook–chill meal, etc.

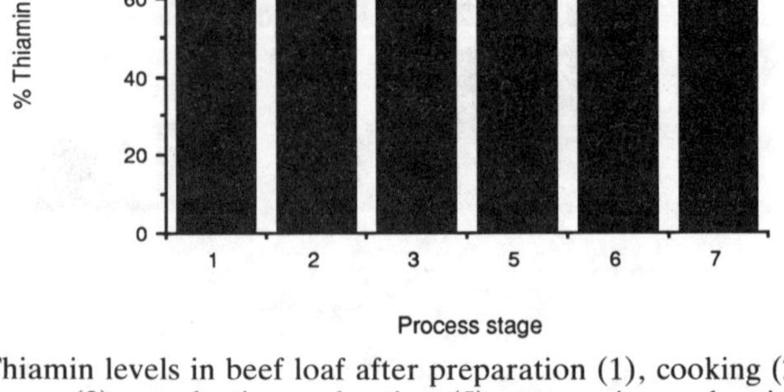

Fig. 2.8. Thiamin levels in beef loaf after preparation (1), cooking (2), 1 day's chilled storage (3), conduction re-heating (5), convection re-heating (6) and microwave re-heating (7).

Some losses of essential amino acids were also noted when protein quality in cooked, chilled foods was considered in that the content of seven essential amino acids in beef, pork, chicken and fish dishes, decreased from 100% in the raw product to around 70–80% in chilled products. Frozen products made from the same raw materials and stored for 2–10 weeks showed virtually no loss of essential amino aids (except in a fish product where there was a greater loss than in the chilled equivalent). These results seem to indicate that some loss in protein quality can also be expected on storage of cooked, chilled foods at 0–3°C but that this is relatively low compared with some of the vitamin losses. Certainly, in the relatively protein-high diets of the developed world such losses could not be considered significant. However, it is fair to observe that far too little work has been done on this important subject to date and that there is room for more research in this area.

Overall, therefore, cook–chill does cause some loss of nutritional quality in foods stored at chill temperatures. However, these losses are relatively low provided the foods are handled well and conditions are optimised. In situations where cook–chill catering is used and diet is critical, such as hospitals, these losses should be accounted for and compensated for with additional foods such as green salads and fruit juices or even, if necessary, supplements in the form of tablets.

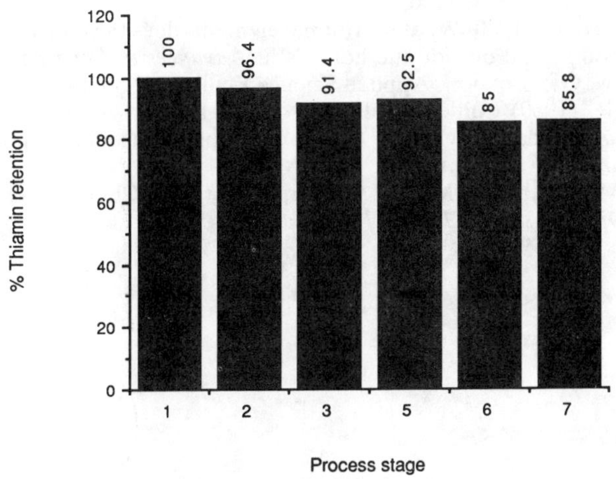

Fig. 2.9. Thiamin levels in peas after preparation (1), cooking (2), 1 day's chilled storage (3), conduction re-heating (5), convection re-heating (6) and microwave re-heating (7).

FURTHER READING

Bognar, A. (1980) Nutritive value of chilled meals. In: *Advances in Catering Technology* (ed. Glew, G.), London, Applied Science Publishers, pp. 387–408.

Dahl-Sawyer, C. A., Jen, J. J. & Huang, P. D. (1982) Cook/chill foodservice systems with conduction, convection and microwave reheat subsystems. Nutrient retention in beef loaf, potatoes and peas. *J. Food Sci.*, **47,** 1089–95.

Fox, B. A. & Cameron, A. G. (1982) *Food Science, a Chemical Approach*, 4th edn, London, Hodder and Stoughton.

James, S. J. & Bailey, C. (1985) Principles and problems in cooling foodstuffs. In: *Proc. Chill Foods*, 7–9th May, Stratford-upon-Avon. Chipping Campden, UK, Campden Food Preservation Research Association.

Jellinek, G. (1985) *Sensory Evaluation of Food, Theory and Practice*, London, Ellis Horwood.

Karlstrom, B. (1982) Sensory quality of peas handled in food-service systems. *Lebensm. Wiss. u. Technol.*, **15,** 326–29.

Mikkelsen, B. E. (1985) The quality of potatoes in four different meals-on-wheels systems. *J. Foodservice Sys.*, **3,** 251–6.

Nicholanco, S. & Matthews, M. E. (1978) Quality of beef stew in a hospital chilled food service system. *J. Amer. Dietet. Assoc.*, **72,** 31–7.

Paulus, K., Piekarski, J. & Leistner, L. (1980) Chilled meals: possibilities and limits. In: *Advances in Catering Technology* (ed. Glew, G.), London, Elsevier Science Publishers, pp. 375–86.

Sawyer, C. A. & Naidu, Y. M. (1984) Sensory evaluation of cook–chilled products reheated by conduction, convection and microwave radiation. *J. Foodservice Sys.*, **3,** 89–106.

van Oosten, H. M. (1980) What nutritional elements does the nutritionist want to see in food served outside the home? In: *Advances in Catering Technology* (ed. Glew, G.), London, Applied Science Publishers, pp. 64–71.

Zacharias, R. (1980) Chilled meals: sensory quality. In: *Advances in Catering Technology* (ed. Glew, G.), London, Applied Science Publishers, pp. 409–416.

Zallen, E. M., Hitchcock, M. J. & Goertz, G. E. (1975) Chilled food systems. *J. Amer. Dietet. Assoc.*, **67,** 552–7.

CHAPTER 3

Food Preservation: the Microbiological Quality of Cook–Chill Foods

INTRODUCTION

All food is contaminated by micro-organisms which live naturally on the food or by organisms which have been introduced onto or into the food, usually from a number of sources, e.g. the atmosphere, from surfaces in stores, during transport or in kitchens (production areas) or by handlers, i.e. human beings. This contamination, as we shall see below, can be in the form of bacteria, moulds or yeasts. Moulds and yeasts can be responsible for extensive damage to fresh foods in the form of food spoilage and once a mould contamination has occurred in a food store it can be very difficult to remove. However, moulds and yeasts are relatively easy to destroy by heating, i.e. the first part of the cook–chill process. The main microbial contaminants of foods encountered in the cook–chill process are bacteria and so the discussion in this chapter will centre on these.

Because of this natural contamination, food needs to be preserved if it is not eaten immediately after harvest whether it is animal or plant in origin. One of the most important principles of preservation concerns the inhibition or complete prevention of the growth of micro-organisms in or on the food during storage. This is because when micro-organisms grow in or on food they can lead to unacceptable changes which include deterioration in the sensory properties of the product ('bad' or 'off' smells, slimy texture, unpleasant flavours) as well as producing toxins (or poisons) in the food or once ingested which can cause illness and even death in people who eat the food. Such toxins can be present in small but dangerous quantities in a food and yet the product may appear to be fit to eat and, conversely, food that appears totally spoiled can be quite safe (if unpleasant) to eat.

The study of micro-organisms is known as microbiology and encompasses all of those organisms that, in general, are uni-cellular (i.e. are composed of only one cell). This includes the bacteria, moulds, yeasts and viruses, although the latter are not often considered in food microbiology. The study of these organisms has led to an understanding of their growth and reproductive characteristics such that it is now largely possible to predict how any one type of organism will grow in a food of known properties under given conditions. Many food microbiologists are advocating the setting up of computer programs which will allow the prediction of shelf-life characteristics in foods with known microbial contaminants. However, there is no substitute for a basic knowledge of the simple principles of microbiology and it is important that all caterers, particularly those engaged in large scale production methods such as cook–chill, should be familiar with these principles.

This chapter, therefore, serves to provide a basic understanding of the principles of food microbiology which underlie the use of any food production and preservation system. They deal with the main types of micro-organisms which cause food spoilage and food poisoning and describe the ways in which these micro-organisms grow and, perhaps more importantly, can be prevented from growing.

A GENERAL INTRODUCTION TO MICRO-ORGANISMS

Micro-organisms (also known as microbes) may be loosely defined as living beings or organisms which are very small (microscopic), usually containing only one cell and which cannot be seen with the human eye. These include bacteria (singular, bacterium), moulds and yeasts, the latter two falling into the larger group known as fungi.

Moulds and Yeasts (Fungi)

Moulds are a group of multi-cellular fungi which grow in thin thread-like strands called hyphae. Each cell can grow alone and so moulds are classed as micro-organisms. Moulds can grow on foods and damage them but some are introduced into foods to give added flavour, e.g. in some cheeses. Most moulds reproduce by forming spores.

Yeasts are single-celled fungi which reproduce by budding, i.e. producing an offspring as a bud off the main parent cell. A wide assortment of yeasts are used in fermented foods such as bread and alcoholic beverages. Many yeasts, particularly those which grow as a bloom on fruit, are able to grow in high acid foods and this can lead to spoilage of fruit and fruit juices, fruit preserves and meat (Table 3.1).

TABLE 3.1
A Summary of the Conditions Under Which a Variety of Micro-organisms are Able to Grow (adapted from Mossell & Westerdijk, Publication 103, Utrecht Central Institution for Nutritional Reseaιcl TND, 1949)

Organisms	*Temperature* (0°C)
	−10 0 10 20 30 40 50 60 70 80 90
Thermophiles	+ + ++++++++++++ ++
Mesophiles	+ +++++++++ + +
Moulds	+ + ++++++++
Psychrophiles	+++ ++++++
	Water activity (as % eq. relative humidity)
	100 90 80 70
Bacteria	+++++
Yeasts	+++++++++++
Moulds	+++++++++++++++++ + + +
	Osmotic pressure (as % NaCl)
	0 10 20 30
Putrefactive bacteria	+++++ + + +
Lactic acid bacteria	+++++++++
Moulds	++++++++++++ + + + +
Micrococci	+++++++++++++++++ + +
Rods (salted fish)	++++++++++++++
	Osmotic pressure (as % sugar)
	0 10 20 30 40 50 60 70 80
Most micro-organisms	+++++++++++++
Osmophilic yeasts and moulds	+++++++++++++++++++ + +
	pH
	9 8 7 6 5 4 3 2 1
Bacteria	+++++++++++++ +
Yeasts and moulds	++++++++++++++++++ + +
	% oxygen in the environment
	100 80 60 40 20 0
Aerobes	++++++
Facultative anaerobes	++++++++++ + + + + +
Micro-aerophiles	+++++++++++++++
Anaerobes	++++++

Both moulds and yeasts are commonly found in the atmosphere, food can, therefore, easily become contaminated by these micro-organisms if it is not kept protected by packaging or other methods. Many foods are spoiled by moulds and yeasts. However, as the major problems associated with the preservation of cooked, chilled foods (i.e. spoilage and food poisoning incidents arising from the consumption of contaminated stored food) generally involve bacteria, the main part of this chapter will deal with this large group.

Bacteria

The bacteria are a large group of micro-organisms which can, in different situations, be both harmful and helpful to man. They are single-celled but contain no nucleus. Some are pathogenic (i.e. cause the illness known as food poisoning) and many cause food spoilage but some live in the intestines of man and provide benefits such as producing vitamins, e.g. biotin, and others are useful in fermenting foods, e.g. yoghurt.

Bacteria can be classified in many ways, e.g. by their shape, by their characteristic uptake of certain microscopical stains, by their resistance to temperature, by their resistance to salt concentration and osmotic pressure and so on (Table 3.1). Later we will consider the various groups of bacteria as defined by their growth characteristics and other parameters but first it is useful to consider the simple factors of shape and appearance.

There are four main groups of bacteria as defined by shape. These are (1) the cocci (sing. coccus), a group of bacteria with a spherical shape, (2) the bacilli (sing. bacillus) which have a rod-like shape, (3) the vibrii (sing. vibrio), bacteria with a short, curved rod-like shape and (4) the spirilli (sing. spirillum) which have a long, coiled shape. Bacteria have also been classified by their characteristics in the light microscope, especially after staining with a material known as the Gram stain. Gram-positive organisms are simply any bacteria which are coloured violet by the Gram stain whilst Gram-negative bacteria do not take up the stain. Thus an organism might be described as a Gram-positive coccus. This would be the first step in characterising it and determining its family and type.

Classification of Bacteria According to Temperature/Growth Relationships

Bacteria can also be classified in terms of their resistance or otherwise to extremes of temperature (Table 3.1). Psychrophilic bacteria, are those bacteria which grow optimally at 20–30°C and even as low as or lower than 0°C. Thus, a psychrophile is merely a psychrophilic bacterium. Psychroduric bacteria are those which can stand the effects of low temperatures, i.e. are not damaged by chilling or freezing.

Mesophilic bacteria are a class of bacteria which grow best at 30–40°C and which will not usually grow below 5°C or above 45–50°C. A mesophile, therefore, is simply a mesophilic bacterium.

Thermoduric bacteria are those which will stand relatively high temperatures whilst thermophiles (thermophilic bacteria) will even grow at high temperatures (e.g. in hot springs).

Reproduction

In terms of understanding the dangers of bacterial contamination of foods and the effects of heat and cold on bacteria it is important to discuss the way in which bacteria grow and reproduce. Firstly, most bacteria reproduce through the simple and rapid process of binary fission (Fig. 3.1). This means simply that one bacterium divides into two daughter bacteria. This process can take place in a relatively short period of time, e.g. 10–20 mins. Thus, the generation time for a class of bacteria living in ideal conditions can be as short as, say twenty minutes. So, in twenty minutes a contamination of just one organism can double to two, in forty minutes it will increase to four, in one hour to eight. By the end of an eight hour shift, at this rate and if conditions do not change, there will be 16 777 216 bacteria on or in the food. In the UK the *DHSS Guidelines on Pre-cooked Chilled Foods in Catering* (see Chapter 4 of this volume) suggest that a bacterial loading of 100 000 (or 10^5) per gram of food is the allowable limit in a cooked, chilled food product. In our example food, if we assume that the original contamination was at a level of one organism per gram, we reached this unacceptable level within 5 h 40 mins.

Fortunately, foods which are chilled do not provide a good environment for most bacteria to grow due to the low temperature. In addition, those foods which have been properly heat processed and not recontaminated by handling after heating should start with a very low load of bacteria anyway. Therefore, the growth rate of bacteria in such foods is usually very low. This is why, under well controlled conditions, microbiological contamination of cook–chill foods should be minimal thus providing a good way of preserving foods over short periods of time.

Bacteria grow in four distinct phases which are known as the lag phase, the logarithmic phase, the stationary phase and the phase of decline. The lag phase is the time before actual increase in cell numbers when cells grow and enlarge. The length of the lag phase varies according to environmental conditions and the numbers of cells initially present. The logarithmic (or log) phase is the period of most rapid growth when cells

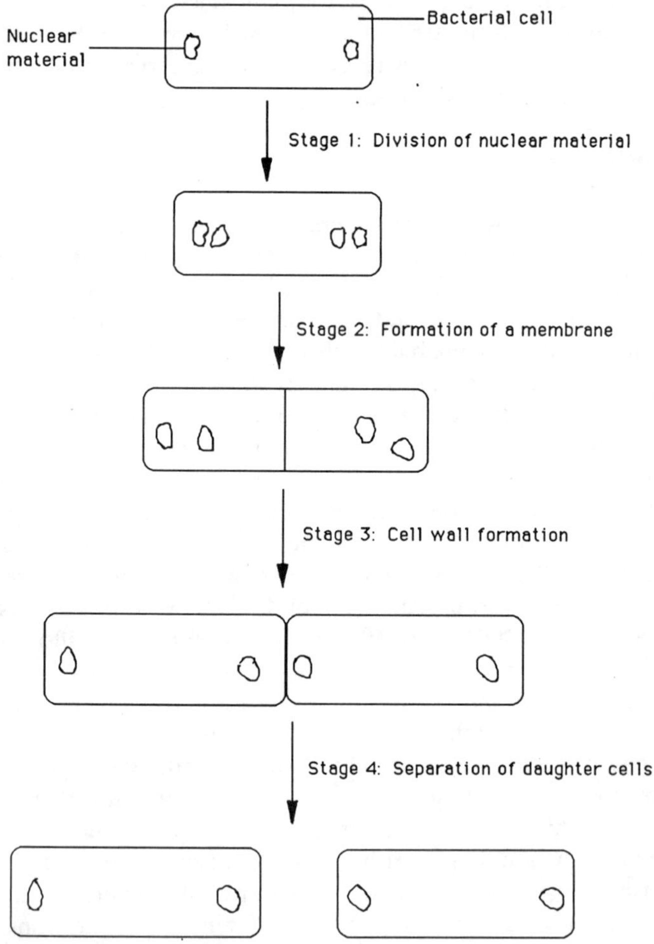

Fig. 3.1. Diagrammatic representation of the reproduction of bacteria by binary fission.

divide at a constant rate. The stationary phase is when, at the end of the log phase, the number of viable cells reaches a maximum and the rate of cell division becomes slower. In this phase cell division is just fast enough to replace dying cells. In the phase of decline cell division stops and the remaining cells die resulting in the overall reduction of cell numbers.

FOOD SPOILAGE AND FOOD POISONING

Food spoilage is any process which leads to the deterioration or spoilage of food such that it eventually becomes inedible due to the production of 'off' odours, flavours and colours and can be caused by chemical reactions, as discussed in Chapter 2, or microbial growth (the growth of micro-organisms). Food spoilage does not necessarily lead to the production of toxins in food. Microbiological spoilage (or microbial spoilage) is any food spoilage caused by the growth of micro-organisms on or in the food, e.g. the growth of mould on stored fruit, the growth of slime-producing bacteria on meat.

Food poisoning is any illness in man caused by eating foods containing harmful compounds. There are three main kinds of food poisoning: 1, chemical food poisoning (caused by chemical poisons, e.g. pesticide residues in food); 2, biological food poisoning (caused by naturally poisonous foods, e.g. uncooked or under-cooked kidney beans which contain a protein toxin which has to be inactivated by heat before consumption); and 3, microbiological food poisoning. Microbiological or microbial food poisoning is caused by the products of micro-organisms, such as toxins, growing in foods. Food poisoning caused by contamination with bacteria is the most common kind.

Food poisoning micro-organisms are called pathogens or pathogenic organisms. These produce toxins which are released into the food the organisms grow on. The toxins can be either exotoxins, which are toxins released into food by micro-organisms during growth and before the food is eaten or endotoxins which are released by micro-organisms eaten with contaminated food. A toxin which is released by bacteria in the alimentary canal is known as an enterotoxin. Some toxins can be destroyed by heat but the more dangerous ones are those which are heat stable.

So far we have discussed the terms food spoilage and food poisoning but it is necessary, before gaining a full understanding of the effects of extremes of temperature on micro-organisms, to understand the nature of micro-organisms themselves.

TABLE 3.2
Some Common Food Spoilage and Food Poisoning Bacteria and Selected Characteristics

Bacterium	*Gram stain*	*Min. growth temp (°C)*	*Spores*	*Oxygen tolerance*
Salmonellae	−ve	5·2	No	Facultative anaerobe
S. aureus	+ve	7·0	No	Oxygen tolerant
C. perfringens	+ve/−ve	15·0	Yes	Anaerobe
C. botulinum	+ve	3·3	Yes	Anaerobe
Y. enterolitica	−ve	< 3·0	No	Facultative anaerobe
L. monocytogenes	+ve	< 3·0	No	Aerobe
B. cereus	+ve	4·0	Yes	Aerobe
Lactobacillus	+ve	< 3·0	No	Facultative anaerobe
Pseudomonas	−ve	< 3·0	No	Aerobe
Brochothrix thermosphacta	+ve	0·8	No	Facultative anaerobe
Bacilli	+ve	—[a]	Yes	Aerobe
Various clostridia	+ve	—[a]	Yes	Anaerobe

[a]Minimum growth temperature dependent on type.

Some common food spoilage and food poisoning bacteria are listed in Table 3.2. The major characteristics of these bacteria relevant to cook–chill catering are also listed. Later in this chapter, after discussing the effects of temperature and atmosphere on micro-organisms, we will review the main types of food poisoning organisms in some detail.

EFFECTS OF THERMAL PROCESSING ON BACTERIA

The growing cellular form of a bacterium is called its vegetative form. In this state most bacteria are easily killed or damaged by heat, whilst others are killed by extremely low temperatures. Damaged organisms might eventually start to grow again and so any heat processing of food should aim to kill as many (if not all) micro-organisms as possible. However, extreme heat treatment can just as easily destroy or, at the very least, vastly reduce the organoleptic (sensory) quality of the food. Therefore, there needs to be a balance between over-heating yet applying enough heat to effectively kill micro-organisms. The thermal death time (TDT) of an organism is a concept used to quantify the killing of microbes in food by heat processing. It may be defined as the time required at any specified temperature to inactivate (kill) an arbitrarily chosen proportion of the organisms, where, obviously the greater the proportion killed the higher the margin of safety. Thus, time can be plotted on a graph against the temperature a food is processed at to give an indication of the TDT (Fig. 3.2). TDT is used in the canning industry to give assurances of safe heat processing but, as yet, has not been widely applied in the preparation of cooked, chilled foods especially in the catering industry.

There are four types of thermal processing used in the food industry as already mentioned; blanching, cooking, pasteurisation and sterilisation. Only the last two can be regarded as controlled processes in terms of the removal or inactivation of bacteria and other micro-organisms.

Pasteurisation

Pasteurisation is a process in which foods are heated to kill only the vegetative forms of the bacteria which contaminate them and, generally,

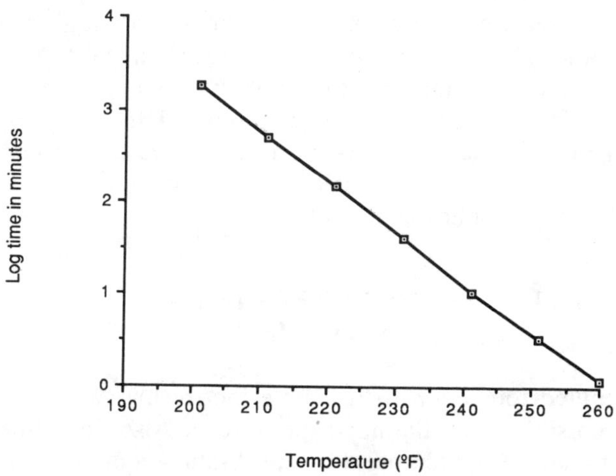

Fig. 3.2. TDT plot for 90% inactivation of *Clostridium botulinum*.

temperatures of less than 100°C (i.e. 65–100°C) are used. For example, milk is pasteurised by an automated process at relatively low temperature and for a short time, i.e. 15 s at 73°C. Although these conditions may seem mild, they have been calculated in order to ensure the destruction of all vegetative forms of the organisms that commonly contaminate milk. It is not merely the temperature to which the food is heated that is critical but the time that it is held at that temperature that also determines the rate (extent) of bacterial kill. Cooked, chilled foods should always be properly pasteurised to ensure that all vegetative (growing) forms of the organisms in the raw materials are properly killed.

Sterilisation

Some micro-organisms form spores. These are bodies produced by some micro-organisms in conditions unfavourable to survival which have a hard coat and can stand extreme conditions. Of the bacteria only the bacilli and the clostridia form spores but these can survive even in cooked (pasteurised) foods and lead to food spoilage or food poisoning. This is because the spores (which are often activated by heat shock) may begin to grow in the cooked food after it is cooled and could reach high enough numbers to spoil the food or produce harmful amounts of toxins.

In order to inactivate spores (as well as the vegetative forms of bacteria and other micro-organisms in food), a heat treatment more rigorous than pasteurisation has to be used. Sterilisation, therefore, is a process in which foods are treated to kill all forms of micro-organisms and spores. Foods can be sterilised with high temperature or with ionising radiation (food irradiation) but in the former case enough heat has to be applied to ensure the total destruction of all micro-organisms and spores. Typically, temperatures in excess of 120–130°C are used and, therefore, much thermal damage is inflicted on the food. Such damage can be restricted by using high-temperature-short-time (HTST) treatments but such methods are not necessarily applicable to all foods.

PRACTICAL USE OF HEAT-TREATMENT IN COOK–CHILL CATERING

Cooked, chilled foods aim to resemble the freshly cooked product as closely as possible given the need for storage. Also, investment in the kind of automated high temperature sterilisation equipment used in the canning industry is impractical for most caterers. Thus, all cooked, chilled foods, at this time, are pasteurised and so will contain a proportion of undamaged spores. Because most spores are activated by heat treatment it is critical that cooked foods are chilled extremely rapidly in order to prevent such germination and growth (see Chapter 4).

It is almost impossible to ensure absolutely the complete destruction of all living organisms in food by most treatments and, therefore, in practical food pasteurisation, the core (centre) of the food should reach an internal temperature in excess of 70°C for enough time to ensure the death of the vast majority of the microflora living in or on the food. As the TDT for any one organism varies with the number of organisms initially present on or in the food, it follows that a food which is more highly contaminated will take longer to pasteurise than one with a low bacterial loading. For this important reason caterers should ensure the highest quality of their raw materials supplies and specify, if possible, acceptable maximum levels of microbial loading. The risk of contamination being carried through with a product is obviously higher for a raw material with a loading which is initially very high.

As we shall discuss later, two of the most dangerous forms of food poisoning are salmonella poisoning (salmonellosis), which is caused by the toxins produced by the *Salmonella* group of bacteria, and botulism, a form of food poisoning caused by the toxins of *Clostridium botulinum*. Botulism often results in death but is very rare whereas salmonella poisoning, which can also result in death, is the most common form of severe food poisoning in the western world. However, bacteria of the *Salmonella* family are relatively easily killed by heat processing, and proper pasteurisation can result in a high degree of assurance of their complete inactivation. The vegetative form of *Clostridium botulinum* is also relatively easy to kill by pasteurisation but this bacterium forms spores which are resistant to such thermal processing treatment. Thus, thermal treatment, particularly pasteurisation should not be seen as a 'stand-alone' method of bacterial inhibition. Fortunately, *Clostridium botulinum* has never been known to grow at temperatures below 3°C and so the combination of proper pasteurisation, very rapid chilling and storage of products at 0–3°C should provide assurance of the inhibition of most dangerous food microbes.

EFFECTS OF LOW TEMPERATURE STORAGE ON GROWTH

Many studies have been conducted over the years on the effect of low temperatures on bacterial growth and some of the general findings of these studies, particularly in connection with the growth characteristics of some food poisoning organisms, are discussed later in this chapter (and see Table 3.1). However, a recent piece of work set out to examine the growth of organisms in three typical catering products prepared under conditions similar to those used in commercial cook–chill catering operations and we shall now consider some of the results of this study as being typical of the many others on this subject.

The initial total viable counts (TVC) of bacteria in three cooked chicken menu items (chicken à la king, minced chicken and chicken drumsticks) were greater in samples prepared at a real catering site than those prepared under laboratory conditions which mimicked cook–chill catering conditions, showing that the care of the food was not as great under commercial conditions as in the laboratory, as might be expected.

Before storage, counts were negligible whereas after storage the main organisms present were pseudomonads (a common class of food spoilage organism). Of 30 colonies isolated from samples before storage all were tentatively identified as staphylococci and micrococci. As these organisms are relatively heat sensitive their presence is likely to have been due to contamination after cooking. The change in the microbial composition demonstrates the selective effect of chilled storage on the nature of the surviving microflora.

Growth rates for bacteria at 0°C were significantly lower than those obtained at 4°C (Figs. 3.3–3.5) and those at 2°C generally lay between the other two. The lag phases for growth on the three menu items were all different but remained the same for food stored at 0 and 4°C for a minimum of two days and a maximum of nine days and was product-dependent. In all cases the TVC had reached unacceptable levels (10^5/g—see Chapter 4) at 4°C much more quickly than at 0°C. The minced chicken TVC had reached the maximum allowable level by day three at 4°C (Fig. 3.4) and by day five the TVC was well in excess of this level although the same product stored at 0 and 2°C was well below the recommended maximal loading.

The initial numbers of micro-organisms were higher in the minced chicken, due to the additional post-cooking handling it received (even though the utmost care had been taken to mince the drumsticks aseptically), and the larger surface area available for growth (Fig. 3.4). The lag phase was also shorter for the minced product (e.g. two days for the minced chicken at 4°C and nine days for the chicken drumsticks—see Fig. 3.3). By the end of the storage period the TVC for the chicken drumsticks ranged between 10^5 and 10^6/g, whereas for the minced product they were at least 10^7/g.

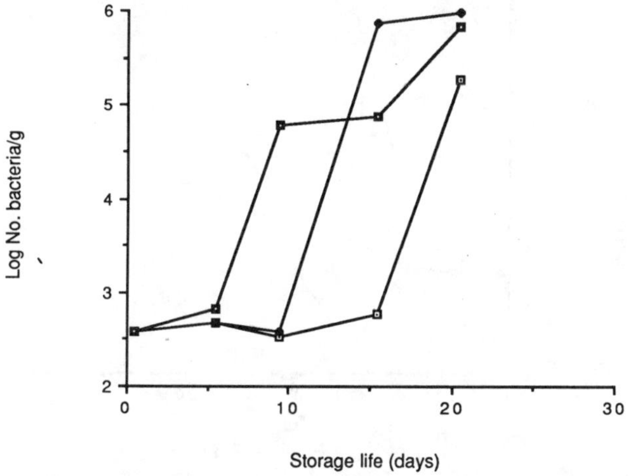

Fig. 3.3. Comparison of bacterial growth in cooked, chilled chicken drumsticks at 0°C (□), 2°C (■) and 4°C (◆). The graphs show total viable counts (TVC) in chicken drumsticks prepared by the cook–chill method and stored for up to 21 days.

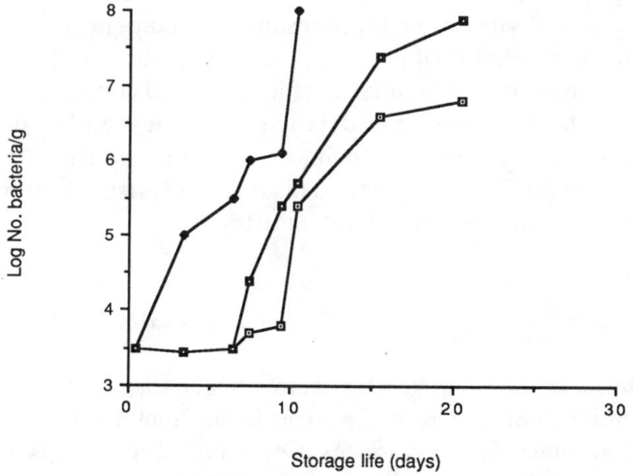

Fig. 3.4. Comparison of bacterial growth in cooked, chilled minced chicken at 0°C (□), 2°C (■) and 4°C (◆). The graphs show total viable counts (TVC) in minced chicken prepared by the cook–chill method and stored for up to 21 days.

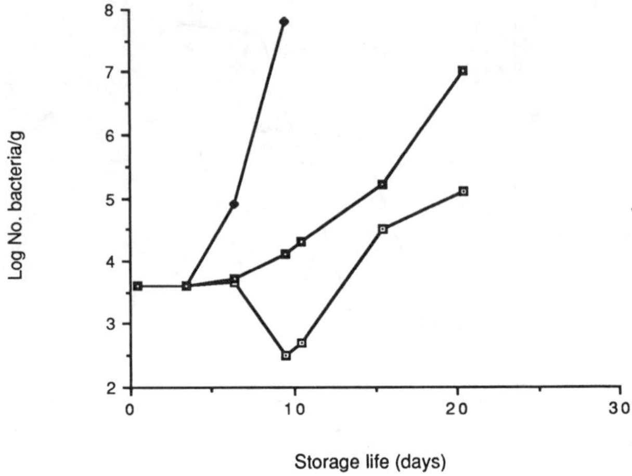

Fig. 3.5. Comparison of bacterial growth in cooked, chilled chicken à la king at 0°C (□), 2°C (■) and 4°C (◆). The graphs show total viable counts (TVC) in chicken à la king prepared by the cook–chill method and stored for up to 21 days.

The growth of microflora in chilled, stored cooked foods is dependent on a variety of factors, some of which are of more concern to the caterer than others. It is obvious from these results that temperature and storage time are two key factors. In practice, the composition of the microflora will be determined by the standards of hygiene and control employed in the kitchen. The microbiologial data from previous studies of catering systems varies enormously, even if the differing nature of the foods studied is taken into consideration. This is not surprising as standards of hygiene are not uniform throughout industry.

EFFECTS OF ATMOSPHERE ON GROWTH

Some bacteria, moulds and yeasts require oxygen for life. The oxygen can be either atmospheric or in a dissolved form. Such bacteria are called aerobes or aerobic micro-organisms. Other bacteria, moulds and yeasts do not need oxygen to live and so these are called anaerobes or anaerobic micro-organisms (Table 3.1). Some bacteria can function in both situations, i.e. with and without oxygen. Thus, bacteria are usually classified in four groups in terms of the influence of molecular oxygen on their development. These are:

1. obligate aerobes which require molecular oxygen for growth,
2. obligate anaerobes which are unable to develop in the presence of molecular oxygen,
3. facultative anaerobes which will grow in the absence or presence of molecular oxygen,
4. micro-aerophiles which will grow in an oxygen tension lower than that of air (Table 3.1).

The main consideration for cook–chill users with respect to these groups is in terms of the ability of the modern food producer to exclude or at least vastly reduce the amount of oxygen in a food and/or its environment by methods in which the food is packed in a container and vacuumised to remove all the air (vacuum packaging) or in which the container is filled with a suitable combination of gases like carbon dioxide and nitrogen (modified atmosphere packaging—MAP). These will be discussed in later chapters (see Chapters 5, 6 and 7) but at this stage it is worth mentioning the general effects of reducing oxygen tension in the environment of a stored, cooked chilled food.

As we have seen above, proper pasteurisation will kill all growing (vegetative) forms of all anaerobes and aerobes. Only spores will survive within the food. If a pasteurised food is placed in a vacuumised bag or if the food is pasteurised within a vacuum sealed bag (see Chapters 6 and 7) the spores of aerobic micro-organisms will not be able to grow because they need oxygen for development. Thus, such foods should be well protected against aerobic micro-organisms. However, anaerobic micro-organisms can grow from spores in pasteurised vacuum-packed foods because they do not need oxygen for development. The same is true for some micro-aerophiles. Unfortunately, some food poisoning bacteria are anaerobic spore-formers (e.g. the *Clostridium* family). Such organisms will not be destroyed by pasteurisation but will grow in anaerobic conditions, e.g. in vacuum packs, under the right conditions of temperature and environmental factors (pH, salt content, etc.). This has an implication for vacuum packed foods as we shall see later (Chapter 8).

Many studies have been carried out on the microbiology of cooked, chilled foods. These have covered such diverse subjects as the influence of temperature and storage time on the growth of food spoilage and food poisoning organisms, the effect of various chemical parameters such as salt content and pH on the growth of micro-organisms, the effect of packaging on growth and development of micro-organisms, the effect of various methods of cook–chill on the type and number of organisms found in chilled foods, the variation in contamination in cooked, chilled foods prepared in the same way by the same caterers and so on. Many of these studies have led to findings which have been mentioned above and, certainly, there are too many to mention all of them here. However, some specific work relevant to the shelf-life of cooked, chilled foods is discussed in Chapter 5 and some key references are cited in the main bibliography at the end of this book.

A REVIEW OF SOME COMMON FOOD POISONING ORGANISMS

The eight most common micro-organisms cited in records of outbreaks of food poisoning in the western world over the last ten years are:

(1) the salmonellae,
(2) *Staphylococcus aureus,*
(3) *Clostridium perfringens,*
(4) *Clostridium botulinum,*
(5) *Yersinia enterolitica,*
(6) *Listeria monocytogenes,*
(7) *Bacillus cereus*, and
(8) *Campylobacter*.

The types of outbreaks reported and the places with which they were associated are reported in Chapter 8. In this section we will consider the growth characteristics of the first six of these. Effective growth in these organisms relies on many factors coming together, e.g. pH, salt content, osmotic strength; however, we will only consider temperature and competition of other micro-organisms. The main points are summarised in Tables 3.2 and 3.3.

Salmonellae

There are many species and strains of the salmonellae (over 1700 serotypes) but all may be considered to be pathogenic for man. The organisms are facultative anaerobes and are motile (i.e. are able to move), Gram-negative, non-sporing rods. The disease resulting from ingestion

TABLE 3.3
The Minimum Temperature for Growth of Several Pathogenic Bacteria

Organism	*Temperature (0°C)*	*Substrate*
Bacillus cereus	4·0	Nutrient agar Vacuum packed bologna
Clostridium botulinum		
group E	3·3	Sterile meat stew
group A	10·0	Meat
group B	10·0	Vegetables
Non-proteolytic:		
group B	3·3	Cooked meat medium
Clostridium perfringens	15·0 5·0	Hartleys digest
group F	18·3	Beef cubes in gravy
Escherichia coli	7·2	Trypticase soy agar
Listeriae	1·0–4·0	
Staphylococcus aureus	6·7 5·0 5·0 8·0	Sterile custard Chicken à la king Bacon Vacuum packed bologna
Salmonella typhimurium	8·0	Trypticase soy agar
Salmonlla heidelburg	5·5	Trypticase soy agar
Salmonella seftenburg *S. enteritidis & S. manhatton*	6·7	Sterile chicken à la king
Salmonella panama	4·0	Buffered glucose tryptone soya peptone agar (GTS agar)
65 *Salmonella* species	7·0	Buffered GTS agar
9 *Salmonella* serotypes	10·0	Beef (not sterile)
Streptobacteria	4·0–8·0	
Atypical streptobacteria	2·0–4·0	
Yersinia enterocolitica	4·0 2·0	GTS agar Vacuum packed bologna

of the organism known as salmonellosis and may be divided into three clinical types, i.e. typhoid fever, paratyphoid fever and gastroenteritis. The first two forms can often be lethal to humans as can the third under certain conditions.

The optimum growth temperature for the salmonellae is 35–37°C. It resists freezing well but is relatively fragile at higher temperatures. Pasteurisation times and temperatures vary for the different forms of the organism (e.g. 5 min at 60°C; 10–15 min at 65°C) but few forms resist temperatures in excess of 65–67·5°C. This means that foods are relatively easily pasteurised with respect to salmonella, as discussed earlier.

The minimum growth temperature is dependent on the type of strain and the substrate upon which it is growing but the lowest recorded temperature under ideal conditions was 5·2°C when growth began within seven days. In fact, salmonellae are sensitive to temperatures around 1–2°C and tend to decline in numbers. In custard (pH 6·6) and ham salad (pH 5·6) various salmonellae serotypes declined in numbers at temperatures from 4·4 to 10°C whereas in chicken à la king (pH 6·2) growth began after four days at 6·7°C. Growth has been shown to be less at these low temperatures in the presence of competing microflora. A review of the available literature shows that salmonellae are unlikely to grow quickly if at all at temperatures below 10°C and that they will not show any growth below 5°C.

In general, the salmonellae do not compete well with other microorganisms. For example, lactic acid bacteria like the lactobacilli strongly inhibit the growth of salmonellae at high temperatures (20–30°C) due mainly to the production of a low pH environment and other chemicals, e.g. hydrogen peroxide and antibiotics. Salmonellae have been shown to grow in sterile food media after inoculation (e.g. in crabmeat). However, they will not grow in the same, unsterilised media because the natural microflora normally 'over-grow' the salmonellae.

Staphylococcus aureus

This organism is found in air, water, milk and sewage but is associated, in general, with the skin and mucosal secretions of human beings and animals. Staphylococci are free growing facultative anaerobes which are Gram-positive, non-motile and do not form spores. About 40% of all humans are said to carry the bacterium and approximately half of these are enterotoxigenic strains. Thus, the main contamination of food by *Staphylococcus aureus* is from food handlers. The form of poisoning caused by the organism is due to the toxin and results in the rapid onset (normally within 2–4 h) of nausea, vomiting, diarrhoea and abdominal cramps. The symptoms normally pass away as soon as the irritant substances are voided from the body.

Staphylococcus aureus is relatively easily destroyed by heat treatment, like the salmonellae, and, as stated above, it has no spore form. However, the toxin is quite heat resistant and, in order to fully inactivate it, needs treatment at temperatures in excess of 120°C for between 5–30 min depending on the medium in which it is carried. This means that a food which carries a high burden of a toxin-producing strain of *Staphylococcus aureus* might have all the organisms killed during pasteurisation (e.g. 80–90°C) but little of the toxin would be destroyed at these temperatures and, therefore, the food would remain hazardous.

The optimum growth temperature for *Staphylococcus aureus* is 37–40°C and for toxin production, 40–45°C. The temperature range for growth of the organism is 7–47·8°C but toxin production only occurs between 10 and 46°C. This means that food contaminated with *Staphylococcus aureus* which is stored below 10°C should remain fit for human consumption with respect to this organism provided no toxin has been produced before storage.

Staphylococcus aureus does not compete well with normal food-borne micro-organisms and competition may increase the minimum growth temperature of the organism. In other words, in situations where there is a mixed microflora on the food the minimum growth temperature of *Staphylococcus aureus* will be higher than in culture on its own and it will not grow well in competition with the other micro-organisms present.

Clostridium perfringens

This organism has been classified in six groups A–F but the food poisoning strains are predominantly in group A (which includes the gas gangrene strain) and, more rarely, group C. *Clostridium perfringens* (other name *C. welchii*) is a spore-forming, anaerobic, non-motile, short thick bacillus which is Gram-positive when young but which can turn Gram-negative with age. Some of the group A strains also produce an enterotoxin which is the cause of food poisoning. It is found in soils, water, most foods, dust, spices and the alimentary canals of man and other animals. The main food-stuffs associated with outbreaks of *Clostridium perfringens* food poisoning are red meat and poultry products. Food poisoning usually starts 12–20 h after the consumption of contaminated food and involves abdominal cramps and diarrhoea without any fever or vomiting. The organism has been implicated as the second largest cause of food poisoning incidents in the UK after salmonellae.

The optimum growth temperature of *Clostridium perfringens* is 32–40°C in which range most forms are able to produce spores. The minimum growth temperature is 15°C in bacterial media under ideal conditions. This has been shown to be similar in real food-stuffs. Low temperatures (1–3°C) and freezing both lead to the destruction of vegetative cells and spores of the organism. One study showed that 1–30% of cells and 41–76% of spores survived refrigerated storage (4–8°C) for 180 days.

The vegetative cells are easily killed by pasteurisation temperatures but the spores tend to be 'heat-shocked' into growth after cooling down from such temperatures. Thus, the majority of outbreaks of food poisoning associated with *Clostridium perfringens* are caused by cooling large joints of meat too slowly such that spores can germinate and grow rapidly, producing toxins which contaminate the meat.

Clostridium perfringens does not compete well with the other natural microflora of foods. Its growth is adversely affected by a range of other organisms such as the lactobacilli and streptococci. In inoculation experiments with egg, total anaerobic counts increased and *Clostridium perfringens* grew at 20 and 30°C but the organism was effectively outgrown by the natural flora of the egg.

Clostridium botulinum

Clostridium botulinum is an anaerobic Gram-positive, spore-forming, rod which is motile and which occurs as seven distinct groups, A–G. Each group can be divided into two sub-groups of proteolytic (protein digesting) or non-proteolytic organisms. The organism is found in both virgin and cultivated soils and so is widely spread throughout nature being largely associated in food with meat and fish. The illness in humans and animals known as botulism is caused by toxins (neurotoxin) produced by certain toxin-forming strains. These are all A strains, proteolytic strains of groups B and F (group 1), group E and non-proteolytic strains of B and F (group 2) which cause human botulism (although F has only been implicated in two incidents) and C and D (group 3) cause animal botulism. Group G has not been associated with human botulism.

Research has shown that the occurrence of *Clostridium botulinum* in meat and meat products is rare and numbers are normally very low. *Clostridium botulinum* group E seems to be largely associated with fish and it appears that poisoning outbreaks are normally associated with very high loadings. When ingested in the detoxified spore condition the bacterium is harmless; only the germinating spores which produce autolysing cells (cells which die and break open) cause illness. Symptoms include vomiting, thirst and difficulty in swallowing, double vision and constipation. The illness usually starts within 24 h of consumption of contaminated food (although it may be delayed as long as 72 h) and the nervous system is attacked, often resulting in death.

Clostridium botulinum group E and non-proteolytic strains of B and F are capable of growth at 3–4°C although the optimum growth temperature of the organism varies, e.g. 37°C for groups A and B and 30°C for group E. In general, group E and non-proteolytic strains of B and F have minimum growth temperatures 10°C lower than those of A and B strains. The higher the initial loading of cells the faster the growth and toxin production. However, growth at 3·3°C, the lowest recorded growth temperature of group E, is very slow, taking an average of 36 days for growth to occur. The toxin survives freezing but can be readily destroyed by heat treatment (e.g. 5 min at 65°C).

Natural microflora which produce acid tend to inhibit the growth of *Clostridium botulinum* but there is some evidence for synergy between lactic acid bacteria and the organism. Some organisms produce inhibitory factors which can be filtered and shown to stop the growth of *Clostridium botulinum* and others produce proteolytic enzymes which break down the toxin.

As *Clostridium botulinum* is a spore-former it cannot be completely removed from contaminated food by pasteurisation. In other words, the treatments used in cook–chill will not guarantee the removal of the organism. Thus, it is critical to maintain temperatures in chill stores below the known minimum growth temperature of 3·3°C and to use raw materials of the highest quality. Also, as the organism is anaerobic, packaging pasteurised high risk foods under vacuum can invite growth if storage temperature is not strictly controlled, especially as it has been shown that the growth of *Clostridium botulinum* in aerobic conditions is normally restricted by over-growth of spoilage organisms which cause the food to be destroyed whereas these organisms will not grow under vacuum and thus, contaminated food may appear safe to eat.

Yersinia enterolitica and *Listeria monocytogenes*

Yersinia enterolitica is a psychrotrophic enterobacteriaceae which occurs as a diverse group of organisms most of which are harmless to man. However, strains 0:3, 0:8 and 0:9 are pathogenic and produce a toxin even at temperatures as low as 3–4°C. It is for this reason that the organism has recently been signalled as being of importance and risk in cooked, chilled, but more importantly, raw chilled foods. The organism grows slowly at 2°C in the presence of low levels of competing organisms but growth at 5°C is fast, reaching, in one experiment, 10^6/g in vacuum packed bologna sausage within 14 days. The organism has been implicated in cases of gastroenteritis and is associated with animals and meat products.

Listeria monocytogenes is a psychrotroph found in soil, meat animals (especially chickens and sheep) and animal manure which is pathogenic and has been recently implicated in outbreaks of human food poisoning resulting in spontaneous abortions in pregnant women and death in some cases. Although the organism was first described in the early part of this century, it has been poorly understood until recently.

The main pathway for transmission of the disease caused by the one known pathogenic form of the *Listeria* family (*Listeria monocytogenes*), listeriosis, remains unclear. Indeed, the incidence of listeriosis in the UK is relatively low (187 cases in 1986 and 259 cases in 1987) and, apart from a few incidents, there is little direct evidence to link the illness with food. However, the organism has been implicated in outbreaks of food poisoning in Europe and the USA associated with unpasteurised dairy products, particularly soft cheeses. A recent report has suggested that an alarming percentage (18%) of retail cook–chill products might be contaminated with the organism (albeit at very low levels— $<$ 100 organisms/g) and one case of food poisoning in the UK has been strongly linked with consumption of soft cheese.

Listeria monocytogenes is relatively easily destroyed by heat, i.e. pasteurisation of milk reduces numbers to levels which pose no threat to human health. However the minimum growth temperature of *Listeria monocytogenes* is below 3°C and so it presents a particular hazard to cook–chill caterers should food be improperly pasteurised. However, once again, recent evidence has shown that it grows only very slowly between 1 and 3°C with a generation time measured in days rather than hours. Even at 4°C the generation time is measured in days whilst at 13°C it is measured in hours. Two per cent of main course items prepared in a range of UK cook–chill catering sites were shown to be contaminated with low levels of the organism. This is a very low incidence but still should be regarded as totally unacceptable as foods made by such units should be fully pasteurised and protected from re-contamination by good quality assurance procedures (see Chapter 9). Because *Listeria monocytogenes* is easily destroyed by proper pasteurisation, grows relatively slowly at temperatures below 3°C and can be destroyed, if still present in stored chilled foods, by proper re-heating, there should be no excuse for incidents of food poisoning resulting from the consumption of cook–chill foods prepared by caterers who use recognised guidelines such as those issued by the UK Department of Health and Social Security.

MICROBIOLOGICAL HAZARDS OF COOK–CHILL CATERING

A further caution should be added here about the phrase 'proper pasteurisation' in connection with organisms that are known to grow at low temperatures, i.e. psychrophiles. Recent controversy has raged in the UK press incited by various reports of *Listeria monocytogenes* food poisoning and its implications for cook–chill caterers. As the organism grows slowly at low temperatures and, in fact, will colonise cooked food held at 4°C (if it has not previously been properly inactivated by complete pasteurisation), it has been suggested that cook–chill methods pose a particular threat to consumers. A similar situation exists with *Yersinia enterolitica* although it has been shown not to survive an internal temperature of 60–62°C in beef roasts. Thus, it is only by the successful and efficient *combination* of the two methods, pasteurisation and chilling, that assurance of the inhibition of the natural microflora of a food product can be gained.

Many other organisms (mainly spoilage organisms) not only survive at low temperatures but can actually grow in these conditions. Thus, as early as the 1930s it was shown that *Pseudomonas, Lactobacillus, Monilia* and *Penicillium* species could all grow at −4°C and *Cladosporium* and *Sporotrichum* species as low as −7°C. These are good examples of psychrophilic organisms. Refrigeration at 0–3°C would do little to halt the growth of such organisms should they be present on a food product.

Environmental conditions in the food (which acts as the substrate for growth of the microbes) can often affect the growth characteristics of contaminating micro-organisms. Thus, it was shown that the presence of salt was enough to raise the minimum growth temperature of *Clostridium botulinum* type E from 3·3°C under ideal conditions to above 8°C by the addition of 4·5% salt to the growth medium. The acidity of the food also plays a role in determining the minimum growth temperature of any particular organism. It was shown that the growth of strains of *Clostridium botulinum* type E was inhibited even at 15·6°C by a pH of 5·2. Therefore, the conditions of the food itself provide another set of variables which help to determine the viability of different organisms in the food at chill temperatures after cooking (pasteurisation).

Finally, it must be remembered that food-borne micro-organisms, i.e. the natural microflora of raw food, are not the only contaminants in food. Micro-organisms are found in the air, on surfaces and on humans as well as in raw food. Therefore, post-processing contamination of foods by any of these sources can and does regularly happen. It is essential that caterers guard against re-contamination of cooked, chilled products and this will be discussed in Chapter 9. Thus, pasteurisation and chilling can only go part of the way to assuring shelf-life in terms of restricting the growth of food poisoning and food spoilage micro-organisms. For this reason, the shelf-life of such products is obviously limited and is normally much shorter than that of cooked, frozen foods (stored at −20°C) and canned foods (heat sterilised).

FURTHER READING

Azadian, B. S., Finnerty, G. T. & Pearson, A. D. (1989) Cheese-borne listeria meningitis in immunocompetent patient. *Lancet*, Feb. 11th, 322–3.

Bryan, F. L. (1974) Identification of food hazards in food service establishments. *J. Environ. Health,* **36,** 537–40.

Bunch, W. L., Matthews, M. E. & Marth, E. H. (1976) Hospital chilled food service: acceptability and microbiological characteristics of beef soy loaves when processed according to systems procedures. *J. Food Sci.,* **41,** 1273–6.

Cremer, M. L. & Chipley, J. R. (1977) Satellite food service system assessment in terms of time and temperature conditions and microbiological and sensory quality of spaghetti and chilli. *J. Food Sci.,* **42,** 225–9.

Cremer, M. L. & Chipley, J. R. (1979) Time and temperature, microbiological and sensory quality of meat loaf in a commissary food service system transporting heated food. *J. Food Sci.,* **44,** 317–21.

Cremer, M. L. & Chipley, J. R. (1980) Time and temperature, microbiological and sensory assessment of roast beef in a hospital food service system. *J. Food Sci.,* **45,** 1472–7.

Dahl, C. A., Matthews, M. E. & Marth, E. H. (1980) Survival of *Staphylococcus aureus* in beef loaves, potatoes and frozen and canned green beans after microwave heating in a simulated cook/chill food service system. *J. Food Prot.,* **43,** 916–21.

Galbraith, N. S. (1988) Listeriosis. *Brit. Med. J.,* **297,** 754–5.

Gilbert, R. J., Miller, K. L. & Roberts, D. (1989) *Listeria monocytogenes* and chilled foods. *Lancet,* Feb. 18th, 383–4.

Hayes, P. R. (1985) *Food Microbiology and Hygiene*, London, Elsevier Applied Science Publishers.

Lacey, R. & Kerr, K. (1988) The enigma of food-borne listeriosis. *Environ. Health,* **96,** 7–9.

Patterson, J. T. & Gibbs, P. A. (1973) Observations on the microbiology of cooked chicken carcasses. *J. App. Bact.,* **36,** 689–97.

Paulus, K., Piekarski, J. & Leistner, L. (1979) Chilled meals: possibilities and limits. In: *Advances in Catering Technology* (ed. Glew, G.), London, Applied Science Publishers, pp. 375–86.

Pinaga, F., Carbonell, J. V., Valles, S. & Ampara Feria, M. (1979) Pasteurization of ready to serve meals: quality changes during refrigeration of beef goulash in air atmosphere. *Revista de Agroquimica Y Technologiade Alimenta,* **19,** 520–32.

Silla, H. & Simonsen, B. (1985) Shelf-life of chilled, cooked and cured meat products. *Fleischwirtschaft,* **65,** 66.

Snyder, P. O. & Matthews, M. E. (1984) Microbiological quality of food service menu items produced and stored by cook/chill, cook/freeze, cook/hot hold and heat/serve. *J. Food Prot.,* **47,** 876–85.

CHAPTER 4

Cook–Chill Systems

INTRODUCTION

Cooked, chilled foods are produced for both the retail and catering food sectors. They may be made by large food manufacturing companies for sale directly to consumers in supermarkets and stores, by equally large or small specialist manufacturers who supply caterers or stock chilled food vending machines or by caterers themselves for in-house use or for sale to 'outside' customers.

It is important to start with the understanding that the production of cooked, chilled foods (which, for convenience sake, we will refer to throughout this book as cook–chill foods) is a relatively simple, if highly responsible, process which can be operated on large and small scales alike. The former may involve the building of a special production unit or factory with the acquisition of many items of specialised and expensive equipment. The latter could merely involve the purchase of a small blast chiller and a good, reliable chill storage cabinet and their installation in a standard catering kitchen. A recent survey of cook–chill units in the UK highlighted this broad variability (Fig. 4.1).

The application of the cook–chill method to the manufacture of products for the retail food industry is an interesting area which has successfully developed over the last few years. However, although sharing many common features with cook–chill catering, manufacture of cooked, chilled foods for retail purposes will not be directly addressed in this book. Later in this chapter we will consider the different types of cook–chill catering unit and their application but first it is important to remind ourselves of the definition of the process itself.

Cook–chill catering is essentially a food preservation system, based upon chilling, adapted to the needs of the caterer. The process, at its

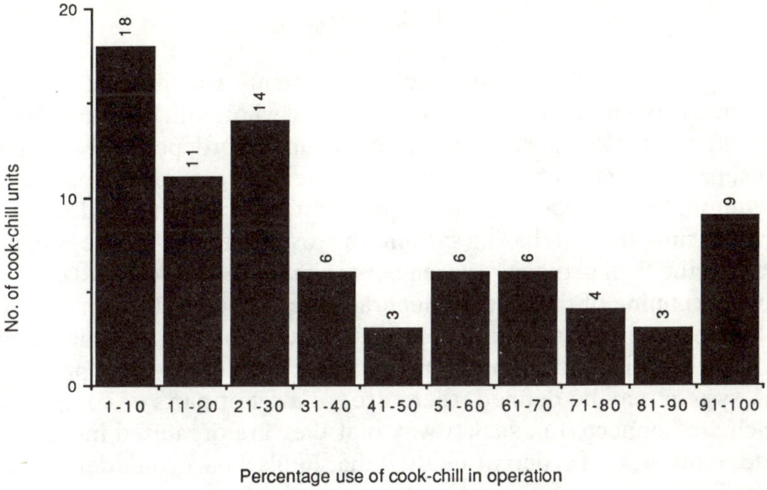

Fig. 4.1. Histogram showing the percentage use of cook–chill in 80 catering units in the UK (from Walker, 1988). Notice that over 50% of the units used cook–chill to provide less than 30% of their total food output. Only 11% used cook–chill for most of their food production (i.e. 91–100%).

simplest involves the: 1, preparation; 2, cooking; 3, rapid chilling; 4, chilled storage; 5, re-heating; and 6, service of food-stuffs. Steps 1, 2 and 6 are identical, in most respects, to the major stages of classical catering (cook–serve); steps 3, 4 and 5 add a new dimension—the time buffer—that allows the rationalisation of food production on a systematic basis. Because of this time buffer, which utilises the holding of food at or near temperatures at which some potentially dangerous micro-organisms can grow (Chapter 3), many writers have stressed that cook–chill contains more inherent risks than other forms of catering. For this reason, guidelines for the use of cook–chill catering have been issued by several authoritative bodies (e.g. government bodies, international health organisations, national food science bodies).

This chapter, therefore, will deal with the general aspects of cook–chill in terms of the theoretical approach, the concepts involved and the actual application. Thus, we will consider the concept of cook–chill as a 'system', the use of chilling as a preservation step in terms of its providing a time buffer in a food service system, models of cook–chill catering, the impact of legislation and quality guidelines on the approach to cook–chill catering and applications of cook–chill in the catering industry.

SYSTEMS THEORY

A systems approach simply requires the treatment of a set of steps, parameters or characteristics as an integrated whole and being concerned with the relationships, structure, function and interdependence of those steps, parameters or characteristics. This means regarding the parts of an operation, be they the staff, the equipment, the buildings and services, the financing, the purchasing system, the raw materials, the steps in production, the final product, as each being vital to the smooth and ever improving running of the whole. General systems theory dates back to the 1940s when scientists tried to formulate a common theory which would allow the identification of problems and offer solutions to those problems.

A system may be defined, therefore, as a set of parts or components which are connected in such a way that they are organised into a complete, inter-related group or method that fulfils a particular defined purpose. The component parts of a system may, each on its own, be viable or complete, but when organised together in the appropriate way, they become a complex whole.

Most systems are not closed but operate in an interactive way with their environment. Thus, a cook–chill catering operation may be able to prepare, produce, store and re-heat foods for service but it has to deal with the outside world in terms of the ordering, receipt and checking of raw materials, reliance on external sources for financing (e.g. banks), arrival of clients and their leaving at the end of the meal. We can construct a simple diagram which illustrates this interaction (Fig. 4.2) but it is the steps within the system 'box' that we are most concerned with here. The system itself or the component parts of the system can be described as a flow chart of operational steps, as in the sub-system of staffing (Fig. 4.3). As can be seen, the feed-back loop, where information about what has occurred is constantly returned, is essential in understanding and controlling the system or sub-system. In the case of staffing, these loops give information on the type of job and its requirements or changes in its requirements, the training needed, or up-dates in training and so on. It is clear that, without the feed-back function staffing issues would soon become a problem. The same is true, of course, of process steps in food production.

In catering, the component parts of a system include the equipment which is designated for specific jobs, the flow of work and work organisation and the people and their skills. All of these parts of the system are in-

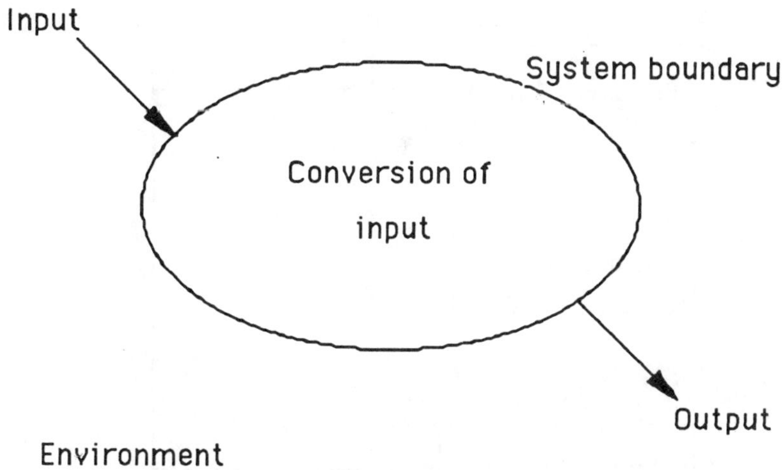

Fig. 4.2. A simple flow diagram illustrating the interaction between a 'system' and its environment.

volved together in taking the raw materials, the food components, and transforming them into the final product, the meal that the consumer eats. The connections between the parts are the organisational control and monitoring of the process, the information which assists and guides that control and monitoring and the flow of food through the production process.

COOK–CHILL AS A SYSTEM

Certain words and terms have come into use over the years that cook–chill has evolved and, before embarking on a brief overview of the system, it would be worth defining a few of the more common and important ones. The *central production unit* (CPU) or *central kitchen* is the area/building where food is prepared, cooked, chilled and, generally, stored before being distributed to end kitchens. The *end kitchen* or *satellite kitchen* is the kitchen to which cooked, chilled food is distributed for final re-heating and service. *Regeneration* is the re-heating of cooked, chilled food before service. A *blast or rapid chiller* is a special device (not a standard refrigerator) for rapidly chilling food, especially hot cooked food, to temperatures between 0 and 3°C. The *chill store* is a room or cabinet designed to

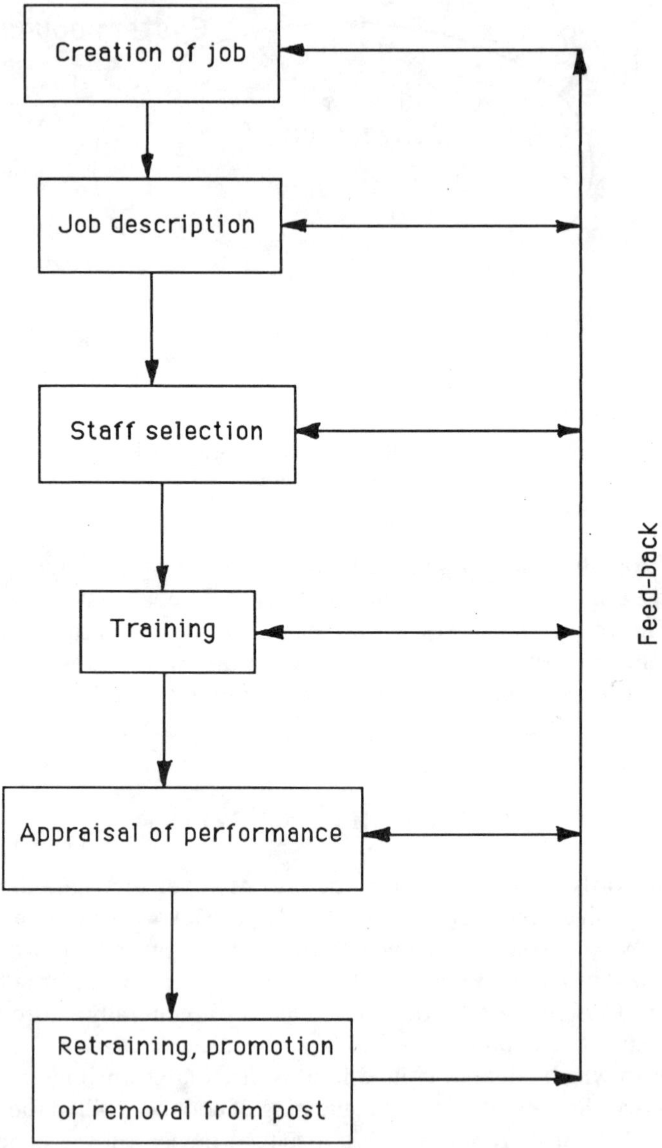

Fig. 4.3. Flow diagram of staffing requirements in an organisation illustrating the importance of feed-back loops.

hold chilled food at 0–3°C until needed for use—this is not a refrigerator and standard refrigerators must not be used for this purpose. Chill stores work to much higher specifications and are normally fitted with accurate temperature monitoring and recording devices (Chapter 5).

In recent years the phrase 'systems approach' has become popular in catering. In cook–chill catering this means adopting a systematic methodology to food production; that is, recognising the various parts of the whole and organising them in a way which leads to optimal efficiency and output. A systems approach can be used in setting up the cook–chill unit, i.e. in bringing together a set of requirements and physical objects, such as the logical layout of a kitchen area to allow maximal efficiency, no cross-contamination of raw and cooked foods, optimal flow of foods through the kitchen and the best possible use of equipment and space. Similarly, as we have seen above, a systems approach can be used to describe the steps of the production process itself. This is a key factor, as we shall see later (Chapter 9) in proper quality assurance.

It is useful to begin by considering the conventional catering unit as a system, i.e. the unit in which food will be freshly cooked and served as soon as it is ready. Figure 4.4 shows the flow diagram of such a system.

In the conventional catering unit, the food is received and stored in some way prior to service. However, to simplify our view of catering as a system, this step can be assumed for the moment. Thus, the raw or pre-prepared food ('convenience' foods—e.g. frozen or dried foods like frozen peas and beans, dried milk or starch thickeners, etc.) is prepared in the kitchen for cooking. The next step involves the cooking of the food and finally, the cooked product is served. This is a simple system which can be adapted to various forms of application from the gourmet restaurant, hotel dining room and coffee shop through to the fast food outlet and the snack bar. As the food flows through the system, information relating to its control and its quality (see Chapter 9) flows back to help maintain standards, timing, quantities and so on.

In the simplest version of the cook–chill system (Fig. 4.5) one more step is introduced into the traditional format laid out in Fig. 4.4, i.e. the time buffer. In other words, a method of storing the cooked food in conditions designed to reduce to a minimum the deleterious effects of storage (in this case chilling), is used to allow some flexibility of production and service of the food. Later we shall consider some of these potential flexibilities. The revised system is not radically different from the classical catering system described by Fig. 4.4. However, a new dimension can be added to the flow diagram shown in Fig. 4.5, as illustrated in Fig. 4.6. In

Fig. 4.4. Simple system diagram of a conventional catering operation.

this case the place of service is not assumed to be the same site at which the food was produced. Now, we have to add to the time buffer box the word *distribution*. In this sense the cook–chill system described by Fig. 4.6 is radically different from the diagram in Fig. 4.4 in that the use of the cooked food is not limited to the site of production. This evidently opens up new horizons for the caterer in terms of planning and optimisation of production.

Also inherent in the revised system diagram is a new dimension of control and monitoring of the process. In terms of food quality, i.e. organoleptic quality and safety (the minimising of microbiological hazard), the caterer must exercise a far greater degree of control in the cook–chill system than in the conventional system. This is facilitated by use of the kind of feed-back of control and monitoring information that is illustrated in Figs 4.2–4.6 and the organisation of the cook–chill unit as a system. Figure 4.7 shows how this is achieved by means of expanding the

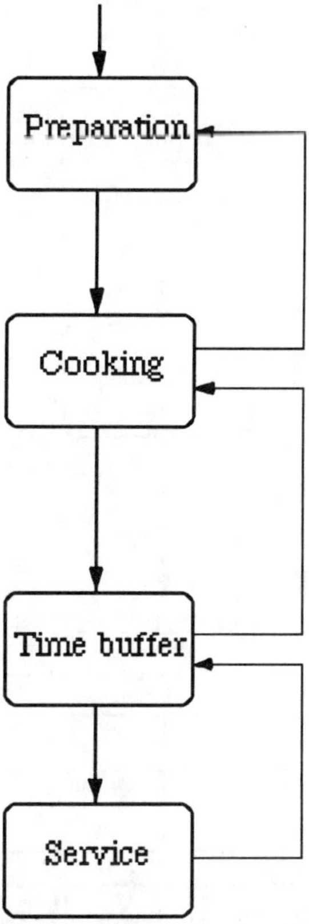

Fig. 4.5. Simple system diagram of a cook–chill catering operation illustrating the addition of the time buffer.

simple diagrams shown in Figs 4.5 and 4.6. The system starts with a well organised store for dry goods, frozen food and chilled foods. The raw materials are then passed on to the preparation area where they are made ready for cooking. The next step is the cooking process which is followed immediately by chilling and portioning. The cooked, chilled products are then immediately sent to chilled storage where they may be held a specified period of time before they are transported in the chilled state to end kitchens which may be on the same site or geographically distant from the production kitchen. At this point they are finally re-heated and served.

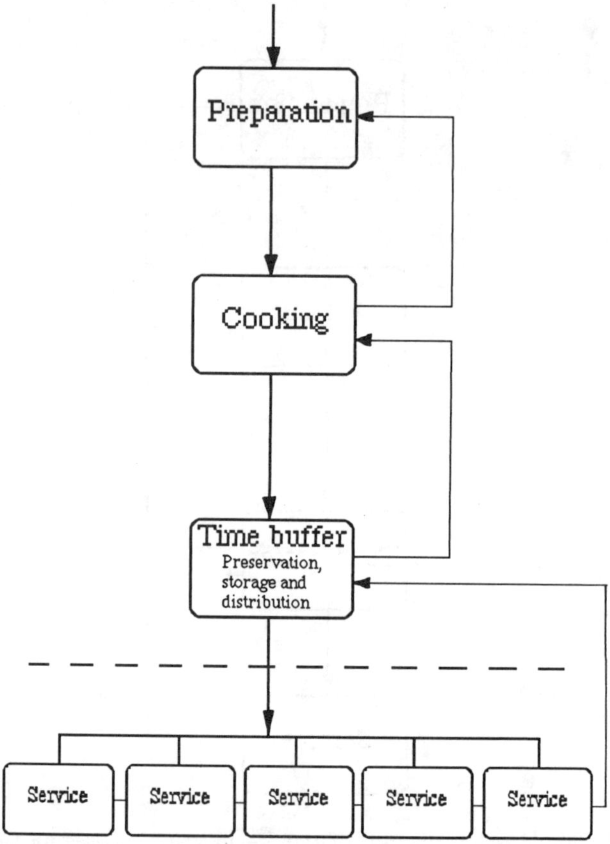

Fig. 4.6. System diagram of cook–chill catering incorporating the distribution element.

THE COMPONENT PARTS OF THE COOK–CHILL SYSTEM

An inseparable part of this system is the equipment needed to carry out the component parts of the process. This is summarised in Table 4.1. For storage of raw materials a good dry store, adequate freezer space and proper chill stores are required. A high quality preparation area with relevant equipment is also needed, as are sufficient cooking facilities. Various ovens and cooking systems have been designed to optimise the cook–chill method and these are discussed in Chapter 5. The two indispensable parts of the cook–chill system and which give it its uniqueness

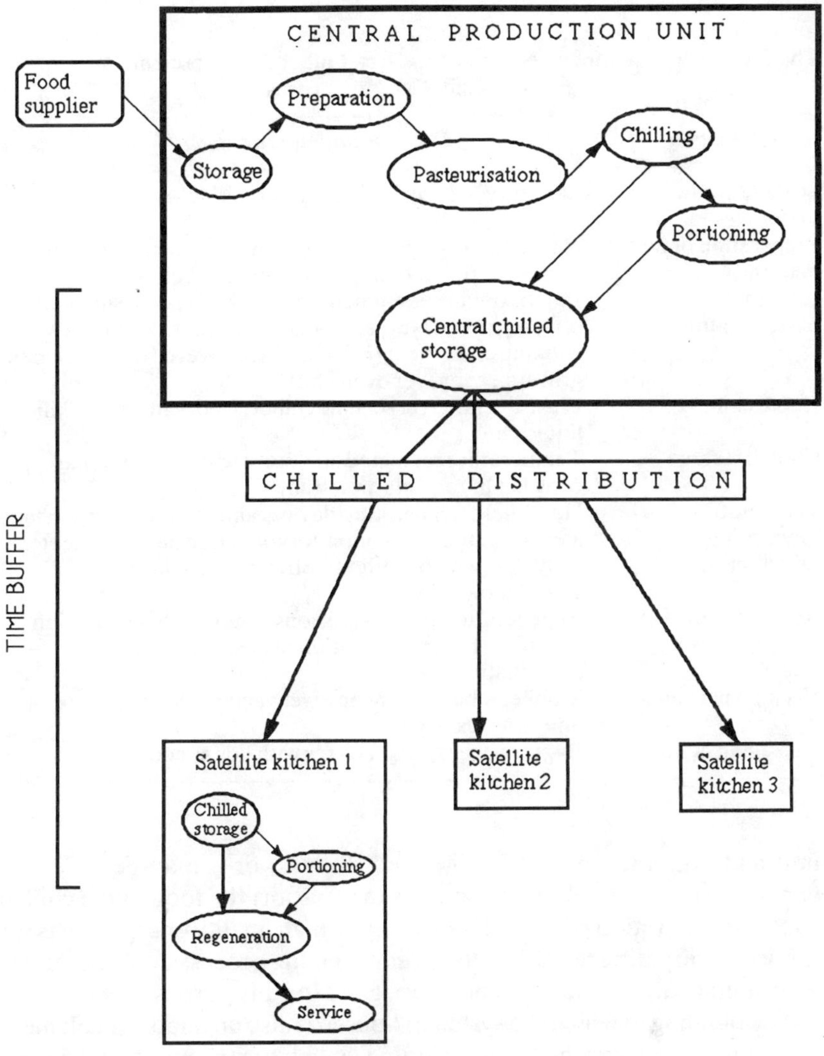

Fig. 4.7. System diagram of a complete cook–chill catering operation incorporating purchasing, initial preparation and regeneration.

are the rapid chiller and the chill store. Both must work to high specifications as will be shown later. Finally, two other unique parts of the cook–chill process are the transport system and the regeneration system. The former may operate within the same building as the central production

TABLE 4.1
The Types of Equipment Needed to Carry Out the Component Parts of the Cook–Chill Catering Process

Operation	*Equipment needed*
Storage of raw materials	Dry store, freezer, refrigeration cabinet/room, chill store
Preparation of raw materials	Normal food preparation equipment. Vacuum packing device for pre-cooking packaging (see below)
Cooking/ pasteurisation	Batch cooking equipment (ovens, bratt pans, steamers, boilers, grills, fryer, combination ovens); continuous cooking equipment (cook-tanks, conveyer fryer/steamer/ grill etc., conveyer oven etc.)
Rapid chilling	Blast air chiller, cryogenic chiller, iced water bath chiller, liquid chiller
Chilled storage	Temperature regulated chill store with read-out, hard copy facility and alarm system
Temperature monitoring	Hand-held temperature devices and various food probes as well as automated monitors on requisite equipment
Distribution	Vehicles (with or without refrigeration), insulated containers
Regeneration	Regeneration trolleys, various conduction, convection or radiant heat regeneration ovens, microwave, combination oven
Vacuum packaging	Mobile, table-top or conveyer vacuum device with sealing unit as necessary
Packaging materials	Bags, pouches, trays, boxes with lids as necessary

unit and take the form of refrigerated trolleys, or it may be a fleet of vans or lorries with refrigeration units to transport the food in the chilled state. Regeneration of chilled foods to give high quality end-products for service is very different from the prime cooking process and, therefore, dedicated heating units have been designed for this purpose.

The building in which the system is housed is also an important element and components such as services, drainage, lighting, proper goods-in and goods-out areas, well-designed clean areas for steps such as portioning, adequate and well-placed changing rooms for staff and so on are all integral to the success of the system.

The final elements in the cook–chill system are the staff through and by whom the production and control and monitoring processes are operated. In any food preservation system it is of paramount importance to ensure the safety of the food in terms of microbiological contamination and, of course, if the final product tastes poor it will be rejected by the consumer. Considerations of good hygienic practice and the evolution of a system, to be implemented by staff, to assure food safety as well as food quality, therefore, complete the cook–chill system. These aspects will be considered in full in Chapter 9.

The other component that the personnel of the cook–chill system contribute is, of course, the traditional skills of the good caterer as well as some new skills that have been learned by cook–chill operators over the years. These also will be discussed in later chapters.

THE USE OF A TIME BUFFER

The concept of food preservation, i.e. the time buffer, in catering arose primarily through the need to produce quality food products in advance of need in a highly reproducible way and the necessity of providing maximal assurance of product safety. Thus, in food manufacturing, the maxims and precepts of large scale, systematised food production grew up from infancy in the use of simple dehydrated products (dried fruits and meats) and cured foods (salted meats and fish) to the use of technologies in the bottling and then the canning industries. Subsequently, with the introduction of reliable refrigeration systems, the frozen cooked food sector began to grow.

The usefulness of cooked, frozen foods was recognised in the catering industry in the 1960s and earlier but some began production of such foods themselves. With the realisation that chilled foods also resisted corruption by micro-organisms (albeit for a far shorter time than frozen foods) and did not lead to the undesirable effects of freezing on texture, flavour and even colour (see Chapter 2), this move was soon followed by the equally rapid growth of the use of cooked, chilled food products in catering. In the early 1960s and 1970s, this new method was often taken up and implemented in an *ad hoc* way. In the last few years, however, with the experiences of early users (and the food manufacturing industry) and the potential hazards associated with the method being more widely publicised, caterers have begun to adopt a far more systematic and rational approach in cook–chill catering.

The extra stage or time buffer in food preservation systems used in catering aims to alleviate the problems of inflexibility inherent in the classical cook–serve system and the cook–hot-hold system (which is known to result in damage to the quality of the food—see Chapter 2) whilst maintaining optimal food quality. Amongst the advantages offered by such a time buffer, used correctly, are the flexibility to cope with changes in demand, ease of maintaining high levels of quality control, a larger range of dishes available, the use of manufacturing techniques to ensure product quality and rationalisation of the operation, better utilisation of equipment and labour leading to higher productivity, better portion control leading to less waste, standardisation of food products, bulk buying leading to economies of scale, more sociable working hours for staff, more even work load during the day and more time for staff to use culinary skills.

Self-evidently, this is an impressive and perhaps daunting list of claims and it would be foolish to proceed without acknowledging that these benefits can only be achieved by close adherence to the principles of good cook–chill catering practice and the building in of hygiene and quality control (see Chapter 9). The management benefits are set into context and discussed more fully in later chapters.

Similarly, there are disadvantages in the cook–chill system when compared to other methods; cook–chill food production is not a panacea and brings with it its own set of problems and pitfalls. Amongst its disadvantages are the fact that on removal from the time buffer, the food has to be regenerated before service, a risk to public health if the system is not managed properly, loss of product quality during storage, problems with the regeneration of a wide range of food products, additional investment in rapid chillers, chilled storage, transport and regeneration equipment, possible resistance to change from staff who see new systems as demanding less skill and eroding jobs and, obviously, the social problem associated with job loss.

LEGISLATION AND QUALITY GUIDELINES

It is acknowledged that there are hazards associated with storing cooked foods prior to sale or service, therefore, various organisations around the world, including governments and international technical and advisory bodies, have laid down specific guidelines and, in some cases, legislation relating to the practice of cook–chill catering. Most experts agree that the ideal conditions under which a cook–chill unit should be run adhere very closely to these guidelines and laws and, therefore, it is important to consider the system in the light of these before progressing any further. In the UK, government advice on technical aspects of the cook–chill system of catering is contained in the *Guidelines on Pre-cooked Chilled Foods* (1980, amended 1983) issued by the Department of Health and Social Security (DHSS). In the USA similar guidelines are issued by the Food and Drug Administration (FDA), whereas in France the veterinary authorities have the say in this regard. The international bodies which lay down codes of practice in relation to food microbiology and food manufacturing, the Food Hygiene Committee of the Codex Alimentarius of the World Health Organisation (WHO) and the Food and Agricultural Organisation (FAO) of the United Nations Organisation are currently working on a draft code of practice on 'Mass Catering' which includes cook–chill and cook–freeze catering systems. Similarly, the Institute of Food Science and Technology (UK) has also issued its own guidelines on the matter.

In the UK the DHSS guidelines are not law although legislation on general practice in the food industry is in place to underpin them. They are intended to ensure that any cook–chill system will be safeguarded against the risk of food poisoning and will provide as nutritious a product as is possible within the limitations of the system. In the UK, the principal statute governing the safety and quality of food is the Food Act (1984). The main provisions of the Act as they apply to the safety of cook–chill meals are:

Section 1, which makes it an offence to prepare or sell injurious food.

Section 2, giving general protection to the consumer against the sale of food that is not of the nature, or of the substance, or of the quality demanded by the purchaser.

Section 8, which deals with the sale, or offer or exposure for sale, or possession for the purpose of sale or preparation for sale, etc. of food that is intended for but unfit for human consumption.

Section 16, which requires premises used for the preparation or manufacture of preserved food intended for sale to be registered with the local authority. Preserved includes preservation by cooking. This section does not apply to premises used wholly or mainly as catering premises. Catering premises means premises where, in the course of a business, food is prepared and supplied for immediate consumption on the premises. Where a cook–chill system involves the preparation of the food in a central kitchen that does not have on-site facilities for the consumption of meals that establishment might have to be registered for the preservation of food by cooking in the same way as any food factory.

Other regulations in the UK, which are relevant to cook–chill catering are the Food Hygiene (General) Regulations 1970 and the Food Hygiene (Markets, Stalls and Delivery Vehicles) Regulations 1966. These regulations are general terms but in relation to catering they do make specific requirements regarding the temperature of all food consisting of meat, fish, gravy or imitation cream, or food prepared from or containing any of those substances of any egg or milk (with certain exceptions).

At present there is no legislation specific to cook–chill catering or preservation systems used in catering in the UK. Usually, specific legislation on food processing has only been made when epidemiological evidence has indicated that there is a strong need for such legislation. Examples are the Ice Cream (Heat Treatment) Regulations 1959 originating in the 1940s following outbreaks of food poisoning and food-borne disease in earlier years and the Liquid Egg (Pasteurisation) Regulations 1963 which followed episodes of salmonella food poisoning associated with contaminated imported liquid egg. There has, at the time of writing, been no report of an outbreak of food poisoning associated with a cook–chill catering system in the UK.

The *DHSS Guidelines on Pre-cooked Chilled Foods in Catering* were produced by a working group and were first published in 1980. These have now been up-dated and revised and were published in 1989. The guidelines were produced in 1980 largely because technological advances, particularly in relation to the rapid chilling of food, had rendered the system practicable and safe provided it is properly controlled and operated in accordance with the recommendations of the guidelines.

The DHSS guidelines, then, like others of the same sort, provide a rigid and safe framework within which the cook–chill caterer can design his or her system. There are some differences between the DHSS guidelines and those issued by other organisations (e.g. in the shelf-life allowances and the exact cooking temperatures) but these are generally concerned with finer points. All of the guidelines recognise the potential hazards of the cook–chill system if not operated properly and within adequate control. They provide, therefore, the springboard for the setting up of all good cook–chill systems.

A MODEL OF COOK–CHILL CATERING

The use of cook–chill technology in catering, especially for large scale food production (more than, say, 1000–2000 meals per day) should be regarded as food manufacture rather than catering. Although, as we have discussed above, the methods used in cook–chill catering share many areas in common with classical catering skills and techniques (e.g. food preparation, prime cooking and presentation) the introduction of the chilling and re-heating technology which allows storage of food for use at a later date, the batch production methods, the rationalisation of shift work and the distancing of the central production unit (where the food is cooked and chilled) from the client make the system considerably different in many respects from conventional catering. For these reasons and, more importantly, for reasons of hygiene and quality control, cook–chill catering should be viewed as a food manufacturing technology.

A survey of cook–chill in the UK indicated that, at the end of 1985, there were approximately 240 cook–chill installations in the United Kingdom in a wide variety of sectors from hospitals to staff catering in large companies (this survey will be more fully discussed in Part 2 of this book). It is likely, given the interest in the technique and the growth in the number of equipment companies and consultants specialising in cook–chill catering in the 1980s, that this number is now considerably higher. There seems little doubt that, world-wide, the method is already very widely used, especially considering its popularity in the travel catering sector, e.g. airline meals and so on. The units producing cook–chill meals can vary in size from kitchens producing only a few meals a day to those involved in providing food for thousands of people a week. Obviously, the differences in scale and approach involved in this broad range of units are extensive but all share common ground in terms of the methods used and the objectives of the system. We can, therefore, lay the foundations of a better understanding of the cook–chill system by discussing the basic principles of the method.

At the outset, it would be worthwhile considering the main points raised by the DHSS guidelines as these provide the working framework for good practice in cook–chill catering as discussed above. The main recommendations contained in the guidelines and which lay down a good procedural guide for the cook–chill system are:

- i the food should be given an adequate cook initially to a recommended centre temperature of 70°C;
- ii chilling should commence within 30 min of leaving the cooker and the food should be chilled to below 3°C within a further period of 90 min (joints should preferably reach 3°C within 150 min—max. recommended size 3 kg);

iii storage should be at a temperature between 0 and 3°C;
iv distribution should be under controlled conditions to minimise any rise in temperature;
v re-heating should follow immediately upon removal of the food from chilled conditions and should raise the temperature to at least 70°C;
vi the food should be consumed as soon as possible after re-heating;
vii if the temperature of the food does rise, a tolerance of up to 5°C is acceptable for short periods of time. If the temperature exceeds 5°C but not 10°C the food should be consumed within 12 h. If the temperature exceeds 10°C the food should be regarded as unsuitable for use and discarded;
viii the food should be consumed within 5 days (including the day of cooking and the day of consumption);
ix the microbiological guidelines should be used when setting up new installations or establishing new resources to provide necessary assurance that a satisfactory standard is reached.

The temperature tolerances laid down above are designed such that all chilled meals should be stored at or below 3°C. Some people have misinterpreted the guidelines as meaning that if the food is to be consumed within 12 h it may be stored between 5 and 10°C. This, of course, is not correct and must be recognised as a very hazardous practice.

In the guidelines there is a strict performance requirement for the rapid chiller. It should be capable of reducing the temperature of a 5 cm layer of food from 70°C to 3°C or below in a period not exceeding 90 min and when fully loaded.

We can now propose a simple model of cook–chill catering built upon these recommendations. Because the cook–chill system uses refrigeration at 0–3°C as the food preservation step which provides the time buffer for food storage the pieces of equipment used to provide this step are a rapid chilling device, a chilled storage room or cabinet (operating between 0 and 3°C) and a method of refrigerated transport and distribution which ensures that the temperature of the food does not rise unduly (i.e. above 3–5°C) before re-heating (regeneration) at the end kitchen. Thus, before setting out procedures there is a requirement for the correct equipment which will work to specification.

Having provided the equipment, the first requirement of the system is that the food should be properly pasteurised before chilling. In other words, even though the food may not be thoroughly cooked in terms of eating quality, the prime requirement is that full pasteurisation (Chapters 2 and 3) to the centre of the food has taken place, ensuring the death of all viable vegetative organisms. This can be generally achieved by heating to a central temperature of greater than 73–75°C but it is crucial to understand that it is the relationship between the temperature and time of treatment that leads to full pasteurisation (see Chapters 2 and 3). The food must then be rapidly chilled to ensure that it passes through the temperature ranges in which rapid growth of micro-organisms can occur, i.e. 10–50°C, as quickly as possible. As we saw in Chapter 2, this also ensures the maximal retention of sensitive nutritional factors like vitamins. This means that the rapid or blast chiller must be built to specifications that allow such rigorous chilling to occur. As we shall see in later chapters, not all blast chillers may be relied upon to perform to such specifications and most vary in their performance dependent on the food type and load. Three main systems exist for chilling: standard blast chilling using pre-cooled air, chilled and driven by electro-mechanical means; cryogenic chilling using gaseous nitrogen derived directly from liquid nitrogen and therefore at very low (sub-zero) temperatures; and iced water bath chilling, utilising direct contact with ice cold water. These three are discussed and compared in Chapter 5. As described above, there are specifications for the maximum time of chilling laid down in various guidelines for cook–chill catering and in the UK, this time is 90 min from cooking temperature to less than 3°C.

The storage temperature for the cooked food is set at 0–3°C because, although some organisms grow slowly at this temperature, the major food poisoning organisms which might survive pasteurisation do not show appreciable growth at these temperatures (Chapter 3), thus the chill storage unit must operate within these set specifications. Recent controversy over the likelihood of food poisoning from *Listeria monocytogenes* in cook–chill foods stored at 0–3°C is discussed in Chapter 9.

Products can be stored at 0–3°C for a set maximum time (which may vary from product to product) and so a careful management system must be inaugurated to monitor products through the process so that storage times for each individual product are always known. Products must be removed from storage when needed, transported to the site at which they are needed and then regenerated to temperatures above 70°C at the centre of the product before service. During the transportation of the food its temperature should not rise above 3°C and it should be re-heated as soon as possible after delivery at the satellite kitchen. If food temperature does rise above 3°C during transport the food should be re-heated and consumed immediately.

All equipment should have temperature read-outs and recording devices which enable the user to constantly check the performance of the unit. This is especially important in rapid chilling and chill store devices. Thermometers should also be available and used for checking the core temperature of cooking food. It is essential in cook–chill catering to check the temperature of chillers and chilled storage units or rooms at least daily and preferably during every production run and to have a continuous record of temperature to enable checks to be made out of working hours.

Figure 4.7 summarises this model of cook–chill catering. Whilst Fig. 4.7 describes the basic requirements of a cook–chill catering system and, therefore, provides a sound foundation from which to build an ideal cook–chill unit, it must be stressed that adaptations are common. Many systems incorporate additional factors and developments, e.g. vacuum packaging or modified atmosphere packaging steps (Chapter 6), the use of freezing and chilling side-by-side, which lead to a more complex process than shown in Fig. 4.7.

APPLICATIONS OF COOK–CHILL CATERING

Cook–chill has been applied in almost every sector of the catering industry. These include the five main categories of:

(1) hospital catering (in the UK, the National Health Service);
(2) education and welfare catering (universities, colleges, schools, old-peoples' homes, childrens' homes, social services units);
(3) the hotel and leisure sector (hotels, restaurants, fast food outlets, coffee shop operations, holiday camps, outside caterers);
(4) industrial catering (company-run canteens, contract catering in industry);
(5) in-flight catering.

In each sector cook–chill catering has been applied in a variety of ways. These include conventional cook-chill catering for 'mass' catering (service at large canteens/restaurants), occasional use in banqueting situations, specialised service, e.g. at ward level in hospitals, in fast food catering, as vended foods. It should be noted that catering on board aircraft is something of a special case. In-flight catering was one of the first areas in which mass preparation of chilled meals was used and today this branch of the catering industry is a major user of the cook–chill methodology. However, the technology and application of the method is similar in principle to standard methods and so in-flight catering will not be dealt with separately in this book.

To a large extent the system grew up, especially in the UK, in the institutional sector and it was in this sector, in the early and late 1960s, that experiments were tried in Sweden and the US with different versions of the cook–chill system utilising packaging methods to increase the quality of the products, lengthen the safe shelf-life of the products and allow more flexible control with higher assurance of hygiene and safety (see Chapter 7). In hospitals and the social services sector cook–chill provides a method of assuring reproducible quality in food-stuffs with maximum flexibility in terms of production and menu planning. However, the nutritional status of food for the sick, young and elderly (as in hospitals, schools, old people's homes and 'meals-on-wheels') and the increased risk of these groups to food-borne disease has led, in recent years to a body of criticism of the method. It is argued that cook–chill cannot provide food of adequate nutritional value for these groups and that the hazards associated with the storage of pasteurised food at chill temperatures are too high to warrant its use in these sensitive sectors. These issues were dealt with in Chapters 2 and 3 but it should be recognised that, at present, they remain contentious.

The application and value of pre-prepared meals in the leisure sector is self-explanatory and a number of top hotels now rely on cook–chill or one of its more sophisticated varients (see later chapters) to cope with the pressure of large functions such as banquets. It is extremely doubtful whether many consumers at such occasions are aware that the food they are eating may have been cooked up to four or five days beforehand. Indeed, the popularity of cook–chill meals in retail food outlets (see Table 4.2) is proof of the acceptance of such goods. In fact, retailed cooked, chilled foods in the UK are well priced being, in general, only slightly more expensive than their frozen counter-parts. Table 4.2 shows that chilled meat and fish dishes were only 13% more expensive than similar frozen meals in July 1989.

After cook–chill catering had been initiated in the UK by National Health Service caterers and for use in school meal production, its value as a system of mass catering began to be recognised in the industrial sector. As explained in Chapter 1, during the recession of the early 1980s, its use began to escalate and proliferate to all of the sectors of the industry

TABLE 4.2
Examples of Prices per Gram of Frozen and Chilled Foods in UK Supermarkets (July 1989)

Product	*Preservation method*	*Price per gram (£)*
Vegetable lasagne	Chilled	0.40
Brocolli mornay	Chilled	0.36
Onion pakora	Chilled	0.30
Pasta in tomato sauce	Chilled	0.33
Vegetable samosas	Chilled	0.54
Mean price for chilled vegetable dishes		*0.39*
Tagliatelle carbonara	Chilled	0.41
Beef bolognese	Chilled	0.55
Moussaka	Chilled	0.48
Beef stroganoff	Chilled	0.52
Beef hotpot	Chilled	0.44
Sweet and sour pork ribs	Chilled	0.55
Chilli con carne	Chilled	0.32
Chicken tikka masala	Chilled	0.74
Beef lasagne	Chilled	0.33
Cannelloni di carne	Chilled	0.38
Meat samosas	Chilled	0.54
Chicken vindaloo	Chilled	0.50
Coronation chicken	Chilled	0.64
Pasta with chicken	Chilled	0.77
Pasta with tuna	Chilled	0.70
Mean price for chilled meat/fish dishes		*0.53*
Beef curry	Frozen	0.37
Cannelloni	Frozen	0.27
Glazed chicken	Frozen	0.83
Beef provencale	Frozen	0.68
Kashmiri chicken curry	Frozen	0.73
Toad-in-the-hole	Frozen	0.49
Beef madras curry	Frozen	0.34
Chicken curry	Frozen	0.37
Ham and cauliflower cheese	Frozen	0.45
Spaghetti bolognese	Frozen	0.39
Shepherds pie	Frozen	0.26
Cod and prawn pie	Frozen	0.44
Mean price for frozen meat/fish dishes		*0.47*

discussed above. An early application was in situations where consumers needed meals out of normal working hours, e.g. during night shifts in industry. In these cases, full-meal vending was often used as a method of storing and dispensing cook–chill foods.

Cook–Chill Catering in Full-Meal Vending

Full-meal vending needs a little more explanation, perhaps, than some of the other areas already discussed. It is an automated food service system in which individual food items or complete meals are stored in a chilled food vending machine (FVM) and are purchased by the consumer who, if necessary, can re-heat the food in a microwave oven. Vended cooked foods have been available for some time although their popularity has been fairly low. Packaging requirements for cooked vended foods in the USA were described as early as 1963 and, in the UK, one of the first chilled food vending systems accompanied by microwave re-heating was opened in 1966. As the potential of this 23·5 h/day system (0·5 h/day are needed for cleaning and loading the machines) has been realised, significant growth in the full-meal vending industry has been seen although it still accounts for only a low percentage of the catering industry as a whole. In addition, full-meal vending represents only a small proportion of the total vending industry in the UK. For example, in a survey conducted in 1982, only about 2% of the population who had used vending machines had ever purchased hot or cold meals through a vending machine and only 3% of that population had purchased sandwiches.

Although changes have been made in the design of some FVMs, a recent study of the temperatures in such machines showed considerable fluctuation, e.g. −1°C to +16°C. If food products are not sufficiently heat processed before chilling and storage or are contaminated after cooking the risks of growth of food poisoning and/or food spoilage organisms are greatly increased.

A survey was carried out of a number of UK sites operating vending services as a part of their catering system. The results of the survey showed that full-meal vending is seen by managers as a convenient form of food service but there were indications that some managers were not aware of the limitations in terms of shelf-life of some of the cook–chill products they were using. This survey concluded that there was room for considerable improvement in the service offered by the chilled food vending industry in the quality, presentation and marketing of foods.

Another survey, which looked at the operating temperatures of chilled food vending machines, showed that there was considerable fluctuation of temperatures in different parts of many of the machines surveyed and that large temperature variations could be observed in working machines at different times of the day. It was not uncommon for machines to exhibit temperatures over 10°C and sometimes as low as −1°C. Different machine types showed variability in maintaining low storage temperatures. It was shown that some machines (45% of those studied) were unable to maintain the temperature in parts of their cabinets between 0 and 5°C for even as much as 50% of their working time although much of this variation was ascribed to consumer use (i.e. through the opening of cabinet doors), as laboratory trials showed that one type of machine was able to maintain an average cabinet temperature of 4·4°C±1·4°C. In view of the known growth behaviour of pathogenic micro-organisms it was considered that there was a definite risk of food poisoning from chilled vended foods on the basis of the large temperature fluctuations observed in this study.

Similar data on the performance in the field of rapid chilling and chill store devices used in main-stream catering is not available. However, consideration of these surveys of the chilled food vending industry highlight the problems and potential hazards of the method when used improperly and lead to the question of proper hygiene and food management which will be discussed in Chapter 9.

NOTE

During the time that this book was in press the Department of Health and Social Security in the UK issued their revised guidelines on cooked, frozen and chilled foods in catering. Although there was no time to revise the text of this book accordingly, the pertinent points of the revised guidelines are included in Appendix 4.

FURTHER READING

Armstrong, R. (1986) Cook–chill catering. *Environ. Health,* **94**(10), 253–5, 257–8.

Collison, R. & Johnson, K. (1980) A general systems theory approach to catering. In: *Advances in Catering Technology* (ed. Glew, G.), London, Elsevier Applied Science Publishers, pp. 327–34.

Cooper, H. (1987) Experts defend cook–chill. *Hotel & Catering Technology,* July/Aug., 14.
DHSS (1980) *The Guidelines on Cook–Chill Catering.* HMSO, London.
Fennema, O. R. (1975) *Principles of Food Science: Part II Physical Principles of Food Preservation,* New York, Marcel Dekker.
James, S. J. & Bailey, C. (1985) Principles and problems in cooling foodstuffs. In: *Proc. Chill Foods,* 7–9th May, Stratford-upon-Avon. Chipping Campden, UK, Campden Food Preservation Research Association.
Mobbs, C. C. (1986) Cook–chill production. *Hospital Caterer,* Sept., 20, 22.
Paulus, K., Piekarski, J. & Leistner, L. (1980) Chilled meals: possibilities and limits. In: *Advances in Catering Technology* (ed. Glew, G.), London, Applied Science Publishers, pp. 375–86.
Sheppard J. (1987) The big chill. A report on the implications of cook–chill catering for public services. London Food Commission. Report no. 15.
Wadey, C. (1987) What future for cook–chill? *Caterer and Hotelkeeper*, 30th July, 16.
Walker, A. (1988) The transfer of technology: A study of UK cook–chill catering operations. PhD Thesis, Dorset Institute, Poole, Dorset, UK.
Young, H. (1986) Factors affecting the shelf-life and acceptability of cooked, chilled foods. PhD Thesis. Dorset Institute, Poole, Dorset, UK.
Young, H., Youngs, A. & Light, N. D. (1987*a*) A qualitative survey of the use of chilled vending machines as a means of food service in the UK. *Food Sci. & Technol. Today* **1**(4), 247–51.
Young, H., Youngs, A. & Light, N. D. (1987*b*) Operating temperatures in chilled food vending machines and risk of growth of food poisoning organisms. *Food Sci. & Technol. Today,* **1**(4), 252–6.

CHAPTER 5

Equipment and Application of the Technology

INTRODUCTION

In Chapter 4 we considered the fundamental aspects of the cook–chill catering system. It is evident from this introduction to the method that it is reliant on a number of well-defined and strongly related steps, each of which is associated with a specific process. Each of these processes (i.e. cooking—or pasteurisation, thermal processing, thermalisation; production—portioning, rapid chilling, chilled storage, chilled transportation, satellite chilled storage and re-heating), is carried out using specific pieces of equipment.

In the early days of cook–chill catering much of the equipment used in the process was traditional machinery adapted to the purposes of the newly designed cook–chill catering system. Even now, to some extent, because the uptake of the technology has been rather haphazard in this somewhat traditional industry, there is a tendency for caterers to associate their needs with the more traditional items of kitchen equipment such as bratt pans.

These are not necessarily the wrong pieces of equipment for cook–chill, rather they can supplement the more specialised items of machinery. However, there has been a trend towards the production of instrumentation which has been specifically designed with cook–chill in mind. This reflects the recognition by catering equipment manufacturers and by caterers alike that the cook–chill catering system is far more akin to food manufacturing than to catering as it has been known in the past.

Indeed, it could be argued that, because of the inherent risks involved in cook–chill catering and the large scale on which food can be prepared and distributed, the method should be treated as a food manufacturing system and not a catering system and relevant machinery, controls and good manufacturing practice (GMP—see Chapter 9) are all essential elements of the system. This assumption would seem to be an excellent starting point for any aspiring cook–chill caterer but a major survey of cook–chill catering in the UK carried out in 1986 showed that more than a large minority of caterers were using cook–chill on a very small scale (see, for example, Fig. 4.1). Thus, many cook–chill users cannot afford large scale equipment and would have no application for it. For small scale cook–chill catering units the adaptation of existing preparation and cooking equipment is, therefore, sometimes a necessity.

Unfortunately, we do not have time to review the huge range of traditional and new standard catering equipment in these categories. This chapter will deal, therefore, with the specialist equipment involved in the cook–chill catering process, i.e. the different system elements of the cook–chill process including large scale cooking (pasteurisation), portioning, rapid chilling, chilled storage, chilled distribution, re-heating (regeneration) and manufacturing control systems. Equipment which is recommended, required or useful for each part of the process will be discussed in turn.

It is inevitable that, given the changing nature of modern technology, specific items of equipment will become quickly outdated; therefore, only the general principles of each piece of equipment will be discussed in terms of the job done by that item, its design and structure, its controls and its advantages and disadvantages.

FOOD PREPARATION AND COOKING

Figure 5.1 shows a diagrammatic representation of the work flow in a standard cook–chill catering operation. In a traditional catering operation (i.e. cook/warm-hold/serve) one of the main requirements of skilled staff is in the choice and preparation of raw food materials. This is not always the case in cook–chill catering. Operations may use the head chef, the catering manager or even a purchasing officer to lay down specifications for raw materials and to choose and buy food. Because the emphasis is on the manufacture of a set of menu items in bulk and to a reproducible standard, a fair proportion of raw materials may be purchased as pre-prepared goods. Thus, much of the preparation equipment of the classical kitchen may not be required.

As can be seen in Fig. 5.1, goods received, the raw materials, may be stored in any one of four ways, i.e. dry store, frozen storage, chilled

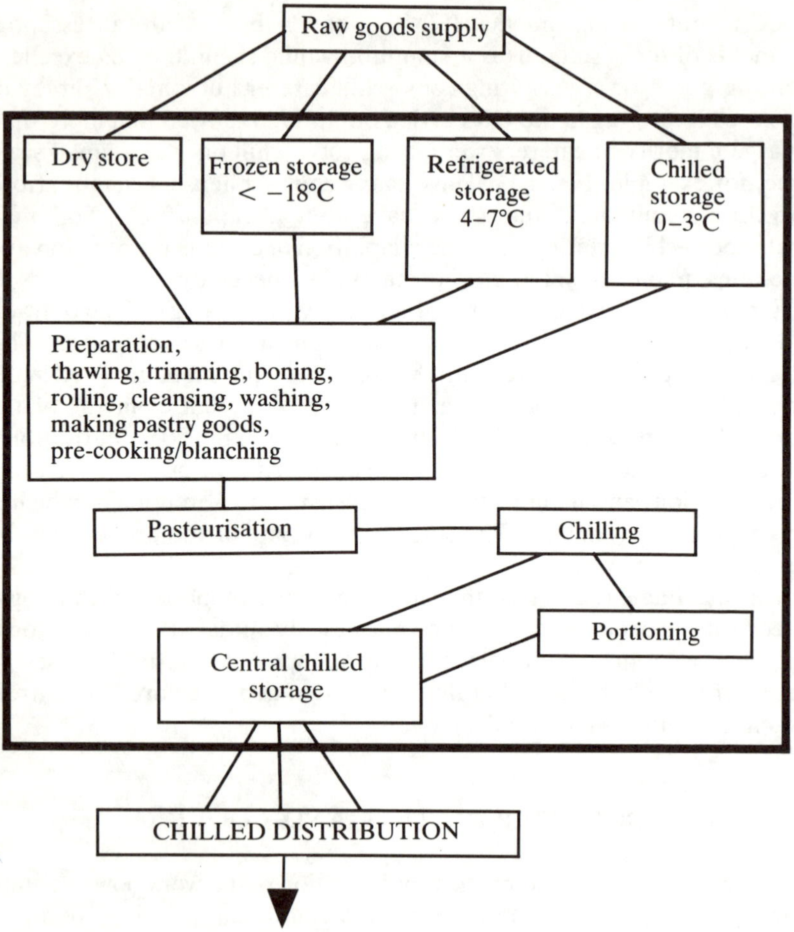

Fig. 5.1. Stylised work flow for a standard cook–chill catering operation.

storage (0–3°C) or refrigeration. In the latter three cases food may have to be prepared before cooking, e.g. trimming, boning, rolling meat; preparation of fish; washing, trimming, chopping vegetables; making pastry, rolling pastry, beating eggs. Raw materials from the dry store are more likely to be ready prepared (e.g. canned foods, dehydrated products) as are many of the frozen products. The same should be possible with many of the chilled (stored at 0–3°C) and refrigerated (stored in a refrigerator, i.e. 3–7°C) goods.

Cooking Equipment

Once the food materials have been assembled and prepared the first step in the cook–chill process is to cook them. It is important to stress several points at this stage. As we saw in Chapter 2, there are several good reasons for cooking food and in terms of preservation the most important of these are:

(a) to change the texture of the food;
(b) to produce flavours;
(c) to inactivate enzymes which might damage the food during storage;
(d) to kill or inactivate food spoilage organisms;
(e) to kill or inactivate food poisoning organisms.

However, for the purposes of cook–chill catering and the optimal retention of quality throughout the storage period, there are certain other requirements which need to be considered.
These are:

(1) Pasteurisation is required in all thermally processed products.
(2) However there is not necessarily a need to thoroughly cook all products before storage, for example, some products, if fully cooked, might become over-cooked and 'stodgy' or 'limp' after re-heating prior to service.
(3) Complex products may need to be assembled after cooking the separate elements individually, e.g. pie fillings and crusts.
(4) Some new methods may require some pre-cooking or blanching of some food components, e.g. in *sous-vide* cooking (see Chapter 8).

What is clear about all thermal processing carried out during a cook–chill operation is the strong requirement for maximal control, particularly of the critical parameters of time and temperature.

In some large scale modern cook–chill catering operations there is a trend to utilise food manufacturing style equipment for thermal processing. Much of this equipment is programmable and self-monitoring with time and temperature settings and recording, cut off and/or alarm devices. There has also been a limited uptake of integrated systems which allow total thermal processing, i.e. pasteurisation and chilling in the same device. However, more traditional pieces of equipment are still common in the industry so that bratt pans, boiling pans, steamers and other items may often be found alongside the more technological breed of equipment. In what may be regarded as 'standard' ovens (dry heat or steam combination) foods are placed into standardised rectangular metal cooking trays (the gastronorm), which are then often loaded onto trolleys designed to take the trays on open shelves and which may be wheeled direct to chilling devices. Many substitutes for metal gastronorms now exist in the form of disposable units and even heat sealable units. Such items of equipment and accessories have a place in the cook–chill catering operation but they will not be discussed further in this book as they are adaptations of existing technologies about which much has already been written.

The use of food manufacturing style cooking equipment in the cook–chill catering industry is increasing. There are four main categories of this type of machinery; cook-tanks for pumpable materials, combined cooker/chillers, continuous (conveyer style) thermal processing equipment and units in which pre-packaged products (generally vacuum packed) are cooked and chilled. These latter devices rely on the concept of in-container pasteurisation to obtain longer shelf-life for products than other methods and include such systems as batch steaming/water heating and cook/chill tanks. These types of equipment can be used to cook a range of vegetable dishes, rice and pasta, meat joints and composite dishes but, as they rely on the relatively recently founded principles of pre-vacuum packaging of the food they will be dealt with later (Chapter 7) when large scale specialist methods will be discussed in more detail. The other three categories of cooking device, however, will be considered here.

Cook-Tanks

This type of machinery is composed of a kettle or cook-tank with some form of continuous, controllable stirring or agitation device for mixing the food within the tank (Fig. 5.2). The food, generally, liquids (e.g. soups, sauces) or semi-solids (e.g. meat pieces in sauce like bolognese, chicken à la king), is continuously heated and mixed within the tank until it has reached the required temperature for the set time (i.e. it has been pasteurised). The tanks vary in shape but are usually hemispherical or cone-shaped, lidded and are either directly heated or heated through a steam jacket. Agitators and 'scrapers' within the tank allow constant mixing and, in some cases, folding of the food.

These devices are normally fitted with controls for the regulation of agitator speed control, temperature, a simmering facility, timer and

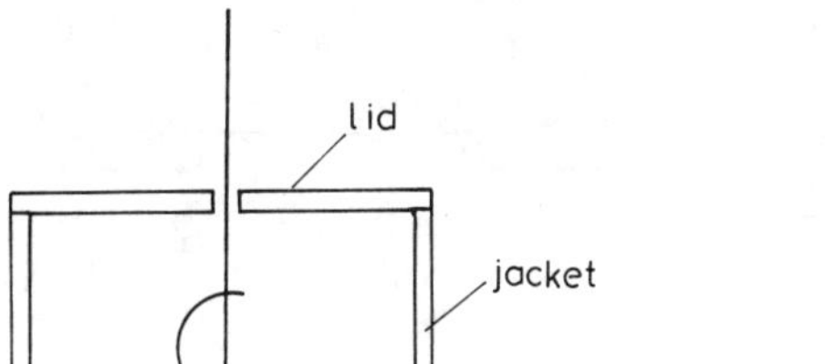

Fig. 5.2. Simple diagram depicting the principle of the cook-tank system.

pumps, allow pre-programming of the pasteurisation cycle and, once loaded and pre-set, do not require attendance or checking. They also usually form a part of a complete system and are fitted with or linked to a pumping station which can pump the cooked food from the kettle to a container (or set of containers, e.g. plastic pouches or bags, trays, etc.). The portioned (packaged) food is then rapidly chilled and stored.

The main advantages of this system are the programmable production methodology and the fact that (it has been claimed) minimal cooking skills are required to operate the devices. The size of food pieces that can be handled by the pumping system is about a 25 mm cube but this is not a major restriction on use.

Combined Cooker/Chillers

Current equipment of this type has a capacity in the range 30–50 litres but it is likely that larger scale equipment will be available in the future. These

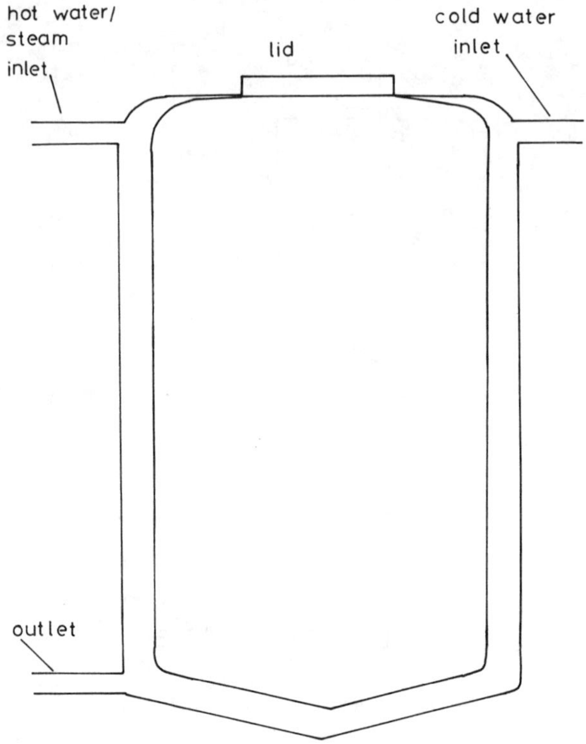

Fig. 5.3. Simple diagram depicting the principle of the combined cooker/chiller system.

items consist of jacketed tanks in which the food is both cooked and chilled by control of the water temperature within the jacket (Fig. 5.3). Existing models require the food to be removed manually but it seems likely that future developments will also include pumping systems as in the cook-tank system. The units are normally used for soups and sauces but, as in the case of cook-tanks, they can be used for meat-in-sauce dishes providing that the size of meat pieces is small enough to allow adequate pasteurisation during thermalisation. Cooking and chilling in one tank reduces handling and movement of the food but the present units are rather small and the transfer of food for portioning is laborious and could be open to microbial contamination because of its manual and multi-step nature.

Continuous (Conveyer Style) Thermal Processing Equipment

Steamers, grills and fryers have been adapted for conveyer style cooking on a continuous production basis (as have microwave cookers although these will not be considered here). The principle of the equipment is simple. Food is loaded at one end of a continuous belt, carried through a cooking chamber (open or closed) suited to the type of cooking required and is un-loaded at the end of the cooking cycle (Fig. 5.4). For example, continuous fryers have a moving metal belt which is loaded with products at the start. The belt carries the food down a slope into a pre-heated deep fat (oil) unit. As in all types of continuous device, the belt speed is adjusted to give the required time in the cooking medium, in this case the oil. In the case of the steamers, cold water is usually sprayed over the vegetables or other products to stop the cooking process as the food leaves the thermal processing chamber.

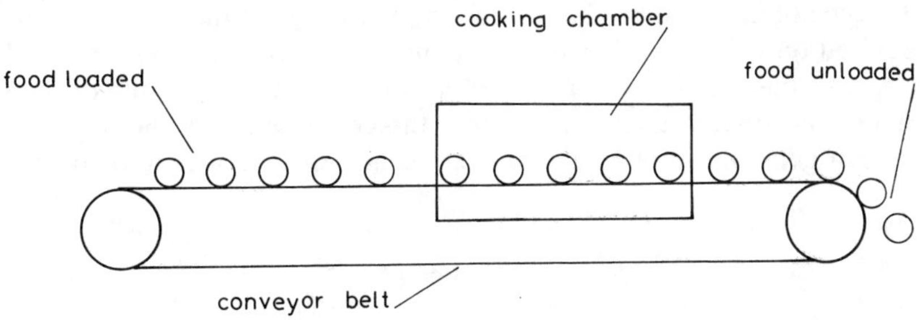

Fig. 5.4. Simple diagram depicting the principle of the continuous (conveyer style) thermal processing system.

Obviously, given the kind of continuous thermal processing equipment available, the types of product which can be cooked using this process include battered fish and other deep fried menu items, many vegetables including potatoes, rice and pasta.

PORTIONING

Portioning is the process by which bulk cooked or stored foods are split down into defined units of size, either single person portions or multiples of the single person portion. Obviously, this process is important in terms of ensuring the correct and reproducible size of portions (which has implications for customer satisfaction and cost control) but it is also important with regard to hygiene and quality assurance (see Chapter 9) because, in the past, portioning has always involved handling by operatives i.e. it has been a manual operation. Automatic or semi-automatic machinery is available to help rationalise portioning procedures. Such equipment should reduce the risk of food contamination during portioning and should also increase the degree of control over portion size/weight.

The types of equipment that fall into this category are filling and container-forming devices, dedicated pumping devices for use with cook-tanks, vacuum pressure transfer systems and vacuum/modified atmosphere packaging systems.

Filling and Container-Forming Devices

This type of machine produces containers, formed to the shape and size required on-line or pre-formed containers on a conveyer system which supplies a filler station at which food is pumped or deposited into the containers and, thereafter, sealed within the containers after lidding (Fig. 5.5). In general, systems which incorporate formers are used for pre-

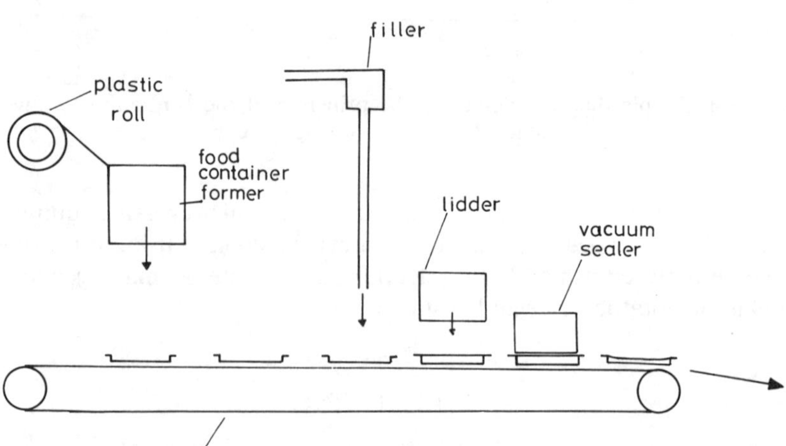

Fig. 5.5. Simple diagram depicting the principle of the continuous container-forming, filling, lidding and sealing system.

chilled or cold products whereas those which utilise ready made containers are applied to the packaging of hot products. The technology is from the food manufacturing industry and is highly flexible, generally being adaptable to any one single user's needs. The incorporation of such devices requires production line type techniques common in the food manufacturing industry but has the advantage of allowing systematic process control (see Chapter 9) and removing, to the greater extent, the human factor.

Dedicated Pumping Devices

In the last section we discussed cook-tanks or kettles which require emptying devices, i.e. pumps (see Fig. 5.2). Some of these are supplied as simple pumping stations which can be adapted for portioning under full manual control but many come complete with bagging or packaging facilities. This type of machine comprises a food pump which empties the cook-tank and a bagging unit which allows the controlled filling of bags or pouches with pumpable foods. The filling unit is also equipped with a clipping or similar sealing device which allows the bag, once full, to be sealed and trimmed. Most units also incorporate a labelling system for the packaged products.

An advantage of this system is that foods at pasteurisation temperature can be bagged and sealed, thereby reducing the risk of microbial recontamination. Most systems do have to be manned, however, and there are inherent risks in handling foods at very high temperatures (over 80°C). However, recent developments include the investigation of fully automated units which would deposit a pre-programmed amount of product into a bag, clip or otherwise seal the bag which is then trimmed before passing the portioned food onto the next stage of the process.

Vacuum Pressure Transfer Systems (VPTS)

VPTS is a variation of the simple pump unit described above with the sophistication of vacuum transfer and in-line chilling (Fig. 5.6). Here, the food is transferred to a holding tank from a cook-tank by a vacuum pump, i.e. it is sucked out of the kettle, through a heat exchanger in which the food is chilled as it passes along the pipework of the system. By controlling the vacuum, the rate of transfer, and thus chilling, can be altered. The chilled food is then pushed out of the holding tank under pressure

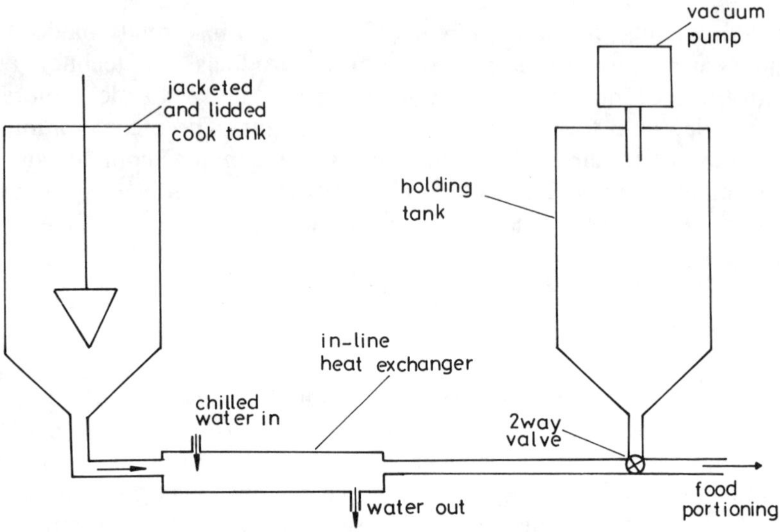

Fig. 5.6. Simple diagram depicting the principle of vacuum transfer and in-line chilling.

into the required portioning system (bags, pouches, trays, sealable boxes).

The main advantages of this method are that it can be fully automated with food passing from a pre-cooked to a chilled, packaged state without exposure to the external environment and with minimum, if any, handling.

Vacuum/Modified Atmosphere Packaging Systems

In Chapters 6, 7 and 8 we will discuss the theory and practical applications of the vacuum and modified atmosphere packaging methods in full. Therefore, the background and definitions of these methods will not be covered at this stage. Also, although these methods may be regarded as part of a portion control system, they are not portioning methods *per se*. Indeed, a pumping or transfer system of some sort is needed to remove the food from the cooker, chiller or combined cooker/chiller and thus, it is this transfer system combined with its controls which should be seen as the true portioning system and should not be confused with the packaging methods of vacuumisation and modified atmosphere packaging (MAP).

CHILLING TECHNOLOGY

As we have seen in earlier chapters, one of the key points in the cook–chill process is the rapid chilling of the cooked food below the effective minimum growth temperature required for the majority of food spoilage and food poisoning micro-organisms. The major requirement of any chilling system is that it should be able to chill a variety of food types (e.g. fried battered fish, meat-in-sauce, roasts, high density products like mashed potato) from cooking (pasteurisation) temperature to below 3°C within a set time (e.g. 90 min is recommended by the UK *DHSS Guidelines on Pre-cooked Chilled Foods* (1980)). What appears to have become clear in recent years is that no one method can be comprehensively guaranteed to cover all needs and thus a combination of methods is being adopted by many cook–chill caterers.

The chilling rate of the food is dependent on many factors which include the thermal conductivity of the food, the initial (cooked) temperature of the food, the density and moisture content of the food, the presence or absence of a lid on the food, whether or not the food is pre-packed in plastic bags and the size and weight of the food units. The integrated methods which allow both heating and chilling in the same device have already been discussed (see above) but, apart from these rather specialised applications, three main methods are used for rapid chilling; these are air-blast (electro-mechanical) chilling, cryogenic chilling and iced water bath chilling. Other, less widely used, methods are vacuum chilling (as discussed briefly above) and cold water spraying within the cooking cabinet (see Chapter 7). Obviously, cryogenic and air-blast chilling (which tend to be cabinet-based methods) can be used on unpackaged and even unlidded products but iced water bath chilling normally requires the food to be pre-packaged.

Air-Blast (Electro-Mechanical) Chilling

This type of machine is usually constructed as a cabinet and food to be chilled is placed inside (often on trays—gastronorms—stacked on pre-spaced rungs in a trolley). Pre-cooled air, sucked into the machine and cooled using a compressor, is pumped into the cabinet at high velocity and (sometimes being deflected by baffles) around the inside of the unit so that it blows more evenly over the hot food, bringing about the most rapid cooling possible. Various sizes of machine are available and range from pilot plant units (which hold 3–35 kg of food) up to tunnels which will hold a number of trolley loads of food. As with the generation of cryogenic cabinets (see below), air-blast chillers are supplied as standard units but also may be constructed on site to the operators' specifications. Many are also designed so that they lead from the production area, through a wall or barrier to the portioning room which is often held at reduced temperature (e.g. 10°C)—see Fig. 5.7.

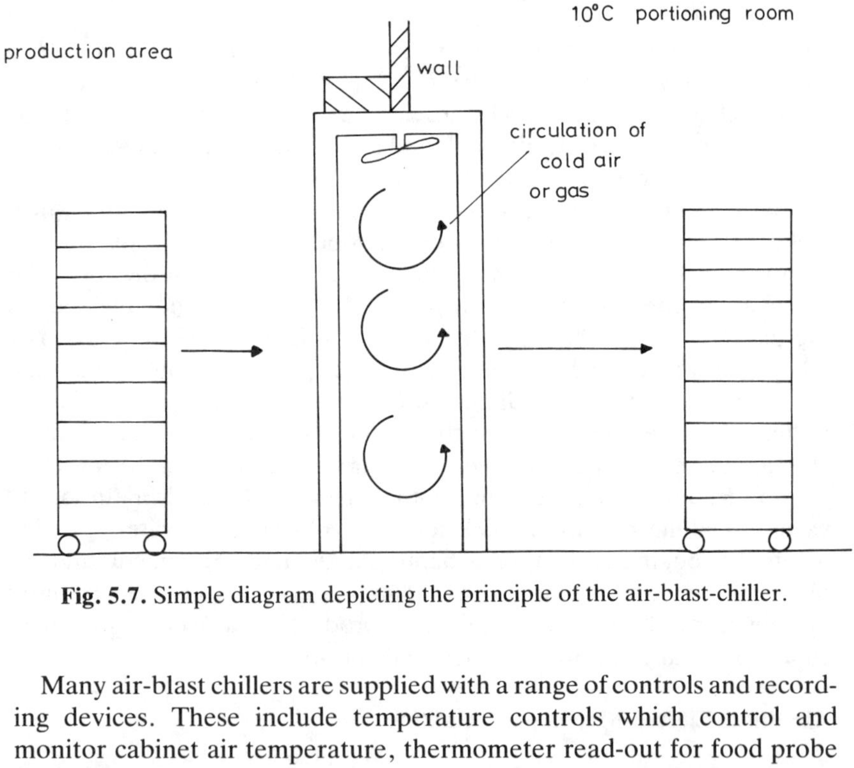

Fig. 5.7. Simple diagram depicting the principle of the air-blast-chiller.

Many air-blast chillers are supplied with a range of controls and recording devices. These include temperature controls which control and monitor cabinet air temperature, thermometer read-out for food probe thermometers, automatic defrost systems with defrost signal lights, pre-programmable chill cycle control, time control systems and alarm systems to signal when the device is faulty or not operating. Most systems also offer, but are not necessarily automatically provided with, 24 h/7 day recording systems which allow constant monitoring of the device. This and the above controls are, as in the cooking systems, vital if proper process control is to be implemented and, thereby, quality assured (see Chapter 9).

Cryogenic Chilling

This form of rapid chilling is roughly analogous to air-blast systems as it normally utilises a cabinet in which the food is chilled although this is simpler than the air-blast chillers having few mechanical parts but, rather, a system of pipes for circulating cold gas around the food stacked in the cabinet. In this case though, instead of air being the circulating gas, nitrogen (N_2) or carbon dioxide (CO_2) are used with the former being by far the more popular. The nitrogen gas is supplied in liquid form (in which state it has a temperature of below −196°C) and as this liquid boils the resultant gas (at −196°C) is pumped over the food in a manner similar to that described above. Carbon dioxide is supplied as a solid and, for food chilling, is converted to a gas which carries a certain amount of the solid as powder ('snow') at about −79°C. As with air-blast chillers, cryogenic cabinets are capable of taking fully loaded trolleys as already mentioned.

The installation cost of cryogenic systems can be high as can the running costs if the size of the unit is too small to make economies of scale. The systems also require an on-site storage plant for the liquid nitrogen which is delivered regularly to the site. Cryogenic systems are generally not recommended in operations under a certain size because of the relatively heavy capital investment needed and the fact that the method is orientated towards larger scale systems. However, because it is a relatively simple system with few moving parts, it rarely goes wrong and maintenance costs are low.

It is also a very efficient method and, because of the low temperature of the gas, very fast. It can also be used as an in-line system for chilling or freezing products coupled with a conveyer belt from an on-line cooking device (Fig. 5.8). It should be stressed that this versatility in the method, i.e. that it can easily be used for both chilling and freezing, is seen by many as a major advantage. It means that a combined chill/freeze operation can be set up with only one piece of equipment for temperature reduction. However, it should be pointed out that the same can be said of air-blast freezers which can also be used to chill foods under carefully controlled conditions. Some cook–chill caterers prefer air-blast freezers used in this dual role but these machines are generally much slower than cryogenic systems.

Some difficulties were experienced on the introduction of the method into the catering world, especially 'crust' or surface freezing. The very low temperature of the gas and inadequate control systems allowed some surface freezing of foods during the chilling process leading to unac-

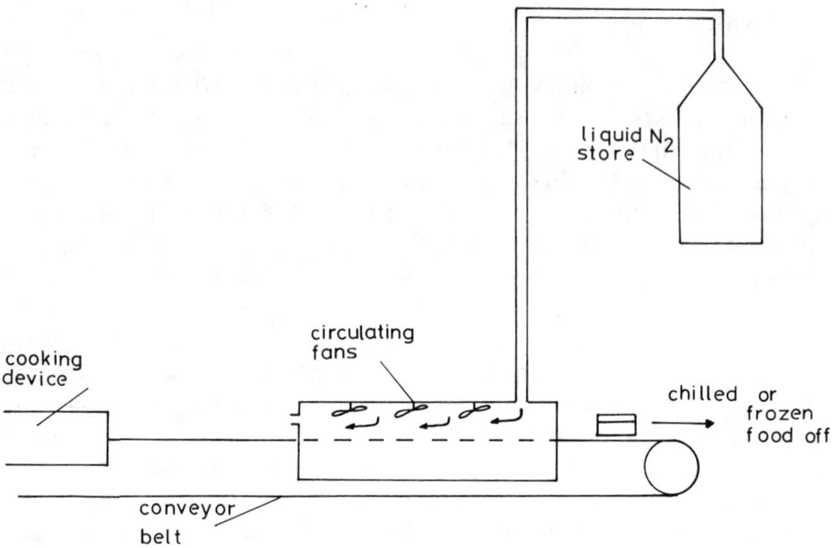

Fig. 5.8. Simple diagram depicting the principle of production line liquid nitrogen chilling.

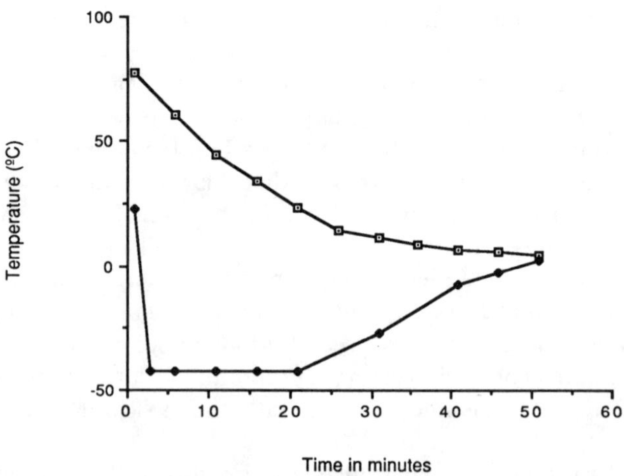

Fig. 5.9. Graphical illustration of the principles of ramp control. (□), Temperature of a food product in the cryogenic chiller; (◆) temperature in the cryogenic chiller cabinet.

ceptable textural changes. This problem has, it is claimed by the manufacturers, been eradicated by the use of sensitive controls which allow the temperature of the cabinet to rise as the food temperature falls (ramp control)—see Fig. 5.9.

Control systems on cryogenic chillers include cabinet and food temperature read-outs, power light, alarm systems to signal the end of a chilling cycle, fan control light. In addition, most of these devices feature a microprocessor control panel which allows pre-programming of a certain type and length of chilling cycle. Many machines are supplied with up to six or more pre-programmed cycles in the memory of the device and in some machines these cannot be over-ridden by manual control.

Iced Water Bath Chilling

In this method food can be very rapidly chilled by the simple expedient of immersing it in ice cold water or brine at sub-zero temperature (−1°C). The method has been shown to be up to four times faster than air-blast chilling but it is somewhat specialised as it requires the food to be pre-packaged to avoid the risk of post-pasteurisation contamination. There is evidence that some operators use cold water systems to chill unpackaged cooked vegetables but this can be risky. A microbiological study of such a practice showed that the immersion baths used (which were supplied direct by unchilled mains water with a temperature of 10–15°C) harboured colonies of psychrophilic bacteria (see Chapter 3) which contaminated the food and led to very high bacterial counts at the end of chilled storage. Although the problem was solved by re-designing the water bath shape and the implementation of regular anti-bacterial cleaning, the practice obviously carries certain risks which may be considered unacceptable by the majority of practioners and food microbiologists.

Iced water bath chilling is used, almost exclusively, for the chilling of pre-packaged cooked foods. As has been described above, pumpable foods cooked in cook-tanks or kettles are often chilled in this way and specialised systems for cooking foods in vacuumised packs also use the same method (see Chapter 7). Most ice bath chillers operate on the same principle, i.e. heat conduction from hot foods pre-packaged in plastic bags and immersed in ice cold water. Two different systems exist. The first uses a specially designed paddle device which agitates the food packed in bags (i.e. the bags are massaged and kneaded during the chill cycle) to maximise heat transfer. The second method provides for the rapid movement of the water around the system and the use of a heat exchanger to retain the low temperature of the coolant water. Another modification of the system utilises the conveyer principle so that cooked food emerging from a packaging plant on a production line is transferred directly to the water chiller where it is immersed for a pre-set time to ensure complete chilling. The controls on iced water bath chillers are much the same as those described above for air-blast and cryogenic chillers and many are fully programmable.

CHILLED STORAGE

Chapters 2 and 3 discussed at some length the importance of maintaining chilled food at a constant low temperature (between 0 and 3°C and, if possible at 0–2°C) in order to inhibit or, at least, minimise microbial growth during storage—a topic that will be taken up again in Chapters 8 and 9. It goes without saying then, that all cook–chill units require first class chill storage units to enable such control. It cannot be stressed enough that these are not refrigerators (which generally operate to much looser tolerances and have average temperatures some 2–5°C higher than proper chill stores). Thus, any caterer who uses such devices, even for short term storage of cook–chill foods, is taking unacceptable risks.

Cook–chill foods can be stored either in cold rooms or chill cabinets. Both utilise enclosed areas which are cooled by compressor units and both are built to specification to retain cold air at a temperature between 0 and 3°C. Cold rooms are often built as part of a continuous system with chillers or cold portioning rooms leading directly into them. Cabinets are available as 'roll in' types to take food on trolleys, free standing or on wheels. Many cook–chill operators utilise banks of cabinets which together can have the capacity of a large cold room but which allow the 'roll-in-roll-out' principle to be used.

All systems must be adequately fitted with controls which should include:

(a) A continuous recorder of some kind which prints out a record of the temperatures maintained within the room or cabinet through each 24 h period. As the most likely time for growth of organisms within the food is during chilled storage it is absolutely vital to be able to know the continuous performance record to the chill store. Should the temperature of the store (and food) rise above 5°C for any amount of time it may be necessary to dispose of the food stocks.

(b) An alarm system for notifying staff if the temperature of the unit falls below or rises above given set levels or if power fails to the unit.
(c) A continuous dial or digital real-time read-out of temperature of the store (and, if possible, a typical product) for visual monitoring.

DISTRIBUTION

Chilled foods may be stored at the central production unit (CPU) before distribution, immediately distributed to satellite kitchens with their own high quality chill stores or stored for a short period at the CPU prior to shipment to end kitchens with high quality chill stores. In each case, there is a need to maintain the temperature of the chilled food between 0 and 3°C during transport to the end kitchen. In order to achieve this reliable and appropriate transportation systems are required. These may be insulated or chilled mobile cabinets for on-site distribution (e.g. from a central kitchen to wards in a hospital), insulated containers for off-site deliveries, insulated vehicles or trailers, refrigerated vehicles or trailers. Another effective method for short-time transport is to place frozen eutectic plates in insulated containers with the food.

Insulated boxes are either front or top loading and generally incorporate runners for the easy loading of trays. The boxes are usually formed from high density toughened plastic, aluminium or stainless steel and have a double wall construction with an insulating material filling the cavity thus formed. Many are fitted with wheels for ease of handling and they come in a range of sizes. Mobile units for use in hospitals and other large institutions tend to be of this type although combined insulated trolleys/regeneration units can be obtained for this purpose (see next section). Obviously, such systems are only applicable in 'on site' situations and have found use in other cook–chill locations such as hotels for banqueting.

Insulated and refrigerated vehicles are self-explanatory and will not be discussed in detail here. However, one of the major problems associated with such systems is the increase in temperature caused by the constant opening and shutting of the doors of vehicles during deliveries. Barrier systems such as heavy duty plastic 'curtains' should be fitted inside the permanent doors of vehicles to limit such rises, and systems should be rationalised to minimise the opening of doors on such vehicles.

It should be stressed that distribution is a key part of the cook–chill process and cannot be considered as an outrider. The conditions used for distribution must match up to the needs of the system, i.e. the food temperature must not rise above 5°C (or, better, 3°C) during this part of the 'cold chain' especially as some distribution networks can involve hundreds of miles and, possibly, hours on the road. Again, it should be axiomatic that proper control and monitoring systems are incorporated so that food temperature can be accurately followed during the transport process.

REGENERATION

Prior to regeneration (or re-heating) foods may be stored at the end kitchen for some time, as mentioned above. Obviously, such chill stores should be of as high a standard and have as efficient a set of controls as those at the CPU. Finally, the chilled food must be re-heated before service. Systems designed to carry out this function need to be quite specialised as the food must be adequately re-heated to properly kill any vegetative forms of micro-organisms that may have grown in or on the food (e.g. at least 70°C at the centre or core of the food as stipulated by the UK *DHSS Guidelines on Pre-cooked, Chilled Foods,* 1980) while at the same time it must not be over-cooked or dried, thereby destroying or, at least, minimising the organoleptic quality of the food. Regeneration, then, represents a fine line between safety and acceptability and, as such, must be carefully controlled.

The four most common pieces of equipment used for the regeneration of foods from cook–chill systems are; bulk regeneration ovens, 'combination' (low pressure steamers/dry heat) ovens, mobile regeneration trolleys and microwave ovens.

Bulk Regeneration Ovens

These ovens are designed to be loaded with trays (gastronorms) of standard size, constructed from steel, aluminium foil, ceramic, or disposable plastic coated, often pre-loaded on a trolley. The units are generally designed to rapidly re-heat products to avoid excessive drying during the process and use direct heat conduction (by placing durable gastronorms on a heated surface), forced air convection or dark infra-red heating. The latter have shelves which are bounded above and below by dark infra-red elements which heat the food by radiant heat. Fans may be incorporated to help equilibrate overall oven temperature. Also, solid shelves on which food is placed can act as heat 'sinks' to capture heat, increase heat conduction to the food and reduce circulation. The operating temperatures of the infra-red ovens are normally in the range 200–250°C but can be as low as 150°C.

The forced air system relies on indirect heating and the circulation of hot air within a cabinet. Oven temperatures are normally in the range 140–185°C depending on the food being heated and the type of oven used. The control systems for this type of oven are variable but most rely on a pre-set programme for control of the heating cycle or are programmable by the operator.

All three types of bulk regeneration oven lead, to some extent, to dehydration of the food as a dry heat source (either direct heat or hot air) is used. Therefore, for optimal recovery of the organoleptic qualities of the food, tight control of the system is required. Many caterers utilise these ovens only for specific items in which dry heating may be a requirement for crisping or browning before service. Other forms of re-heating may be used alongside this method for other products which are more sensitive to drying.

Combination Ovens

These devices are simple oven cabinets which can either use convected dry heat alone (forced hot air) or low pressure steam injection in addition to convection heating at temperatures below 100°C. The equipment is supplied by a wide range of manufacturers who each incorporate specialist design features. However, most follow the same basic principles and are designed to hold standard gastronorm type containers (or packaged products on shelves) and have capacities from 5–40 gastronorms. Larger ovens can be loaded with gastronorms pre-loaded onto trolleys as described for the bulk regeneration ovens above. Regeneration with dry heat usually uses a temperature range of 120–150°C although temperatures of up to 300°C can be attained by some ovens for browning purposes. 'Wet' heat is normally used below 100°C (40–98°C) and provides a quick way of re-heating as steam is such a good conductor of heat.

These ovens can be used for (generally) bulk regeneration although some offer plated regeneration facilities and many can be used to re-heat individual portions of *sous-vide* packaged foods (see Chapter 7). Although designed for prime cooking these ovens offer a useful alternative to bulk (dry heat) regeneration ovens as they allow for re-heating in a humid environment and thus lead to less drying in foods that would be sensitive to such an effect.

Mobile Regeneration Trolleys

This type of equipment falls into two broad groupings, the mobile zone regeneration ovens and split zone regeneration systems. The former comprise a trolley with an integrated re-heating system which provides for the re-heating (often in transit) of plated, chilled meals. These units can also carry cold (e.g. salad) meal accompaniments, condiments and trays for service and provide a complete service system. Not surprisingly, they find their main use in the hospital service. The latter type, which is also used in hospitals, comprises two units—a regeneration section for re-heating plated, chilled meals and an ambient temperature section for the distribution of meals on trays. The benefits of this latter system are that the regeneration unit can be retained locally and only the ambient unit used for distribution (say, at ward level) thus reducing difficulties arising from having to move regeneration equipment from plating to distribution areas.

Both types, like the bulk regeneration equipment, utilise one of three main heating sources, i.e. convected air, conduction plates or dark infrared radiant heat. Like most regeneration machinery, this equipment relies on the skill of local staff to get the process of re-heating right at the point of service. However, in an end kitchen where food can be regenerated in bulk and then plated, trained catering staff can be involved and control the process. One of the criticisms of the trolley system is that it is designed for use by untrained or non-catering staff (e.g. nurses in hospital wards) and that foods can be abused by such systems if the operators do not understand the proper use of them.

Microwave Ovens

Virtually all vended cook–chill food is regenerated by the purchaser using a microwave oven. Many industrial cook–chill sites employ microwave regeneration for certain menu items and for foods served to shift workers at night. Also, microwave re-heating has been used in the USA and UK as a ward regeneration method for cook–chill foods in hospitals. This system has some advantages but again, often relies on untrained operators, i.e. the consumers, 'to get it right'. Food must be clearly labelled with instructions on regeneration times but, in the event, the supplier has to rely on the consumer to follow the instructions. The results of microwave re-heating are reasonably good if used properly although there is some difficulty in regenerating foods of different densities because heating may be very uneven. However, there is little or no control, using this system, of the end temperature of the re-heated food which should, as mentioned above, attain at least 70°C at its centre before consumption.

Perhaps the main advantage of the method, in terms of re-heating single portions for individual distribution, is the speed of the method. Portions of beef loaf, mashed potatoes, frozen and canned beans were re-heated from either chilled temperatures (7–8°C) or frozen (0°C) in an experiment to test the efficiency of the method. Within 85 s all portions had reached 70°C although the mashed potato was slowest (69°C at 85 s). All of the other products reached 70°C within 60 s (Fig. 5.10). These results prove the speed of the method for single portions and thus its usefulness for certain site regeneration purposes and in whole meal vending but the method has found little use as a bulk regeneration method to date.

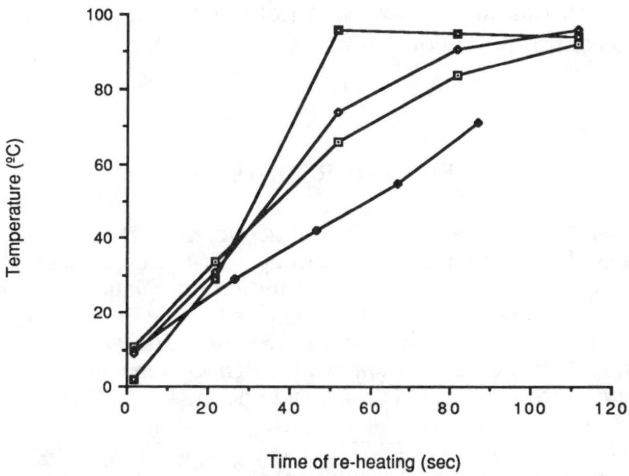

Fig. 5.10. Microwave re-heating of cooked, chilled portions of beef loaf (⊡), mashed potatoes (◆), frozen (■) and canned beans (◆).

This may be due to the difficulty outlined above of heating foods of different densities and water activities on the same plate. Indeed, a customer survey of cook–chill users revealed low satisfaction with microwave reheating amongst consumers who had to use the method.

CONTROL SYSTEMS

As mentioned above and discussed at some length in Chapter 9, it is essential to enforce strict hygiene management and, thereby, quality assurance. Temperatures in the chillers, chill storage rooms and in transport must be monitored and recorded. However, although equipment is improving all the time and, indeed, the systems approach demands that monitoring and control should be as automated as possible, the human factor should not be forgotten. Double checking on systems with the humble hand-held electronic thermometer is an additional control which should not be ignored and, for a small investment, can prove an invaluable tool in reminding staff of the importance of temperature control and checking on the performance of automated machinery.

As computer systems have fallen in price, several are available for managing cook–chill catering systems, providing the benefits of easier stock control, producing labels for all the dishes which pass through the system, storing the recipes and instructions for preparing dishes, storing nutritional information on a wide variety of dishes and all the requirements of a central production unit.

FURTHER READING

Anon. (1987) Computerised cook–chill. *Catering*, May, 117.
Anon. (1988) The cook–chill system. An appraisal of equipment and consumables. Trent Regional Health Authority and the National Health Service Procurement Directorate in association with Greene Belfield Smith.
Armstrong, R. (1986) Cook–chill catering. *Environ. Health,* **94,** 253–8.
Bengtsson, N. (1980) Catering equipment design and food quality. In: *Advances in Catering Technology* (ed. Glew, G.), London, Applied Science Publishers, pp. 121–34.
Boltman, D. B. & Shaw, V. A. (1977) A computer simulation program for establishing equipment and storage requirements of a catering system. In: *Catering Equipment Systems & Design* (ed. Glew, G.), London, Applied Science Publishers, pp. 463–72.

Dahl, C. & Matthews, M. (1978) Cook–chill foodservice systems: Microwave heating pre-cooked portions of beef loaf. *J. Microwave Power,* **13,** 87–93.
Dahl, C. A. & Matthews, M. E. (1980) Effect of microwave heating in cook–chill foodservice systems. Some data on temperature, yield, moisture and fat content of typical foods. *J. Amer. Diet. Ass.,* **77,** 289–95.
Guise, B. (1986) Exchanging gases to extend life. *Food Processing*, Sept., 35–7.
Millross, J., Speht, A., Holdsworth, K. & Glew, G. (1973) *The Utilization of the Cook–Freeze Catering System for School Meals*. University of Leeds.
Mobbs, C. C. (1966) Cook–chill production. *Hospital Caterer*, Sept., 20, 22.
Morris, N. (1987) Designing for hygiene? *Design 460*, April.
Porsdal Poulsen, K. (1980) A system for continuous production for ready made meals and distribution in aseptically filled containers. In: *Advances in Catering Technology* (ed. Glew, G.), London, Applied Science Publishers, pp. 265–7.
Sheppard, J. (1987) The big chill. A report on the implications of cook–chill catering for public services. London Food Commission, Report no. 15.
Skjoldebrand, C. (1980) Re-heating of minced meat products in a convection oven. In: *Advances in Catering Technology* (ed. Glew, G.), London, Applied Science Publishers, pp. 232–46.

CHAPTER 6

Shelf-Life of Cook–Chill Products—Hazards and Potentials

INTRODUCTION

The period of time during which food is both organoleptically acceptable (i.e. from a sensory point of view) and safe (i.e. from a microbiological point of view) is known as the shelf-life. Shelf-life has been variously defined as the period between manufacture and retail purchase of a food product during which the product is of satisfactory quality, and the period between manufacture and consumption of a food product during which the product is of satisfactory quality. The latter is the broader and more useful definition and certainly fits the needs of understanding the subject with respect to catering.

The use of standard cook–chill procedures leads to the production of foods which can be safely stored for short periods of time, i.e. in the UK, up to five days (including the day of production and consumption), as stipulated by the Guidelines on Pre-cooked Chilled Foods issued by the Department of Health and Social Security (see Chapters 1, 3 and 4). Even foods prepared under the strictest conditions of hygiene and stored in first class chill units cannot be guaranteed as being safe for longer than this due to the risk of the growth of food poisoning micro-organisms. Also, oxygen in the atmosphere leads to inevitable deterioration of foods during storage through the oxidation of flavours and colours in the food (see Chapter 2). The subsequent production of off-flavours and unsightly colours as well as the growth of food spoilage organisms can also render the food inedible.

The advent of refinements to the cook–chill method such as vacuum and modified atmosphere packaging has suggested that packaged cooked, chilled foods may have a potentially longer shelf-life than unpackaged foods. Vacuum and modified atmosphere packaging may be used after cooking, after cooking and chilling, or in combination with cooking, i.e. cooking raw or partially cooked foods sealed inside vacuum packs. The research work carried out to date suggests that the application of such methods, when used properly, can lead to substantial extensions to the shelf-life of normal cook–chill foods. However, it has also been suggested that such methods are open to abuse. Short shelf-life at low temperature does not, in general, provide enough time for the growth of dangerous organisms in large numbers but a longer shelf-life obviously increases the risk of this happening. In addition, the recent focussing of attention on food poisoning organisms like *Listeria monocytogenes* which may grow quickly at 4–6°C (see Chapter 3) has highlighted the need for vigilance in the use of standard as well as extended shelf-life cook–chill techniques.

In recent years there has been an international effort to increase awareness of food producers in both retail and catering markets to the dangers, in terms of the increased risk of food poisoning, of lengthening the shelf-life of cooked, chilled foods without adequate safeguards. These warnings have come in the form of detailed investigations of the shelf-life of a range of products published in scientific journals, the publication of guidelines on the production and storage of such foods and a debate in the popular and scientific press on the merits and dangers of storing chilled foods. Some of these topics have already been discussed in previous chapters but others will be detailed here.

Finally, the question should be posed, do caterers want or require more than only a few days chilled shelf-life for their cooked, chilled foods? It may be assumed that such a provision could be of benefit in terms of time and resource management but would the need for extra chilled storage space and more complex date labelling and product monitoring routines counteract these apparent advantages?

This chapter deals with the concept of shelf-life and the actual practice of chilled storage of cooked products in the catering industry, the use of the new technologies of packaging as well as their potential benefits and dangers. The potential hazards of extended chilled storage of cooked foods in terms of the risks of food poisoning will also be considered.

SHELF-LIFE OF COOK–CHILL FOODS

The term shelf-life means the time between production and consumption of the food (Chapter 2). The shelf-life of any stored food depends on how quickly the process of chemical and microbiological degradation of the product takes place (Chapters 2 and 3). These changes take the form of:

(a) oxidative reactions which impair or change the flavour, odour and coloured chemicals in the food;
(b) the interactions of reactive food components to produce unpleasant products;
(c) the breakdown or irreversible alteration of various nutrients (although this is not often regarded by many as a reason for rejecting a food product); and
(d) the production, by micro-organisms, of unpleasant or poisonous products in or on the food.

Many of these changes can be noticed (by sight or smell) because off-colours and off-odours are often produced during most of these processes. However, the changes that take place in a food must be seen as dynamic, i.e. they begin when the food is first chilled after cooking and progress during chilled storage. Thus, food spoilage organisms may grow on the product from the moment it is cooled but it need not be rejected until the point at which enough by-products of the spoilage organisms are produced to make the food inedible due to changes in texture, flavour, odour or appearance. The same is true of chemical changes associated with oxidation or other reactions. These may begin early after cooking in the life of the food but only when they have reached a certain point need it be rejected. Thus, a food might be discarded for sensory reasons brought about by food spoilage or by oxidation or by both.

It is important to realise, however, that the presence of food spoilage organisms in a food does not necessarily signal the spoilage of that product. For example, even though the DHSS guidelines on pre-cooked chilled foods (see Chapter 4) specify a maximum acceptable level of 100 000 viable organisms per gram of food (10^5/g) a study of vacuum packed sliced luncheon meats inoculated with lactobacilli and *Brochothrix thermosphacta*, showed that a count of 800 000 000 organisms per gram (8×10^8/g) multiple food portions are prepared each day (e.g. 10 000 meals per day) age depended on the conditions of storage, the actual nature of the experimental product, the strains of bacteria chosen (or present in the food), and, in sensory analysis, the taste panellists' expectations.

In a similar study, in which the shelf-life of cured cooked meat products was examined, it was shown that the end of shelf-life was not closely related to a particular bacterial count. Therefore, we may conclude that limits of spoilage bacteria are not valid unless combined with organoleptic criteria in determining the actual shelf-life of a product.

In the case of food poisoning the situation is far more complex. Here, small amounts of toxin might build up in the food without being detected. Similarly, levels of organisms which, when ingested will cause poisoning, might be high enough to be dangerous to human beings even though the food appears and tastes absolutely satisfactory. In an organisation where multiple food portions are prepared each day (e.g. 10 000 meals per day) it would be impossible to test each portion for food poisoning organisms before it is eaten (especially because such tests often take too long). Thus, the end of product shelf-life brought about by the growth of food poisoning organisms can only be assessed by predictive methods based on a sound knowledge of the organisms, the food environment and other such criteria. This method forms the basis of most systematic food quality management techniques which will be discussed in detail in Chapter 9.

POTENTIAL HAZARDS OF STORING COOKED, CHILLED FOODS

The main hazards in storing cooked, chilled foods arise from the possible presence of food poisoning organisms and/or their toxins (see above and Chapter 3). It is self-evident that a cooked food-stuff that contains pathogenic organisms which can grow under the conditions of storage of that food presents a danger to human life and health. However, as previously discussed, if good manufacturing practice (GMP) is followed in cook–chill production, i.e. the food is properly pasteurised during the thermal processing process, is rapidly chilled and is stored at 1–3°C for a short time (generally less than 5 days) then all the evidence indicates that such foods will be safe to eat after adequate regeneration.

Unfortunately, such GMP relies on the goodwill and efforts of humans who are, of course, fallible. Thus, the risk of food poisoning will be increased substantially if any of the following errors are made in production:

(a) a food is not thoroughly pasteurised during thermal processing;
(b) a food is not chilled rapidly enough after heat processing;
(c) a food is contaminated with food poisoning organisms after cooking and/or chilling;
(d) a food is not stored at a rigidly controlled low (0–3°C) temperature;
(e) a food is inadequately re-heated before service.

If a combination of these errors is made then the risk of food poisoning will be compounded and the likelihood of an incident of food poisoning will be greatly increased.

Many food poisoning organisms are heat sensitive and will be killed by adequate pasteurisation (Chapter 3). The spore-formers, like the clostridia, will not grow at temperatures below 3°C. Any low level contaminations with mesophiles will be resolved by proper re-heating. However, the psychrophiles recently identified and associated with serious food poisoning problems such as *Yersinia enterolytica* and *Listeria monocytogenes* are, as yet, poorly understood and could pose a threat to cook–chill catering practice if rigid GMP is not always observed.

VACUUM PACKAGING

Food vacuum packaging is the removal of all air and/or other gases from the local environment of a food product. Vacuum packaging can be used to enhance the shelf-life of both fresh and cooked foods because it removes the oxygen in the air which leads to the oxidative deterioration of foods and which sustains the life and growth of aerobic micro-organisms (see Chapter 3). Raw meats have been shown to benefit from vacuum packaging and stay fresher and safer for longer, freshly cooked speciality biscuits and cakes can be kept fresher for longer and their flavours can be sealed in by vacuum packaging, cooked, chilled complex meat dishes (which normally undergo oxidative rancidity) can be preserved longer at chill temperatures because oxygen is excluded. These are just a few examples of how the method works. In addition, cooked food can be packed hot into heat-resistant bags, pouches or containers and then chilled and stored at 0–3°C prior to use. Such a method has been adapted for large scale cooked, chilled food production (see Chapter 7). Most cooked foods suit this method of storage except, possibly, some bakery goods. It is especially applicable to most entrées.

Let us examine the actual process of vacuum packaging. Foods can be sealed in containers under vacuum. Such containers can be pouches, bags of various sizes (e.g. from individual portion size to 5 kg containers), plastic boxes, trays, etc. The vacuumising process is simple and can be carried out using a specially designed piece of equipment. The food product is placed inside the container (bag, pouch, box, etc.) and a lid is applied if necessary. The whole is then placed in the vacuum machine and air is withdrawn from the package until either no more air-space is visible between the food and the container or a reading on the vacuum device indicates that a good vacuum has been achieved. It should be remembered that ultimate vacuums, in which no gas molecules remain in an environment, are almost impossible to achieve. Thus, what is sought in the vacuum packaging of a food-stuff is merely the removal of all or most of the air from around the food such that the packaging fits tightly around the outer surfaces of the product. In addition, with some solid foods it may be possible and advantageous to continue the vacuum procedure until some of the air trapped within the food is withdrawn. However, this latter procedure may cause the food to physically collapse or distort if taken too far and, therefore, must be closely monitored. Most vacuum machines have a facility which allows the visualisation of the packaged product during the vacuuming process.

When the appropriate level of vacuum has been reached the bag, pouch or container must be hermetically sealed (i.e. sealed against further invasion of the container by air or other gases). This is normally achieved by heating the edges of the bag, pouch or container until a welding together of the plastic materials is achieved. The material from which the container is constructed must not, obviously, allow the ingress of air. This means that plastic bags or pouches and plastic boxes must be made of gas-impermeable materials or, at worst, materials with a very low permeability to air. As such materials may not be compatible with contact with food (i.e. they may be, or may contain, poisons or undesirable chemical agents), these active materials may have to be sealed as thin films between layers of acceptable plastics. Thus, over the last few years a battery of composite materials consisting of co-extruded films of plastics have grown up for use in the vacuum packaging of foods. Some of these have even been developed to withstand high temperatures and still remain gas-impermeable (see Chapter 8).

Finally, air can be simply excluded from a packaged food by physical means, i.e. manipulating the package. This is not true vacuumising but, potentially, has the same effect and can be very efficient and safe provided the package is closed off at pasteurisation temperature. When foods are heated, the dissolved gases in them form bubbles and escape to the atmosphere. After prolonged heating (cooking or pasteurisation) much of the gas may have been driven off (especially from fluids or semi-solids such as stews). At this point the food product can be inserted at high temperature (80–90°C) into a plastic bag. If the top of the bag is then immediately closed off with the exclusion of a small amount of the fluid, i.e. the deliberate exclusion of all air, and the bag sealed, then there will be little or no oxygen present to affect the food during chilled storage. This simple method has also been applied in some large scale methods with successful results (see Chapters 5 and 7) and may be considered to be equivalent to vacuumising.

MODIFIED ATMOSPHERE PACKAGING

The method of controlling post-harvest changes in fruit and vegetables during storage in a gas-tight storage room which is regulated by means of monitors and whose environment (temperature, gas content and humidity) is closely controlled is known as controlled atmosphere packaging or storage (CAP). This method has been available for some time. A related method, known as modified atmosphere packaging (MAP), is also used in extending the shelf-life of certain food products. This method involves packing a food into a container and then replacing the air present in the container with a mixture of pure gases specially formulated to inhibit or retard chemical changes in the food, affect the rate of ripening or the colour of a fresh product and reduce or stop the growth of micro-organisms.

MAP has been used successfully in the presentation of fresh meat products in retail outlets where colour is a major factor influencing the consumer's decision to buy a particular product. In this case, various gas mixtures can be used to retain the red colour and 'sheen' of meat to improve its appearance. Poultry and fish products can also have their shelf-life extended and, in some cases, their appearance enhanced by use of MAP. Table 6.1 shows typical gas mixtures for some meat and fish products. Fresh vegetable products can have an extended shelf-life in the

TABLE 6.1
Various Gas Mixtures and Their Recommended Application in Modified Atmosphere Packaging of Fresh Foods

Fresh Food	*Gas mixture*
Red meat	20% carbon dioxide, 80% oxygen
Fish (trout)	20% oxygen, 66% nitrogen, 14% carbon dioxide
Fish (white)	30% oxygen, 30% nitrogen, 40% carbon dioxide
Fish (fatty)	40% nitrogen, 60% carbon dioxide
Poultry	20% oxygen, 30% nitrogen, 50% carbon dioxide

retail outlet by being packaged in a container within which the local atmosphere is designed to prolong the normal metabolic processes of the live plant material. In all of these cases oxygen is present in the gas mixture.

However, it is also possible to use the method to exclude oxygen from a package containing a cooked, chilled food by replacing it with other gases. Carbon dioxide has been shown to be a powerful inhibitor of the growth of many micro-organisms. Also, most micro-organisms do not use nitrogen and other inert gases in their metabolic processes. Thus, mixtures of carbon dioxide and nitrogen have been used to replace air within containers. In addition, carbon dioxide can be added to enrich the air already present in the container. The former method, which requires prior vacuumising of a packaged food, removes all oxygen and thus inhibits oxidative changes whereas the latter does not.

MAP using carbon dioxide/nitrogen mixtures can be profitably used with cooked, chilled products which are delicate or will expand or be damaged under high vacuum. Partial vacuums can be applied to such foods followed by gas flushing. Re-evacuation and re-flushing can ensure very low levels of oxygen and corresponding protection against aerobic microbial growth and oxidative deterioration. It should be noted, however, that high concentrations of carbon dioxide can give food a tainted or acid taste after extended storage.

MICROBIOLOGICAL ASPECTS OF SHELF-LIFE EXTENSION

In the following chapters we will examine the benefits and disadvantages of packaging foods prior to pasteurisation but in the present chapter we will consider the application of vacuum and MAP techniques to pre-cooked products and their effects on the shelf-life of cooked, chilled products. Much of this section will be concerned with the microbiological acceptability of food products and so a brief introduction is necessary. In Chapter 3 we considered the fundamental principles of food microbiology as they affect cooked, chilled foods. Now we will consider the sort of bacteria which occur naturally in cooked, chilled foods.

Research into the microflora of cooked, chilled foods prepared in catering has taken two forms—those studies which concentrate on the micro-organisms associated with foods prepared under normal cook–chill conditions in the laboratory and those actually prepared at a catering site. One of the most striking aspects of these studies is the wide variation in the results obtained. These variations are reflected within a single

TABLE 6.2
Summary of Some Studies Carried Out on Cooked, Chilled Foods Between 1972 and 1985 (from Young, 1986)

Menu item	*Chilled storage conditions*	*Analyses carried out*	*Results obtained*
Bakery items, meat and entrée items, salad items.	Rapid chilling to 7·2°C in two hours. Storage at < 4·4°C for < 96 h	TPC, sensory analysis, consumer tests	After cooking; TPC < 200/g. After 9 days storage; meat TPC increased to $1{\cdot}6\times10^5$/g, potatoes > 10^6/g
Roast turkey	Cooled at room temp. for 1–3 h. Stored whole at 7·2°C for 18 h	*Cl. perfringens*, staphylococci, salmonellae	After cooking; all tests negative
Spaghetti and chilli	Chilled in a mechanically stirred chilled tank at −12·2°C for 1·2 h. Stored between −1·1 and 1·6°C for 100–130 h	TPC, clostridia, yeasts and moulds, staphylococci, coliforms	After cooking; TPC spaghetti $2{\cdot}1\times10^3$/g, chilli $1{\cdot}1\times10^3$/g. After storage; TPC spaghetti $5{\cdot}7\times10^3$/g, chilli $2{\cdot}1\times10^4$/g. Clostridia numbers increased during storage from 0·34 to $2{\cdot}04\times10^3$/g in spaghetti
Beef stew	Chilled and stored at 5·5–10°C for 22 h	TPC, total coliforms, sensory analysis, vitamin retention	After cooking; TPC $4{\cdot}6$–$8{\cdot}1\times10^4$/g. After storage; TPC $7{\cdot}5$–$18{\cdot}0\times10^4$/g
Eleven menu items; entrées, vegetables and desserts	—	TPC, coliforms, *Escherichia coli*, coagulase-positive staphylococci	No system was significantly different in the microbiological quality of the food delivered
Meat loaves	Chilled and stored for 25–80 h at 4–18°C	TPC, yeasts and moulds, staphylococci, coliforms, clostridia, sensory	After cooking: TPC, $2{\cdot}3$–$3{\cdot}7\times10^3$/g; staphylococci; 1–3×10^2/g; coliforms; 2–4×10^2/g clostridia; 6–12×10^2/g
Roast beef	Chilled and stored for 45 h at 1·7–6·7°C (at room temp. for 3·5 h)	TPC, yeasts and moulds, staphylococci, coliforms, clostridia, sensory	After cooking: TPC; $1{\cdot}2\times10^3$/g; staphylococci; 10/g; coliforms;— clostridia; $2{\cdot}7\times10^2$/g; yeasts and moulds; 40

TABLE 6.2—*contd.*

Menu item	*Chilled storage conditions*	*Analyses carried out*	*Results obtained*
Chicken and noodles	Packed in plastic casings, chilled in a water bath at 8·7±3°C for 2 h. Stored at 0·2°C for 3–31 days	TPC, coliforms, staphylococci, anaerobic spore count	After cooking: all counts < $3{\cdot}0\times10^2$/g. After storage: all counts < $3{\cdot}0\times10^2$/g
Beef loaf Potatoes	Chilled and stored for 24 h at 7°C	TPC	After cooking: TPC; 1×10^4/g. After storage: potatoes; $1{\cdot}5\times10^3$ $4{\cdot}7\times10^6$/g Beef loaf; < 3×10^2 $6{\cdot}8\times10^5$/g
Beef loaf	Chilled and stored at 1±1°C for 24±1 h	TPC, yeasts and moulds, coliforms, staphylococci, clostridia	After cooking: coliforms, yeasts and moulds, and staphylococci all negative TPC; $7{\cdot}8\times10^1$/g; clostridia; $1{\cdot}7\times10^1$/g; After storage: TPC; $1{\cdot}3\times10^2$/g; clostridia; $3{\cdot}3\times10^1$/g
Roast chicken	Chilled at 1°C for 1 h. Stored at 1–3°C or 16°C for 7 days.	TPC	After cooking: TPC of breast skin<3–14×10^3/g. After storage: TPC of breast skin; at 1–3°C 9–$4{\cdot}5\times10^3$/g; at 16°C < 3×10^2–$2{\cdot}4\times10^5$/g
Ground beef gravy	Chilled in water bath at < 10°C for 1 h. Stored at 4°C for 16 h and a further 5 h at 5·5°C	TPC, *Cl. perfringens*, coagulase-positive staphylococci	After cooking: TPC; $5{\cdot}0\times10^1$/g. After storage: TPC; $4{\cdot}4\times10^2$/g. *Cl. perfringens* negative, *Staphylococcus aureus* positive in some samples
Meat loaf	Chilled and stored at either 0 or 5·5°C for 9 days	TPC, cooking losses, TBA test, sensory	After cooking: all loaves but one TPC<$1{\cdot}0\times10^2$/g. After storage: all loaves but two TPC<$1{\cdot}0\times10^2$/g

(*continued*)

TABLE 6.2—*contd.*

Menu item	*Chilled storage conditions*	*Analyses carried out*	*Results obtained*
Beef soy loaf	Chilled and stored at 5·5±3°C for 24, 48 or 72 h	TPC, consumer evaluation	After cooking: TPC corner of loaf; 2·1–4·2×10^3/g; centre of loaf; 5–7×10^4/g. After 24 h storage: TPC centre of loaf; 1·3–1·5×10^5/g.
Roast chicken	Chilled and stored at either (1) 4°C or (2) 2–13°C for 10 days	TPC, McConkey agar, manitol salt agar, trypticase soy agar, identification of isolates	After cooking: TPC; 1–1·0×10^3/g. After storage: (1) 4·2–10·6×10^6/g (2) 2·1–3·9×10^7/g
Brussels sprouts Minced beef in stock	Storage 8 days ([a])	TPC, coliforms, vitamin retention	After cooking: TPC sprouts; 4·8×10^3/g; meat; 6·3×10^4/g. After storage: TPC sprouts; 7·8×10^3/g meat; 5×10^5/g after 6 days.

([a]) Chilling times and temperatures not given.
TPC = total plate count (of aerobic+anaerobic bacteria).

sample of a menu item, between similar samples produced at the same time and between samples produced at different times but by the same cook–chill system. It has been shown that the composition of the microflora of a cooked, chilled food and their numbers are determined by the microflora of the raw food, the characteristics and parameters of the processing system (cooking, chilling, storage, etc.) and the numbers and types of post-processing contaminants.

Examples of the types of organisms found by research workers in these studies in foods as various as entrées, bakery items, salads, roast meats, vegetables, noodles and beef gravy are *Clostridium perfringens, Staphylococci*, salmonellae, coliforms, yeasts and moulds and *Es-*

cherichia coli. Table 6.2 summarises some of the studies carried out on cooked, chilled foods between 1972 and 1985.

It must be remembered that, after proper heat processing of foods (i.e. pasteurisation), only heat-resistant spores will remain in the food. In a study of chicken, cooked to an internal temperature of 85°C for 50 min, it was shown that only spores of *Bacillus subtilis* and *Clostridium bifermentans* remained. However, most cook–chill operations include a portioning step (during which post-cooking contamination can occur) and, of course, the food is generally open to the atmosphere, handling and thus contamination, during chilling, portioning and chilled storage. Compounding these potentially contaminating steps is the human element. Handlers can themselves introduce contaminants to the cooked, chilled food in the form of pathogens or spoilage organisms that are carried on their hands and bodies. Therefore, cooked foods can contain a variety of micro-organisms which may be naturally occurring and left after heat processing or external contaminants.

A study of vacuum packed bologna-style sausage wrapped in plastics of different permeabilities and stored at 2°C revealed that the aerobic bacterial counts and counts of *Brochothrix thermosphacta* (a strict anaerobe) rose to the highest levels in products packaged in the most permeable films but bacterial counts in all samples exceeded 100 000 per gram (10^5/g) after two weeks. This shows quite clearly, as might be expected, that aerobes are well inhibited by vacuum packing foods in plastic films with high gas-barrier properties (i.e. which restrict the influx of air) but that the method is not absolute proof against the unacceptable growth of micro-organisms as anaerobes (and aerobes living on small amounts of residual air left in the food or package) will eventually reach high levels even at chill temperatures.

When oxygen is entirely eliminated from a food only anaerobes or facultative anaerobes can grow. However, the presence of high tensions (levels) of CO_2 can also inhibit bacterial growth. Strict anaerobes and facultative anaerobes are not affected by CO_2 in the environment of the food. Examples of such organisms are the Gram-positive food spoilage bacteria *Enterobacter, Brochothrix thermosphacta, Lactobacillis, Aeromonas* and *Yersinia*. In contrast strict aerobes are well inhibited by CO_2. For example, *Pseudomonas*, a Gram-negative food spoilage organism, is not inhibited in low oxygen tensions (e.g. down to 0·8%) but its growth rate can be halved by the addition of 10% (v/v) CO_2. However, the situation is more complex than these simple statements might suggest. For example, in one study it was shown that the microbial composition of meat inoculated with aerobes, anaerobes and facultative anaerobes and stored at 3°C was the same whether it was stored under air or 80% air/20% CO_2. This was because the inhibited growth rate of the aerobes (*Pseudomonas*) was still higher than the unaffected growth rates of the anaerobes.

The shelf-life of raw MAP foods (particularly meat) has been shown to be extended by the addition of CO_2 to the environment of the product. In experiments, shelf-life could be extended by 25–30% if the growth of aerobic spoilage organisms was similarly inhibited by the addition of the gas. Also, this effect has been shown to be enhanced if the gas is added to the environment of the food before microbial growth commences because its presence seems to lengthen the lag phase, i.e. the period of latency before active growth of micro-organisms.

Until recent years, very few studies had been carried out which examined the microbiology of foods prepared under the conditions used in the 'cook–chill' system, either in laboratory or actual operating conditions. Similarly, very little research work has been carried out on the effects of vacuum and MA packaging on the microbiological shelf-life of cooked, chilled foods. Vacuum packing has been used by caterers for the preparation of cooked foods, such as entrée dishes, which are pasteurised after packing, i.e. in the pack (e.g. large scale techniques such as the

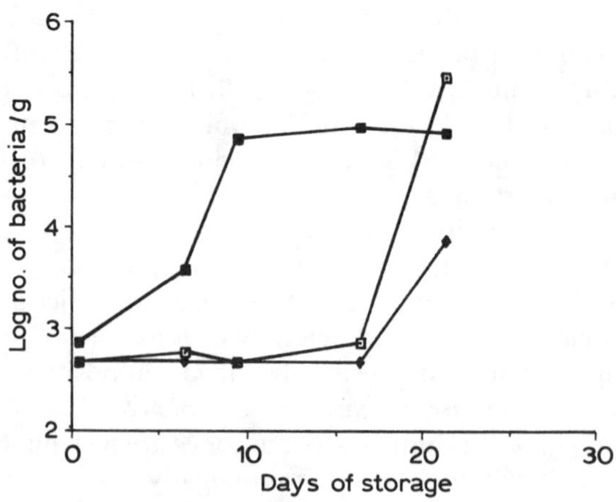

Fig. 6.1. Comparison of the effects of cling film wrapping (□), vacuum packaging (◆) and modified atmosphere packaging (■) on bacterial growth in cooked, chilled chicken drumsticks stored at 0°C. The graphs show total viable counts (TVC) in chicken drumsticks prepared by the cook–chill method and stored for up to 21 days.

Nacka method—see Chapter 7—or smaller scale methods such as *sous-vide* cooking—see Chapter 8).

Some studies of pasteurised cook–chill foods held under chilled storage conditions have shown that when such products are vacuum packed after cooking they have a longer shelf-life than untreated foods. Differences in the effectiveness of vacuum and MAP systems have also been shown. For example, it was found that storage of pre-cooked beef roasts in a 100% CO_2 environment produced more effective inhibition of bacterial growth than vacuum packing.

In another study two products, fried chicken drumsticks and chicken à la king, were selected as being representative of the type of cooked, chilled food used in the catering industry. The growth of microflora in these products at 0 and 4°C was studied when the products were stored wrapped in cling film, after vacuum packaging and after modified atmosphere packaging (see Figs 6.1–6.4). In addition, two representative species (*Pseudomonas* and *Lactobacillus*) of the naturally occurring microflora of these food items were introduced into the pasteurised products by means of direct inoculation and their growth characteristics were studied in cling film wrapped, vacuum packed and MAP products (see Figs 6.5–6.8). The findings of the research clearly showed that both vacuum packaging and modified atmosphere packing, when used at storage temperatures of

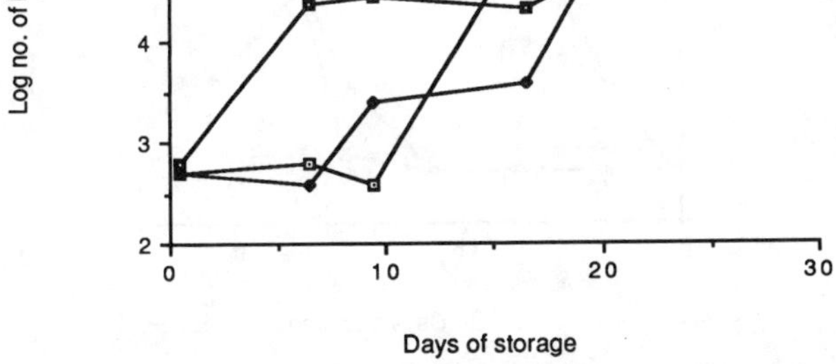

Fig. 6.2. Comparison of the effects of cling film wrapping (□), vacuum packaging (◆) and modified atmosphere packaging (■) on bacterial growth in cooked, chilled chicken drumsticks stored at 4°C. The graphs show total viable counts (TVC) in chicken drumsticks prepared by the cook–chill method and stored for up to 21 days.

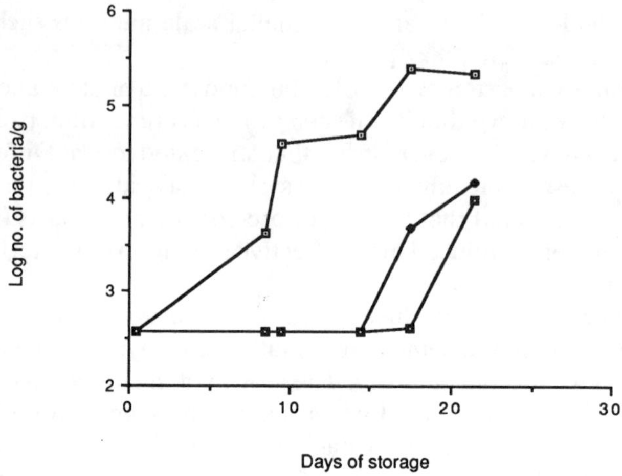

Fig. 6.3. Comparison of the effects of cling film wrapping (⊡), vacuum packaging (◆) and modified atmosphere packaging (■) on bacterial growth on cooked, chilled chicken à la king stord at 0°C. The graphs show TVC in chicken à la king prepared by the cook–chill method and stored for up to 21 days.

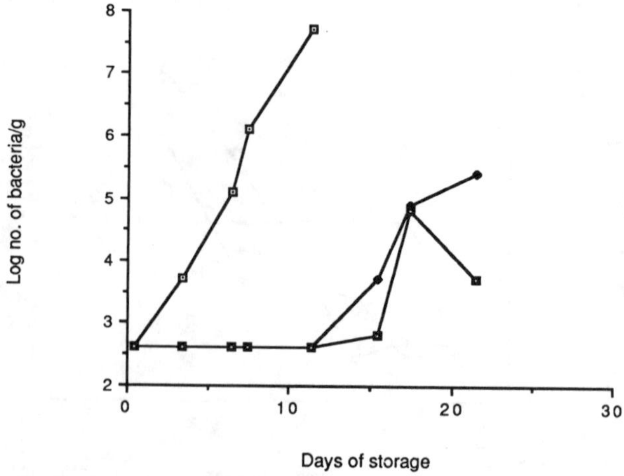

Fig. 6.4. Comparison of the effects of cling film wrapping (⊡), vacuum packaging (◆) and modified atmosphere packaging (■) on bacterial growth in cooked, chilled chicken à la king stored at 4°C. The graphs show TVC in chicken à la king prepared by the cook–chill method and stored for up to 21 days.

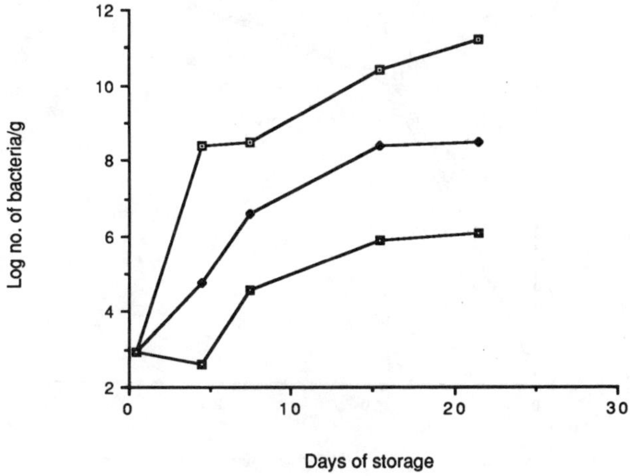

Fig. 6.5. Comparison of the effects of cling film wrapping (⊡), vacuum packaging (◆) and modified atmosphere packaging (■) on TVC in cooked, chilled chicken drumsticks inoculated with *Pseudomonas* and stored at 4°C.

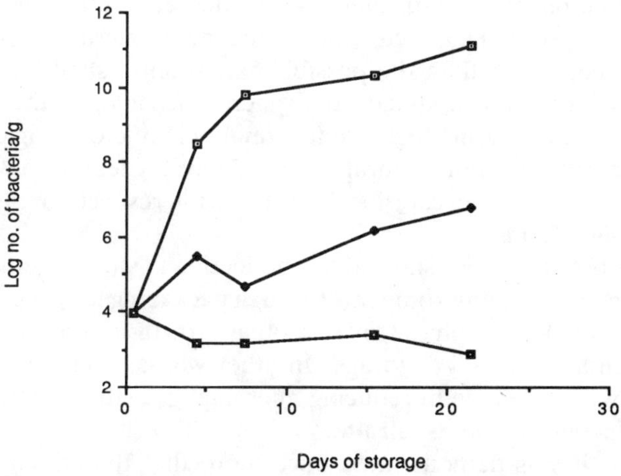

Fig. 6.6. Comparison of the effects of cling film wrapping (⊡), vacuum packaging (◆) and modified atmosphere packaging (■) on the growth of *Pseudomonas* in inoculated cooked, chilled chicken à la king stored at 4°C.

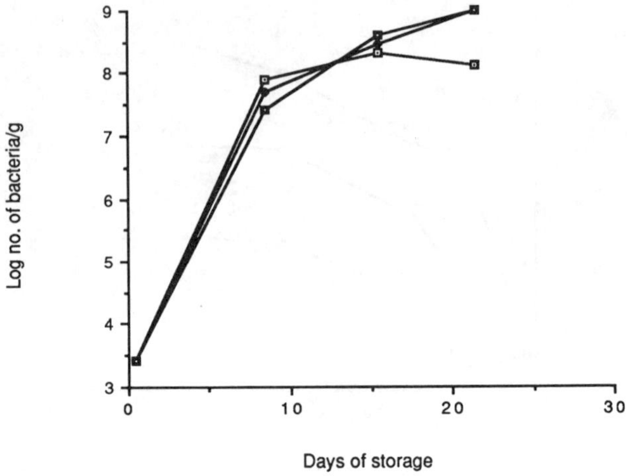

Fig. 6.7. Comparison of the effects of cling film wrapping (□), vacuum packaging (◆) and modified atmosphere packaging (■) on TVC in cooked, chilled chicken drumsticks inoculated with *Lactobacillus* and stored at 4°C.

0±1°C and 4±1°C, inhibit bacterial growth in cooked chicken menu items to levels below 10^5/g for up to 17 days (Figs 6.1–6.4). The direct direct inoculation studies indicated that the degree of this effect was dependent on the growth characteristics of the initial microflora (Figs 6.5–6.8). More important than the possible extension to shelf-life brought about by vacuum and modified atmosphere packaging is the value of these methods in ensuring high product quality in microbiological terms over short periods of chilled storage (up to 5 days). Clearly, these results have importance in the catering industry with respect to the use of cooked, chilled foods.

The results of this work showed that reduction in storage temperature and use of controlled environments through packaging led, in general, to lengthening of the lag phase for microbial growth and a limitation of growth even after 21 days' storage. In other words, vacuum packaging and MAP were effective in reducing microbial growth in cooked menu items and increasing their shelf-life.

However, it was demonstrated once more that the effects of these techniques were not as simple and predictable as might have been expected. For example, the effect of storage at 0°C in increasing product shelf-life in chicken à la king was greatest in the cling wrapped products

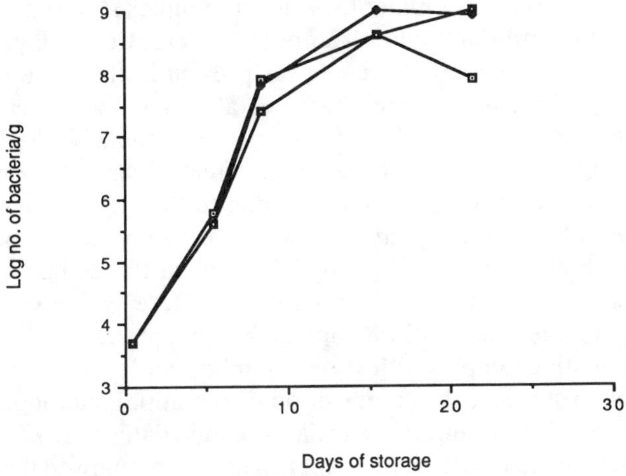

Fig. 6.8. Comparison of the effects of cling film wrapping (□), vacuum packaging (◆) and modified atmosphere packaging (■) on the growth of *Lactobacillus* in inoculated cooked, chilled chicken à la king stored at 4°C.

and less in the MAP and vacuum packed products. The lag phase of organisms growing in the cling wrapped product was extended from 5 to approximately 10 days and the growth rate reduced whereas in the MAP and vacuum packed products the difference in lag phase and growth rate was relatively lower. This may have been due to the aerobic strains of bacteria which grew in the cling wrapped samples being more susceptible to changes in temperature than the different bacterial strains which grew in the MAP and vacuum samples.

These results agree with data obtained in other studies in which it was shown that a temperature increase from 2 to 5°C resulted in a greater increase in growth of psychrotrophic organisms on pork chops in air than in CO_2 atmospheres. However, another study showed that aerobic plate counts and counts of *Brochothrix thermosphacta* from vacuum packed bologna sausage at 2 and 5°C were similar. These contradictory results further consolidate the view expressed earlier that it is difficult to predict the effect of treatments on the growth of micro-organisms in foods stored at chill temperatures.

The microflora of vacuum packed cooked chicken was shown to be dominated by *Lactobacillus* after storage at 7, 24 and 37°C. In the study of chicken à la king and chicken drumsticks mentioned above lactobacilli were shown to be the dominant species in samples stored under all conditions for both products investigated. However, after 60 days' storage at 4±1°C *Brochothrix thermosphacta* was present in the same numbers as lactobacilli in vacuum packed samples although this species was not present in MAP samples. In the latter case lactobacilli dominated the microbial population. The differences in effect of packaging on microbial growth in the two products are partly due to the different nature of the initial microflora. Had *Pseudomonas* species been present on the cling wrapped whole drumsticks initially, the TPC at the end of storage may have been greater, thus showing a greater difference between the cling wrapped and the MAP and vacuum packed samples.

By inoculating samples with known numbers and types of micro-organisms prior to storage, the composition of the initial microflora could be standardised. Two strains of bacteria were inoculated, *Pseudomonas* and *Lactobacillis*. The results of this experiment clearly showed that the effect of packaging in restricting growth is dependent on the type of bacteria present on the packaged food. In the presence of a mixed microflora the inhibitory or stimulatory effects of modifying storage atmospheres may be masked. If, however, the two organisms studied are taken as representative of common contaminants of cooked foods then microbial growth may be predicted to be least in MAP products, followed by vacuum packed products and greatest in cling wrapped or unwrapped products.

The results of this study of cooked, chilled chicken à la king and chicken drumsticks showed that control of storage temperature to between 0 and 3°C and the application of methods such as vacuum packing and MAP could extend the microbiological shelf-life of cooked, chilled foods used in the catering industry but that this extension was variable and hard to predict. Perhaps a more important result of the study was that the use of controlled temperature storage (0–3°C), combined with packaging techniques resulted in a high degree of confidence in the microbiological safety of cooked, chilled foods over the storage life recommended by the UK DHSS guidelines on pre-cooked chilled foods (5 days).

SENSORY ASPECTS OF EXTENDED SHELF-LIFE

In this section we will consider the effects of packaging on the sensory characteristics of cooked, chilled foods stored for extended periods. This has also been the subject of research over recent years although few authoritative studies have been published to date.

The occurrence of rancidity in cooked meats, with the development of a characteristic flavour, within a few hours of cooking, is well documented as a result of both sensory and chemical analyses. Cooked chicken is especially vulnerable to pronounced flavour changes during refrigerated storage but little work has been undertaken on a composite chicken dish, such as chicken à la king, where the addition of other ingredients may have an effect on these changes.

In cooked meats rancid flavours are frequently described as 'warmed over flavour' and are organoleptically different from rancidity in uncooked meats. This may be because of the differences in the side reactions after hydroperoxide decomposition (see Chapter 2). It was reported that vacuum and N_2 packed chicken and turkey meats showed less lipid oxidation than did the meats packaged with CO_2 and air.

A previous study also found that changes in odour and flavour of cooked poultry stored at 4°C were more pronounced than changes in other sensory properties. This work showed that storage of fried chicken and duck increased the perceived intensity of gluey, fishy, rancid and stale flavours. In another study it was found that vacuum packed beef roasts retained the 'healthy bloom appearance' that they had acquired immediately after cooking, whereas those packed under CO_2 or a mixture of gases lost this bloom after 7–14 days. Also the vacuum packed precooked beef roasts were significantly less juicy ($P < 0{\cdot}05$) than MAP samples after 7 days' storage but after 21 days' storage they were significantly more juicy, at which time the vacuum packed samples were also lower ($P < 0{\cdot}05$) in moisture loss.

In a study of cooked, chilled noodles clumping after two weeks of refrigerated storage was attributed to gelatinisation of starch. It was also found that the colour of freshly prepared sauce appeared close to white but was significantly ($P < 0{\cdot}01$) more yellow after two and four weeks of refrigerated storage and re-heating.

In a recent study of cooked, chilled chicken à la king and chicken drumsticks, vacuum and MA packaging were both shown to significantly increase the sensory quality of the stored cooked, chilled products. It was shown that storage time had a significantly ($P < 0{\cdot}05$) deleterious effect on all the sensory attributes of chicken drumsticks whilst just over half the sensory characteristics of chicken à la king were significantly affected and these were mainly concerned with appearance.

The use of vacuum and MA packaging was shown to increase the sensory quality of both products in terms of chilled storage time, that is, both products could be stored for longer with less loss in quality than controls wrapped or covered in cling film. Packaging was effective in reducing the rate of deleterious change in the chicken drumsticks but did little to prevent colour changes and the consistency of the sauce in the more complex dish. MAP was the most effective method for preventing deleterious changes in the stored drumsticks. Although vacuum and MA packaging were both effective in extending the shelf-life of the chicken à la king in sensory terms, it was not clear from the results which method was preferable.

In an extension of this work, consumer panels were used to establish preference for fresh and stored products. In this case, only the chicken drumstick product was studied. Overall, as may be expected, consumers preferred the fresh drumsticks (day 0) and found the cling wrapped chicken drumsticks stored for 7 days to be least preferred. Two separate consumer groups were studied. When the mean scores for samples were ranked in the first study, the most preferred drumsticks was the day 0 sample, followed by vacuum packed samples stored for 4 days, followed by MAP samples stored for 4, 7 and 11 days, followed by the vacuum packaged sample stored for 11 days and lastly the cling wrapped drumsticks in order of storage life (4, and 7 days). In the second study the main effects of storage were seen between 0 and 4 days of storage, after which time there was little change in acceptability scores.

These results support the taste panel study and point to the ability of vacuum and MA packaging to inhibit or slow down the unwanted sensory changes which take place in cooked, chilled foods. Like the results of the related microbiological study discussed above, these data suggest that cooked, chilled products (made under highly hygienic and controlled conditions and stored at strictly regulated temperatures), can have extended storage lives when packaged under vacuum or in MAP conditions. However, it must be remembered that such methods of lengthening storage life are not foolproof and that malpractice and abuse of such methods could lead to food poisoning.

SHELF-LIFE NEEDS IN THE CATERING INDUSTRY

The discussion above assumes that caterers need and/or want an extension to the current storage life expectancy of their products (5 days in the UK and 6 days in France). The methods of vacuum and MA packaging seem to offer the hope of safe shelf-life extension but is this really a necessity?

In 1988 a survey was carried out in the UK to provide quantitative information regarding equipment usage, cook–chill user problems and the views of cook–chill users on the need for the extension of shelf-life of cook–chill food products. A total of 250 cook–chill users representing all sectors of the catering industry were polled as to their needs and opinions (42·3% replied to the survey). Of the replies, 50% were from the service sector, 39% from industrial caterers and 9·6% from the commercial sector. Over half of all sites surveyed had operated cook–chill for over three years, with only 5% starting their unit within the previous year.

Just under half of all institutions surveyed used some form of secondary packaging to improve product quality, 54% using cling film and 40% using modified atmosphere and/or vacuum packing. The largest proportion of cook–chill users stored their products for 3 days (43%), but 16% stored their products to the full 5 days, i.e. the maximum recommended storage life for cook–chill products (DHSS, 1980).

When asked whether they would welcome an extension of shelf-life for their cook–chill products, 32·6% of the respondents thought they would benefit from an increased storage life. However, 50% of these only required an extra 2 day's storage, i.e. a maximum of 7 day's shelf-life. Most

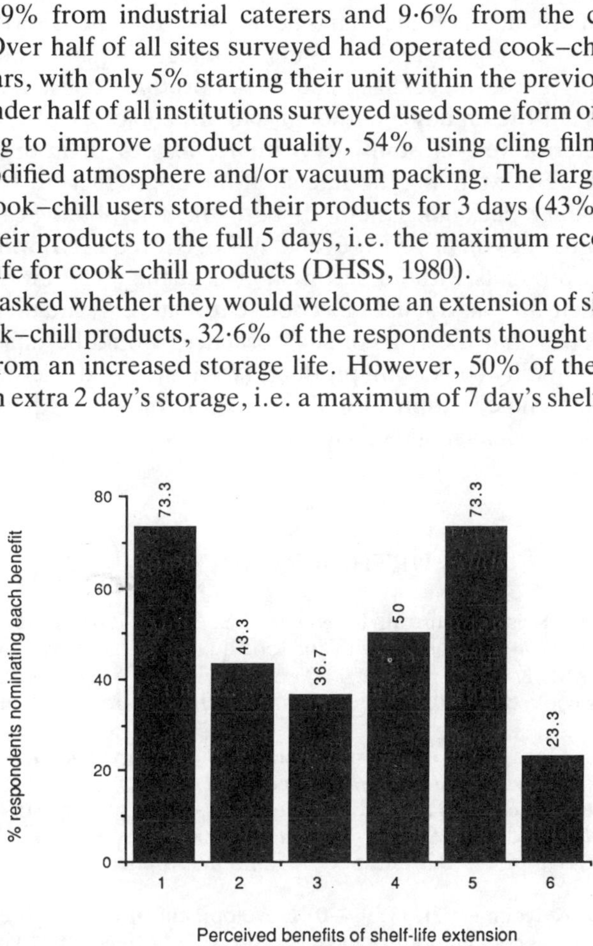

Fig. 6.9. Reasons cited by UK cook–chill caterers for requiring extension of shelf-life of cook–chill foods to greater than 5 days. 1, Increased flexibility in production; 2, increased product flexibility; 3, staff savings; 4, decreased waste; 5, convenience at holiday times; 6, to facilitate outside sale of goods.

respondents felt that an extension of shelf-life would decrease wastage (50%), increase flexibility in production and product range (73 and 43% respectively) and add convenience at holiday times (73%). Other perceived benefits included staff savings and facilitating outside markets for the sale of cook–chill food products (see Fig. 6.9).

It would appear then, that some cook–chill caterers would welcome an extension of product shelf-life. It has been shown that there is a direct correlation between organoleptic and microbiological quality factors and storage life. Hence, unpackaged cook–chill products consumed at their maximum shelf-life on day 5, although safe microbiologically, are likely to have deteriorated substantially with respect to their organoleptic acceptability. By extending the safe shelf-life of products to 10–15 days using secondary packaging methods both the microbiological and organoleptic qualities on day 5 (or even day 7 or 8 if required) will be much higher than products without secondary packaging. Therefore, even if extended shelf-life was not used by cook–chill caterers (and there is evidence to suggest that the risks of such practices would be unacceptably high) higher quality cook–chill products (with increased microbiological safety) could be made within normal production rotas by the judicious use of post-cooking packaging techniques.

FURTHER READING

Daniels, J. A., Krishnahmurthi, R. & Rizvi, S. S. H. (1985) A review of the effects of CO_2 on microbial growth and food quality. *J. Food Prot.*, **48,** 532–7.

Davidson, C. M. & Webb, G. (1973) The behaviour of *Salmonella* in vacuum packaged cooked cured meat products. *Can. Inst. Food Sci. Technol. J.*, **6,** 41–4.

Etheridge, I., Graff, N., Grant, A. & Rudge, T. (1988) Cook–chill under the microscope. *Environmental Health,* **196** (1), 21–8.

Glew, G., Berg, T. & Yin Yam (1980) Some effects of the presence or absence of oxygen on stored precooked chilled chicken and pork meat. In: *Advances in Catering Technology* (ed. Glew, G.), London, Applied Science Publishers, pp. 417–23.

Jacobson, M. & Koehler, H. H. (1970) Development of rancidity during short term storage of cooked poultry meat. *J. Agr. Food Chem.*, **18,** 1069–72.

Lewis, A. & Light, N. (1988) A survey of cook–chill catering in the UK and the need for the extension of product shelf-life. *Food Science & Technology Today,* **2,** 214–17.

Light, N., Schafheitle, J., Williams, R., Barrett, J. & Hudson, P. (1988) A pilot study of *sous-vide* (vacuum) cooking. *Int. J. Hospitality Man.*, **7,** 21–7.

McDaniel, M. C., Marchello, J. O. & Tinsley, A. M. (1984) Effect of different packaging treatments on microbiological and sensory evaluation of precooked beef roasts. *J. Food Prot.*, **47**, 23–6.
Nielsen, H. J. S. (1983) Influence of temperature and gas permeability of packaging film on development and composition of microbial flora in vacuum packed bologna type sausage. *J. Food Prot.*, **46**, 693–8.
Sheppard, J. (1987) The big chill. A report on the implications of cook–chill catering for public services. London Food Commission. Report no. 15.
Silla, H. & Simonsen, B. (1985) Shelf-life of chilled, cooked and cured meat products. *Fleischwirtschaft,* **65,** 66.
Spahl, A., Reineccius, G. & Tatini, S. (1981) Storage life of pork chops in CO_2-containing atmospheres. *Appl. Environ. Microb.*, **41,** 816–18.
Walker, A. & Light, N. (1988) Is cook–chill being put to its best use? *Hospitality,* Feb., 15–17.
Young, H., Youngs, A. & Light, N. (1988) The growth of naturally occurring microflora in cooked, chilled foods used in the catering industry. *Food Microbiol.*, **4,** 317–27.
Young, H., MacFie, H. & Light, N. (1989) The effect of packaging and storage on the sensory quality of cooked chicken menu items served from chilled vending machines. *J. Sci. Food Agric.*, **48,** 323–38.

CHAPTER 7

The Use of Vacuum Packaging in Cook–Chill Catering—the Large Scale Approach

INTRODUCTION

The concepts of the cook–chill system and of vacuum packaging and shelf-life extension (which we have discussed in Chapters 1, 4 and 6), were embodied in some of the earliest work in the development of large scale cook–chill catering systems, i.e. in the evolution of methods which would allow the large scale controlled production of cooked, chilled foods with low microbiological risk and good shelf-life. We have also suggested that large scale cook–chill catering should be regarded and operated as a food manufacturing system rather than merely an adaptation of normal catering. This was also the approach taken in these early large scale systems.

Ironically, in a 1986–87 survey of UK cook–chill units many were shown to be simple, small scale 'add-ons' in a conventional operation; quite the opposite of the manufacturing style operations designed to optimise the concept of the system in the early 1960s. This chapter will consider these pioneering systems which utilised vacuum packaging on a large scale as well as examining some of their successors.

Three large scale manufacturing style cook–chill systems evolved in the 1960s and 1970s in Sweden and the USA. All three methods used the principles of vacuum packing (either before cooking or of vacuum packing hot foods) as a means of assuring product quality and extended shelf-life. These three systems were the Nacka and AGS systems and, much more recently, the cook-tank system. The advantages of using vacuum packing for cooked products should include the partial or complete inhibition of growth and proliferation of aerobic micro-organisms (Chapter 3). In addition, there should be handling benefits in that packaged food is not as susceptible to re-contamination as un-packaged food stuffs because packaged products may be handled directly without the need for additional containers. Finally, direct labelling of bags or packages can allow better and more efficient stock control.

The rationalisation of large scale processes in food production requires a systematic approach and it may be argued that pre-packaged foods offer higher potential for systems rationalisation than foods stored randomly or in rigid tray-like containers. Thus, the large scale vacuum packaging cook–chill systems may claim some advantages in terms of management and control over their non-packaging counterparts. However, the one major benefit that should be obtained from such systems, given tight control of all other parameters in the production and storage process is the potential for longer shelf-life. Extended shelf-life may not actually be demanded or needed by many caterers (although there is an obvious benefit to the retail food manufacturer) but it can offer an additional assurance of microbiological quality (and, to some extent, organoleptic quality) over the shorter shelf-life already safely achieved by the standard system.

This chapter, therefore, will describe the main large scale vacuum packed cook–chill food production systems and their advantages and disadvantages. The following chapter will focus on the more specialised and potentially more beneficial method of vacuum cooking that has been called '*sous-vide* cuisine'.

BENEFITS, DISADVANTAGES AND REQUIREMENTS FOR VACUUM PACKAGING

Vacuum packing can considerably extend the viable shelf-life of many cooked foods, although it should be stressed that it must be used under strict conditions of hygiene and control. It follows that this method could be applied to cook–chill food systems to increase food quality and microbiological assurance. This section will first briefly examine the benefits and disadvantages of vacuum packaging and the main requirements of the packaging materials (for a more detailed discussion see Chapter 6). We will then discuss systems which utilise vacuum packaging combined with cook–chill methods for large scale food production in catering.

Use of vacuum packaging (in gas-impermeable, heat-stable materials) for hot, pre-pasteurised foods or for pasteurising foods within the vacuum pack has advantages which include:

(1) no or low risk of post-pasteurisation contamination;
(2) ease of handling;
(3) inhibition of growth of aerobic organisms;
(4) in-pack chilling can include iced water bath chilling;
(5) inhibition or slowing of deleterious oxidative reactions in the food during storage due to the oxygen barrier properties of the packaging material.

The disadvantages of the method include:

(1) added expense of equipment for vacuum packaging and purchase of non-reusable packaging materials for each food item;
(2) growth of anaerobic organisms, normally restrained by competitive growth of aerobes (see Chapter 3), could be stimulated or, at least, allowed by the conditions within vacuumised food packs;
(3) whilst in-pack, foods cannot be properly inspected for spoilage or unwanted changes without breaking the bag seal and thereby curtailing the specific sample's shelf-life;
(4) without specially designed devices it is difficult to measure the core temperature of vacuum packed foods;
(5) added training necessary for food production staff.

There are a number of criteria required of the films used for vacuum packaging in large scale production methods, especially those which require in-pack pasteurisation (i.e. the pasteurisation of the food after it has been packaged). In fact, in the original publication on the first large scale cook–chill catering system to use vacuum packaging as an integral part of the methodology for food production (the Nacka system—see below), various of these criteria were laid down for the packaging film used in the process. Today we can list even stricter requirements. These include:

(1) high durability, i.e. able to withstand considerable mechanical stress during packaging, handling and transport;
(2) retention of flexibility even at low temperatures (−2°C to 4°C) to enable satisfactory handling in the packaging and refrigeration rooms;
(3) ability to withstand heating to at least 100°C without structural damage, leaching of potentially toxic plastics or plasticisers or changes in any other properties (e.g. gas permeability—see below);

(4) impermeability to liquids, including oils and molten fats, and macromolecules such as proteins;
(5) impermeability to gases, in particular oxygen, so that oxidative deterioration of the packaged food-stuffs is limited or inhibited;
(6) manufactured from non-toxic, food-acceptable, odourless materials;
(7) must allow air-tight durable heat seals to be made to close packs.

Many of these criteria have been met by a range of materials, mostly multi-laminate plastics. A wide range of materials are now manufactured by a number of companies throughout the world which are suitable or even specifically designed for use in large scale vacuum packaging/cook–chill operations. Much has been written on these materials and their properties so this topic will not be discussed here.

THE NACKA SYSTEM

In 1960 the authorities in Stockholm, Sweden commissioned an inquiry into the rationalisation of food production for the 43 existing city and county hospitals. Simultaneously, the Lowenstrom Hospital began trials of a new method for cooked, chilled food production that had been introduced by a local mining company. These trials were on a very small scale but were promising enough to be transferred in 1963 to the Nacka Hospital which had a newly built kitchen facility. In 1964 the committee appointed in 1960 reported back and effectively gave the go-ahead for the optimisation of the system that was, by now, called the Nacka system. Since then, the Nacka system has been widely used in Sweden and other parts of Scandinavia and Europe.

The Nacka Method

According to the first reports on the method, food is prepared and cooked by various traditional cooking methods but temperature in the food-stuffs is monitored to ensure a core temperature of at least 80°C. Cooked products are then transferred in their hot state to plastic bags, air is extracted and the bag is sealed to make it air-tight. Sealed bags are then placed in boiling water (100°C) for three minutes. The hot bags are then immediately transferred to a conveyer which takes them through a cooling tunnel with running water, the temperature of which is 10°C at first, then

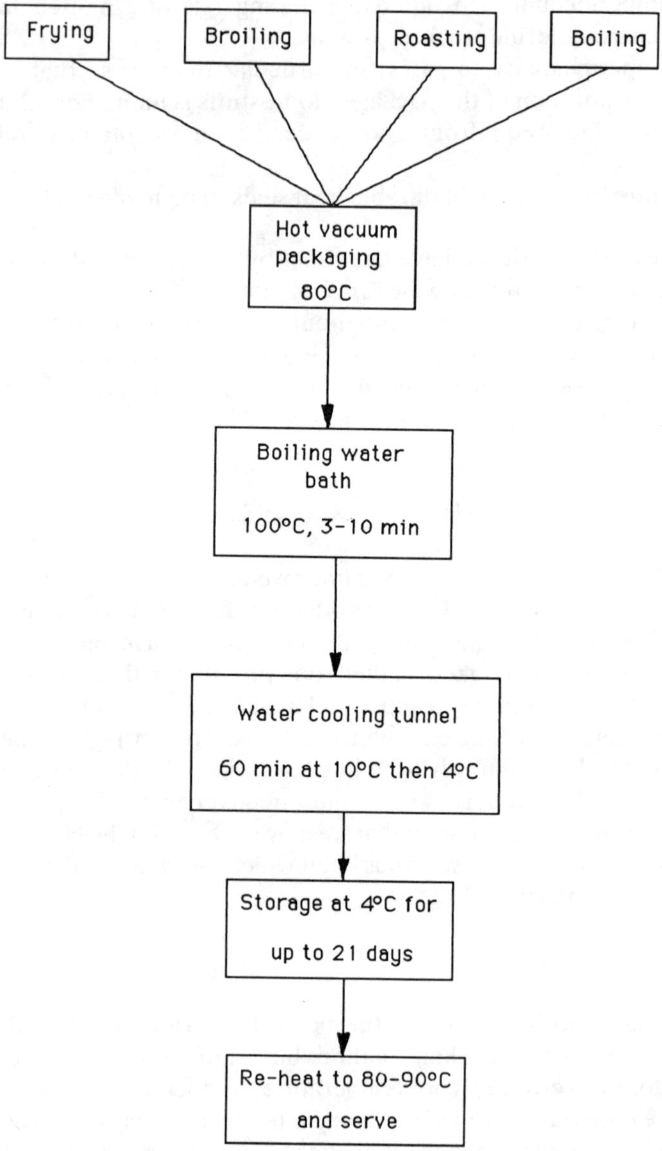

Fig. 7.1. Flow diagram illustrating the original Nacka system of cook–chill production.

reduced to 4°C. The packaged food is dried and stored in a refrigerated room at a temperature of 4°C or less. After the required time (normally 2–21 days) the food is removed from chilled storage and re-heated to at least 80°C (food core temperature) prior to service. The original Nacka food system is illustrated in Fig. 7.1.

Some foods proved hard to adapt to the new system and so were given slightly altered treatments. These included omelettes, some other solid foods, sliced meats and boned meats. All such foods either required cooking temperatures of less than 80°C or cooled unacceptably during slicing or boning. Such products were also packaged and heated at 100°C for three minutes or longer (up to 10 min) to ensure microbiological safety.

In the original Nacka Hospital system, the kitchen was fitted as an entirely conventional unit with fryers, standard roasting ovens, boiling pans and tilting steam kettles. In addition, the unit had a special packaging room in which two machines were used to package and seal food in bags. A heating tank was used to hold the packages at 100°C for the required time (for full pasteurisation) prior to their transfer to a 7-m cooling tunnel which carried food to the refrigeration area. The system allowed the Nacka Hospital caterers to increase production from 1200 to 7500 meals per day and to supply food to ten other Stockholm hospitals as well as a mental hospital 220 miles away.

Microbiological and Organoleptic Quality

There have been several publications on the use of the Nacka system but very few have dealt quantitatively with the problems of microbiological, nutritional and sensory quality. In the original publication on the method it was stated that preparation and packaging should be done under the strictest of hygienic conditions with full temperature and time monitoring. Under these conditions, it was suggested, the food posed little or no microbiological risk. In the first five years of operation at the Nacka Hospital over 5 000 000 food portions were prepared and over 10 000 microbiological tests were done (on site). In no instance was a product found to present a risk to consumers. It was stated that:

> All fully developed bacteria have then been killed by means of this heat treatment (3–10 minutes at 100°C) . . . No development of bacteria from any surviving spores can take place at refrigerated room temperatures of 4°C or lower . . .

Also, the use of taste panels, composed of both 'experts and laymen',

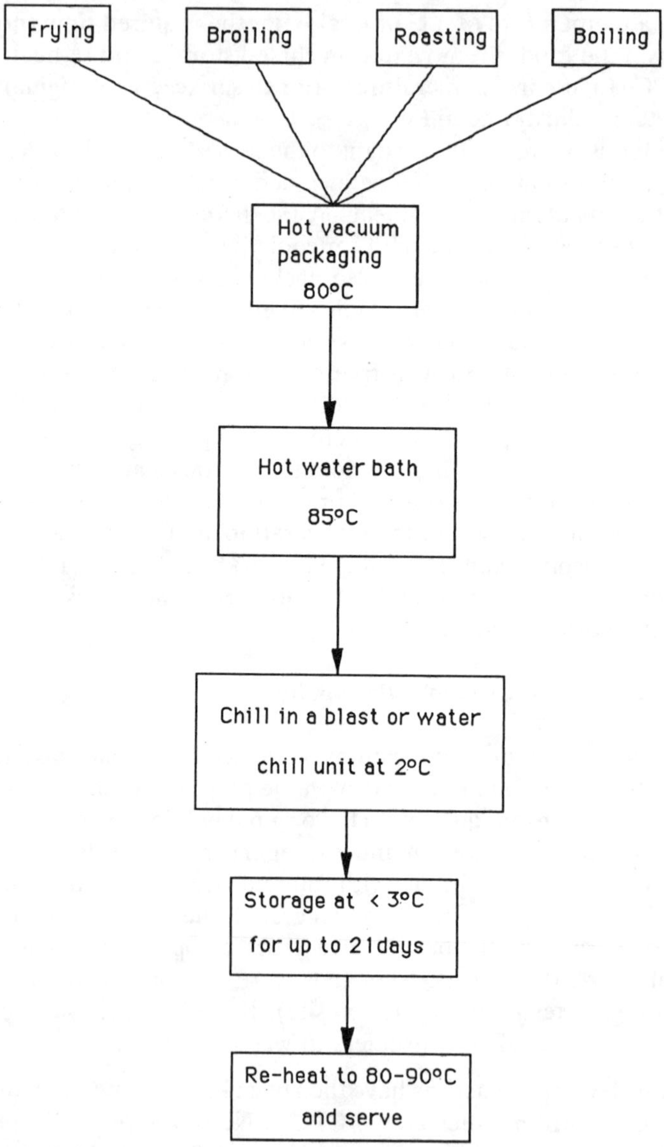

Fig. 7.2. Flow diagram illustrating a modification of the original Nacka system of cook–chill production.

showed the food to be as acceptable as that produced in a standard hospital kitchen under normal conditions. This does not necessarily mean, of course, that the quality of the food was good. Later publications claimed a shelf-life of 2–3 weeks for food produced by the Nacka system although it was recommended that food would have normally reached its limit in terms of organoleptic acceptability by the end of 14 days. Several writers have commented on the 'tired' taste of foods produced by the Nacka system and this almost certainly relates to the triple thermal processing used (i.e. initial cook, 100°C processing prior to chilling, 80°C re-heat prior to service). In other words, the food is over-thermalised in this system.

The point to make about most of the publications which have arisen on the Nacka system, however, is that, unlike most rigorous food science/microbiology research papers, these were somewhat less rigorous and presented very little if any quantitative data. This probably relates to the low level of publicity and interest accorded to catering systems amongst food scientists in the 1960s, 1970s and early 1980s.

Variations on the Nacka System

The Nacka system was not taken up on a world-wide scale but it did attract attention, notably in Germany and Switzerland and was used, especially by professional companies producing foods for institutional caterers. In this form, the method was modified in order to be used with either bags, pouches or rigid trays (aluminium or plastic). Figures 7.2 and 7.3 show diagrams which illustrate two of the modifications of the Nacka system which have found wide application in Europe.

Perhaps the most unusual variant of the Nacka system utilises an extra step after initial cooking (Fig. 7.3). In this case, the food is allowed to cool, in-pack, to room temperature (20°C) where it is held for a period of 24 h. At this temperature, heat-shocked spores of anaerobes will begin to germinate and grow. The next step, therefore, uses a second high temperature pasteurisation (85–90°C) to kill these germinated organisms. Theoretically, after this second heat-processing step, all organisms will be killed and the food will be, effectively, sterile. However, the technique has several major drawbacks:

(1) it is difficult and lengthy to use;
(2) it requires considerable control and understanding to avoid contamination of the food with viable germinated anaerobes;

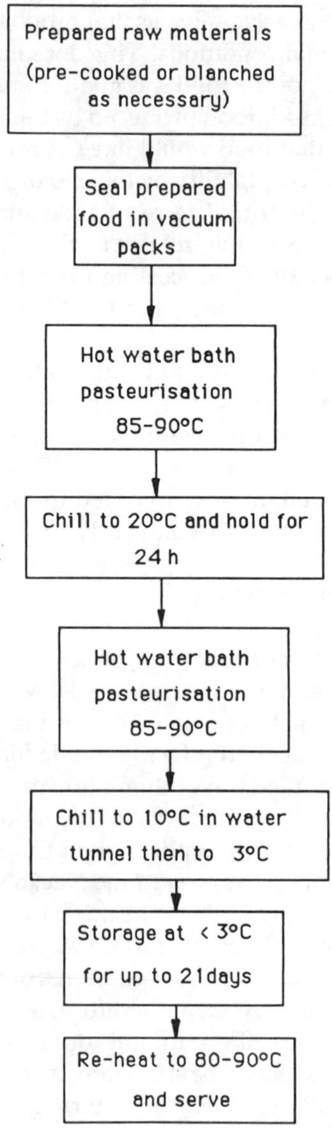

Fig. 7.3. Flow diagram illustrating a modification of the original Nacka system in which spore germination is allowed in a short 20°C incubation before final pasteurisation.

(3) it involves three high temperature treatments as well as holding the food for 24 h at 20°C, i.e. enough thermal treatment to completely destroy the organoleptic quality of the product.

THE AGS SYSTEM

In 1968 the Duke Endowment and three South Carolina hospitals in the USA sponsored a study to produce a new food preparation, storage and service system that could be utilised by hospital caterers. The system that was subsequently evolved (the AGS system) was named after the three hospitals initially involved in its development, i.e. Anderson, Greenville and Spartanburg. Like the committee which brought about the Nacka system, those involved in the AGS study decided that vacuum packaging should be used as an integral part of their new system and that freezing should be avoided due to considerations of higher cost, more time in tempering and thawing frozen foods, difficulties in handling frozen foods and more damage incurred by foods due to ice crystal-induced changes in structure and, thereby, texture.

The AGS Method

The original AGS system (see Fig. 7.4) involved the preparation of high grade ingredients and their assembly in either the raw state or partially cooked as needed. That is, some items might need to be part-cooked or blanched prior to prime cooking (pasteurisation), e.g. some vegetables, sealing of meat. The ingredients of a menu item were then packaged under vacuum in special plastic packs and thermostatically sealed. The in-pack foods were then pasteurised (prime cooked) for a specific time (depending on the menu item) before being transferred to a large iced water tank where they were chilled to −2 to +2°C. The packs were then transferred to a chill store which maintained a temperature of −2 to 0°C although food was not allowed to freeze. The shelf-life of the food under these conditions was claimed to be 60 days. Chilled foods (stored in crushed ice) were delivered to satellite kitchens where they were re-heated in a second hot water bath to a core temperature of 70°C before being plated, given a short microwave heating (10–20 s) to ensure a high service temperature and then served (Fig. 7.4).

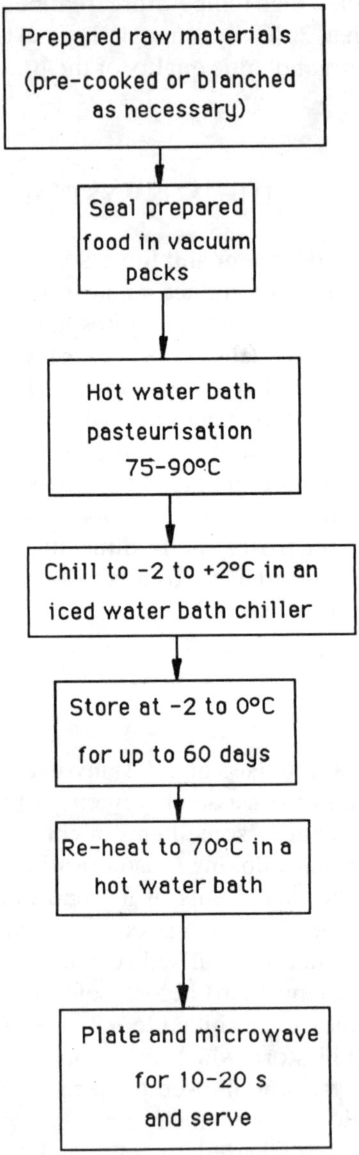

Fig. 7.4. Flow diagram illustrating the AGS system of cook–chill production.

Quality of Foods Prepared by the AGS System

An interesting point to note is that, in original trials the AGS system was very similar to the Nacka system but the investigators found the food to have a 'tired' unpalatable flavour which they attributed to the triple cooking necessitated by this procedure. Instead, they opted for a production process with only two heating procedures, i.e. the cooking in-pack and the subsequent re-heating, thus improving end-product quality somewhat.

There is little or no scientifically accredited literature about the AGS system, in particular, there is no quantitative study of the microbiological, nutritional and organoleptic quality of food prepared in this way, nor is there any report of consumer acceptance of this type of food. However, the method is similar in many respects to the later *sous-vide* method which will be discussed at some length in Chapter 8. This latter method has been investigated to some extent in terms of food quality and these findings could be equally applied to the AGS system.

THE COOK-TANK SYSTEM

The Nacka and AGS systems have been adapted and modified over the years and now there are a number of commercially available systems which are similar to or based on these original methods. These involve:

(1) cooking foods in large scale tanks and then pumping them into heat-stable packs at high temperature (80–85°C—see Chapter 5);
(2) the pre-packaging of foods, their pasteurisation in a large scale jacketed tank at pre-determined temperatures and their subsequent chilling in a chill tank.

Like the AGS system, manufacturers of cook-tank systems have made considerable claims for the possible shelf-life of foods prepared in this way and stored at 0–3°C. In some cases, a shelf-life of 45–60 days is claimed (indeed, in the USA a 45 day refrigerated shelf-life is allowed for these products), whilst in others even longer periods have been quoted. Major advantages quoted by the manufacturers are the high control offered by such systems and the consequent guarantee of food safety through proper pasteurisation and chilling. Most cook-tank systems allow two forms of food preparation. These are the kettle-cooked product method and the cook-tank product method.

The Kettle-Cooked Product Method

Liquids or semi-liquid foods, for example pie fillings, scrambled eggs, soups, sauces, stews, etc., are cooked at pre-programmed and recorded times and temperatures in a jacketed or direct heated tank as described in Chapter 5. When the cooking cycle is complete the cooked product is pumped at 80–85°C into multi-laminate, heat-stable plastic casings with similar low gas permeability properties to those previously discussed. Any air spaces are sucked out (although a true vacuum pump is not used) and the casings are sealed with a metal clip closure. The product is then placed in a bath of water chilled to 1–3°C in which the packs are gently agitated by a paddle system which also helps to 'fold' the food within the packs and thereby hasten cooling. Chilling to less than 4°C is normally achieved within 45 min even in large (5–10 kg) packs. The food is then placed in cold storage at 0–3°C for up to 45 days. Products may be re-heated in-pack or removed to be re-heated in a variety of ways at satellite kitchens.

Cook-Tank Product Method

Products which cannot be produced in a kettle, e.g. whole joints of meat, mixed or large vegetable items or assembled products which could easily break up, are cooked ready packed in vacuumised, heat sealed bags in a temperature-controlled water bath. After initial preparation the products are placed into the same type of casings as above and the air is drawn out in a vacuum machine prior to sealing with metal clips or heat-sealing. The hermetically sealed casings are placed into wire baskets and lowered into a jacketed or direct heated water bath (Fig. 7.5). The tank is pre-programmed for the specific items to be cooked in terms of time and temperature. When the cooking cycle is completed the hot water is replaced by chilled water to bring the product temperature rapidly down to 2°C. Throughout its cooking and cooling cycle, time and temperature of the cooking and cooling waters and the internal food core temperature are recorded on a chart. These systems place considerable emphasis on their ability to record the production details of every food product on a hard copy.

Few studies of any scientific note have been published on this system. Therefore, it is extremely hard to meaningfully appraise the method or to gain any quantitative insight into its viability or the quality of the food produced. Despite this, however, these systems are accredited in Europe

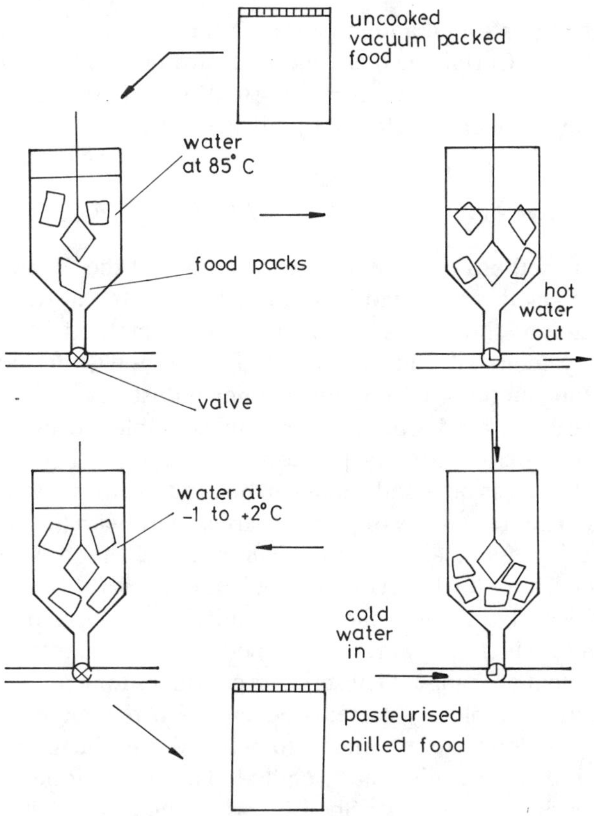

Fig. 7.5. Diagram illustrating the principle of the cook-tank production method of in-pack cooking.

and the USA where they have been evaluated by the regulatory authorities. It is left to the user to assume, therefore, that the degree of assurance given by the method concerning microbiological safety is reasonable for the very long shelf-life quoted. There is no doubt, therefore, that there is a pressing need for an in-depth investigation of the microbiological risks involved with the use of these methods and the tolerance or otherwise of such systems to abuse.

In theory, the combination of two cook-tank methods for pumpable and non-pumpable foods would seem to allow the production of a very wide range of goods on a very large scale with a minimum of equipment and processing provided a high degree of time and temperature control can be exerted. Certainly, these methods are much more akin to food manufacturing technology than catering and, as such, should be viewed as major advances in large scale cook–chill systems.

OTHER METHODS USING VACUUM TECHNOLOGY

Some other methods exist which are variations of the themes discussed above. Perhaps the most widely used of these is the in-tray or in-pack pasteurisation method. This is a batch steaming method for foods which have already been sealed in a container. The method is very similar to the retort cooking methods used in the canning industry.

In this case, the food is pre-packaged in assembled form, either as the raw materials or as partially pre-cooked materials (e.g. convenience frozen foods such as peas and beans can be used), in formed trays or other containers. The trays, bags or pouches are sealed (with lids if necessary) against the ingress of air or water. The packaged food is loaded onto trolleys which have finely perforated shelves to allow the passage of water between the shelves. The trolleys are loaded into a large pasteurisation chamber into which hot water is pumped from above after it has been heated in a heat exchanger. This process continues for a pre-set time until the foods are completely pasteurised at which point the hot water is replaced by chilled water which is also pumped over the food from above until the products are adequately chilled. The chilled foods can then be stored in-pack until needed. Shelf-life is claimed to be 20–40 days at 0–3°C although products prepared by this method in 'hard' containers like cans can have a life of up to two years.

The fundamental principles upon which the batch steaming method are based are the same as those underlying all the other methods so far discussed in this chapter, namely the combination of the beneficial effects of pasteurisation, rapid chilling, controlled chilled storage and vacuumisation. However, in this case foods which require stirring during cooking cannot be made as the packs are held on static shelves but very large batches can be prepared simultaneously and under identical conditions.

THE APPLICATION OF LARGE SCALE VACUUM PACKAGING METHODS

In this chapter we have discussed the main applications for vacuum packaging in large scale cook–chill food production systems. In Chapter 5 we also mentioned continuous (conveyer) cooking in which packs of pre-cooked food can be vacuum sealed on-line. All of these methods rely on two basic principles:

(a) Properly pasteurised food which is vacuum packed whilst held at pasteurisation temperature (or above) will not be contaminated with vegetative forms of any micro-organisms when packed. As long as the packs remain undamaged, the food cannot become re-contaminated by micro-organisms before re-heating and service.
(b) High quality raw materials which are properly pasteurised inside vacuumised containers (boxes, trays, bags or pouches) will not contain any vegetative forms of micro-organisms and cannot be re-contaminated provided the packaging material remains intact.

Both methods also utilise the precept that if foods are very quickly chilled to temperatures below 3°C, then the germination of spores remaining in the food after pasteurisation will be inhibited and the growth of the vegetative forms of spore-forming anaerobes will be slow or even unlikely to occur at all. These principles are well founded and concur with our knowledge of the vast majority of food spoilage and food poisoning organisms but it is interesting and, perhaps, disturbing to note that very little scientific study has been carried out on how these methods are actually being used in practical situations in the catering industry. Only one early study attempted such research but this work set out to compare different menu items under different conditions and, thus, little of any relevance can be drawn from the data produced. Although the experiment was poorly constructed, the data does show that use of vacuum packing and restriction of storage temperature to less than 3°C leads to a high degree of inhibition of growth of micro-organisms in such products.

In the next chapter we will discuss a variation on the systems described above called the *sous-vide* method, in which much finer control of the in-pack cooking temperatures of complex meal items is exercised and very high quality, gourmet-style menu items can be prepared.

FURTHER READING

Anon. (1970) Pressure cooking vacuum packs. *Food Processing Industry*, **39**, 460.

Bjorkman, A. & Delphin, K. A. (1966) Sweden's Nacka hospital food system centralises preparation and distribution. *Cornell Hotel and Restaurant Administration Quarterly*, **7**, 84–91.

Leistner, L., Hechelmann, H. & Vukicevic, Z. (1972) Mikrobiologische Risiken bei industriell hergestellten Fertiggerichten zur Gemeinschaftsverpflegung. *Die Fleischwirtschaft,* **8,** 993–6.

Lott, G. (1973) Mikrobiologische untersuchungen von fleisch fertiggerichten, die nach dem Nacka—Verfahren hergestellt wurden. *Praventivmedizin,* **18,** 371–4.

McGuckian, T. (1969) AGS food system—chilled pasteurised food. *Cornell Hotel and Restaurant Administration Quarterly,* **10,** 87–99.

Porsdal Poulsen, K. (1980) A system for continuous production of ready made meals and distribution in aeseptically filled containers. In: *Advances in Catering Technology,* (ed. Glew, G.), London, Applied Science Publishers, pp. 265–7.

Schwebel, W. (1977) Pasteurised cold stored prepared meals, with shelf life of up to 3 weeks. *GV-Praxis mit Tiefkuehlpraxis,* **17**(2), 69–70.

Tändler, K. (1972) In Folienbehältnissen pasteurisierte Fleischfertiggerichte fur die Gemeinschaftsverpflegung. *Die Fleischwirtschaft,* **52**(7), 845–50.

CHAPTER 8

Cooking in Vacuumised Containers— Sous-Vide *Cuisine*

INTRODUCTION

This chapter deals with a relatively new development of classic cook–chill catering—namely, *sous-vide* cooking. The idea of cooking foods in sealed containers, or, at least, pasteurising foods within hermetically sealed bags, pouches or other containers was successfully used in the 1960s in two large scale systems (see Chapter 6). However, neither of these systems achieved the kind of quality that might be associated with gourmet cooking or haute cuisine.

In the late 1970s a French chef, George Pralus, set out to solve the problem of weight and juice loss in paté de foie gras. He began, following the principle of cuisine *en papillote* (a classical method, mentioned by both Mrs Beeton and Escoffier, of cooking foods wrapped in oiled paper pouches to retain flavour and moisture), by wrapping the foie gras in a plastic layer and cooking. The result was not successful but he persevered and eventually, by using three layers of plastic to enclose the foie gras during cooking, he was able to reduce weight loss dramatically from about 40% to around 5%. This extraordinary result led to a collaboration with a plastics manufacturer and the evolution of a multi-laminate plastic material that was both heat resistant and able to retain significant gas impermeability even after high temperature treatment.

So was born the method of *sous-vide* cooking, or cooking foods sealed under vacuum in multi-laminate plastics. Since then, numerous developments and changes have led to the general acceptance of the method in France and the use of the method as an haute cuisine adjunct to standard cook–chill procedures. Hand-in-hand with these developments has grown an apprehension in the minds of food scientists and food microbiologists concerning the safety and reliability of the method when used as a cook–chill variant. Claims made for the method in terms of shelf-life extension and quality retention may have obscured the more important aspect of microbiological safety and the vulnerability of the method to abuse.

In this chapter we will review the *sous-vide* method itself and its role as a variant of the cook–chill system. We will also consider the drawbacks of the method as a cook–chill variant as well as its claimed advantages, especially in the light of recently expressed concerns such as:

(1) the fact that, in early 1988 the Food and Drug Administration in the USA outlawed the method for retail use pending an enquiry to establish the safety of the system; and
(2) the possibilities for in-pack anaerobic growth of organisms such as *Clostridium botulinum* (Chapter 3) and the consequent threat of food poisoning.

Finally, we will consider the small amount of research that has been carried out to date on the method and evaluate its potential as yet another tool for the cook–chill caterer.

THE *SOUS-VIDE* PROCESS

The literal translation of *sous-vide* is 'under-empty' and merely describes the process of placing food-stuffs in evacuated containers. In fact, it is important to note, at this stage, that, in France, the average chef might refer to the *storage of a cooked food in a vacuum pack* as *sous-vide*. This causes some confusion as the method to which we are referring in this chapter is somewhat more specialised. The key phrase in this context is *sous-vide* cooking. *Sous-vide* cooking involves the pre-packaging of foods in plastic bags or pouches, forming a vacuum, and cooking the food within the bags at controlled temperatures. There are several variations of the method which allow the pre-packaging of part-cooked foods, raw foods and the holding of raw foods at chill temperatures for some time before cooking (see later). However, the system of *sous-vide* cooking in its simplest form involves the following stages:

(1) *Preparation*
The preparation of the raw ingredients e.g. washing, peeling, trimming, boning, seasoning, etc., prior to cooking. These methods follow the traditional approach and merely involve the preparation of the food for cooking as in standard cook–chill methods.

(2) *Pre-cooking or browning*
In *sous-vide* cooking the food is heat processed inside a plastic container. As the food is not directly exposed to radiant heat (in fact, radiant heat should be avoided because of the risk of damage to the plastic containers) the browning effect is minimised. Thus, it may be necessary for some products to pre-brown them by flash frying for example. It is also possible to mark foods to mimic certain cooking methods, e.g. branding a 'grill shape' on the surface of meat. Similarly, some vegetables have to be blanched before packaging because of their strong flavour. A side effect of the *sous-vide* method is the sealing in of flavours and aromas into foods (see later). Products such as chicory which can be very bitter or Brussels sprouts, which can have a very powerful flavour and aroma, have to be blanched to reduce their flavour and aroma before being sealed in the bag or pouch.

(3) *Vacuum packaging*
The prepared food is placed into special heat resistant, air impermeable plastic bags or pouches. These containers are the key to the *sous-vide* process and are produced from multi-laminate plastic materials. The barrier films, such as polyamide (i.e. nylon) which give the multi-laminate its high gas impermeability, tend to be toxic and not suitable for use next to food-stuffs. They are, therefore, sandwiched between layers of heat-resistant plastics approved for use with foods to provide a packaging film with the desired heat resistance and barrier qualities. Once the food is in the pack, a vacuum is produced, using a specially designed vacuum cabinet, to remove all residual air from the bag. The pack is then heat sealed immediately before any loss of vacuum occurs. As discussed in Chapters 6 and 7, the use of a vacuum greatly inhibits the progress of oxidation reactions which generally lead to the deterioration of food-stuffs during storage either before or after cooking and inhibits the growth of aerobic micro-organisms. Also, the use of a vacuum serves to pull down the plastic film tightly onto the food surface. This allows maximal heat transfer into the food through the plastic during cooking. If an air or gas space was left in the bag or pouch this would act as a layer of insulation in much the same way as the air layer between glass sheets in double glazing. In the vacuumised condition the plastic film merely acts as a skin on the surface of the food and does not unduly inhibit heat transfer.

(4) *Pasteurisation*
The packaged food is cooked (pasteurised) for a set time and at a set

temperature appropriate to the individual characteristics of the food to be cooked (or until proven microbiologically safe levels have been achieved). Cooking can be carried out in a water bath, i.e. so that the vacuumised bags or pouches are immersed in water of a controlled temperature. Another popular method is the use of the steam combination oven (Chapter 5). This device has recently been modified to allow more accurate temperature control of the kind needed in the *sous-vide* process. Pasteurisation is one of the most critical steps in the *sous-vide* process. As the method relies on the strict control of temperature to produce highly flavoured, tender products (i.e. often using low temperatures for long times) it is essential that the user is confident that the procedures adopted are sufficient to fully pasteurise all products.

At this stage the food may be served immediately. In this case the method is merely a refinement of the classical catering approach and may be seen as a modern version of the '*en papillote*' system. Figure 8.1 illustrates this method and refers to food dishes which are found in high-class restaurant outlets where high quality expensive ingredients need to be kept at optimum conditions for a short period and maintain the intrinsic gastronomic food qualities expected of such establishments. Cooking a food within a sealed high-barrier container means that the natural juices of the food as well as its flavours and aromas are sealed in during the cooking process. Thus, when the bag is opened just prior to service, the full flavour and aroma of the food is released. This is seen as a distinct advantage by many caterers. This method is currently used very widely in Europe in certain high-class restaurants and food outlets.

After the initial stages prior to cooking have been completed (discussed above), the raw or part-cooked food can be held at controlled chill temperatures between 0 and 3°C. This is a method used as a convenience by some caterers, i.e. the pre-packaged, unpasteurised food (normally in single portion packs although multi-portion packs are equally viable) can be taken from the chill store and immediately pasteurised prior to service or, as we shall discuss below, re-chilling. Usually, only one component of a dish, with some garnish, is vacuum packed, and the sauce is made separately, vacuum packed and added to the dish prior to service when both have been cooked.

This practice may appear to be a viable method for preserving a complex dish prior to cooking and serving but it obviously carries risks of microbiological hazards resulting from the contamination of food (during harvesting or handling) with psychrophiles which may then grow to unacceptably high levels before pasteurisation. The cooking stage may then not be sufficient to make the food safe for consumption.

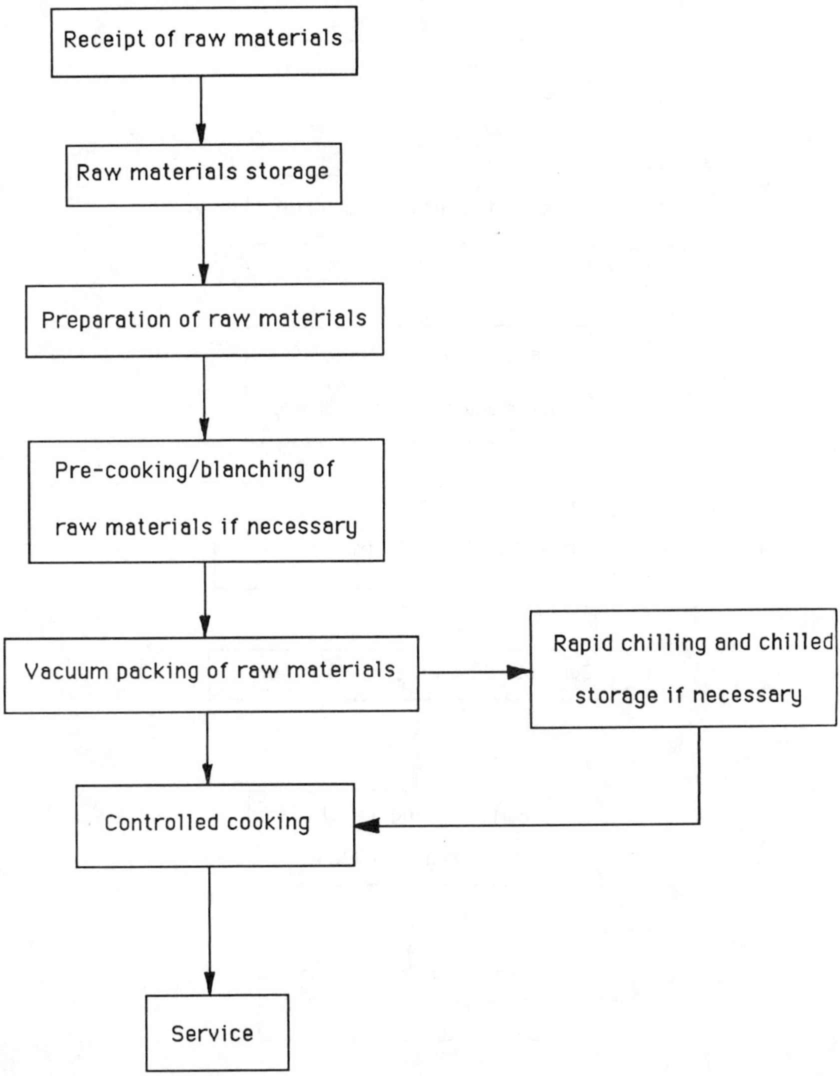

Fig. 8.1. Illustration of the use of *sous-vide* production in a conventional catering process.

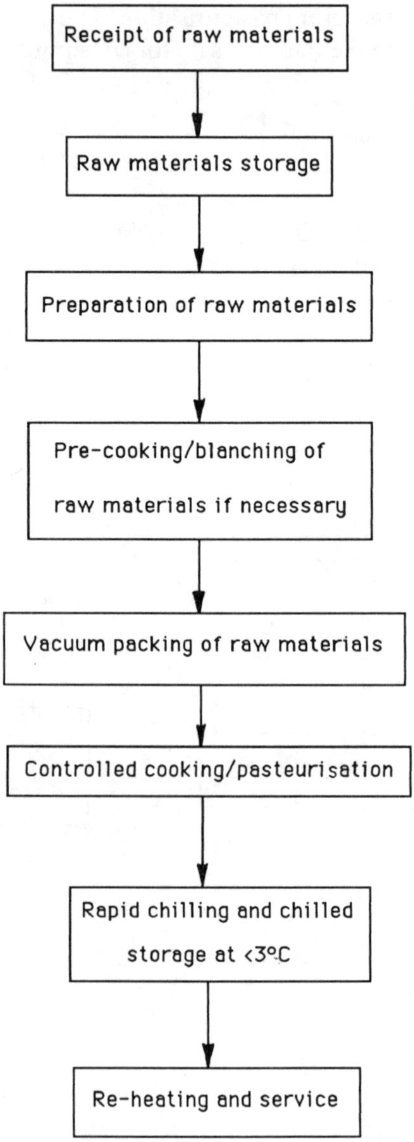

Fig. 8.2. Illustration of the adaptation of the *sous-vide* process as a variant of the cook–chill system.

Sous-vide can also, as we have already mentioned, provide a variation of the cook–chill method. Figure 8.2 shows the work flow of a *sous-vide* system adapted for use as a cook–chill technique. As can be seen, there is very little difference from the standard cook–chill method except that the raw materials may have to be pre-processed prior to pasteurisation and they are, of course, pasteurised in vacuumised bags or pouches. In this case, the procedures following steps 1–4 (above) would be as follows:

(5) *Rapid chilling*
After cooking, which itself must take the form of a proper pasteurisation step, the packaged food is chilled to between 1 and 3°C as rapidly as possible (e.g. within 90 min as stipulated by the UK *DHSS Guidelines on Pre-cooked Chilled Foods,* 1980). Direct chilling in an iced water bath (an option that is normally not possible with most standard cook–chill menu items) is made possible because food is pre-packaged. This method is both cheap and efficient and, from the limited research carried out to date, appears to be a better choice than blast chilling (see below).

(6) *Chilled storage*
Storage at 0–3°C for a period of time prior to re-heating and service. Storage conditions must be as rigorous as those laid down for standard cook–chill menu items. However, the fact that the food is stored under vacuum in sealed containers must lead to certain advantages in slowing oxidation reactions and in inhibiting the growth of aerobic micro-organisms (Chapters 3 and 6). These points are discussed below.

(7) *Regeneration*
Re-heating and service of the product. This procedure may take place in the bag with the product still under vacuum, i.e. in a heated water bath or a combination steam oven, or after the contents of the bag have been emptied into a separate container for heating on top of the cooker or inside the oven in a heat-proof dish. Alternatively, after puncturing the bag, the product could be re-heated in a microwave oven. In any event, whatever re-heating method is chosen, the food must be re-heated to at least 70°C at its centre as advised for standard cook–chill menu items.

In one of the first guide books published on the method the following maxims were suggested for maximum safety and quality of products prepared by the *sous-vide* cook–chill process:

(a) Use only fresh, good quality ingredients from reputable suppliers. *Sous-vide* cannot compensate for poor quality raw materials.
(b) Use only high quality impermeable plastic pouches in conjunction with a perfect vacuum and seal.
(c) Strictly ensure correct times and temperatures for pasteurisation, rapid chilling, refrigerated storage (and refrigerated transport, if used) and re-heating.

Other points to observe in the process in terms of optimisation and safety include ensuring the full and proper training of all staff in the *sous-vide* technique and its application in cook–chill catering, preparing the raw materials in a clean, cool kitchen environment as quickly as possible, maximising the use of manpower and avoiding cross-contamination of foods (raw/cooked/stored), strictly adhering to the standard recipes, using an efficient date labelling system to ensure accuracy of records on stored products.

THE ADVANTAGES AND DISADVANTAGES OF THE *SOUS-VIDE* SYSTEM

It is interesting to note that very little true research has been carried out on the *sous-vide* method to date. This is probably because the method evolved in the catering sector of the food manufacturing business, a sector which has had a poor history of either sponsoring or carrying out scientific research. For this reason, the often quoted advantages and disadvantages of the method have been based, usually, on the working practices and opinions of leading chefs who have helped to develop the method and companies using the method on a commercial scale. Thus, any stated advantages and disadvantages may be coloured by bias, subjectivity and the rhetoric of commercial need. It is therefore important to examine both the advantages and disadvantages of *sous-vide* in some detail, using the results of objective research studies carried out to date. This section will first describe the claims which are based on opinion and qualitative observation and then will consider the potential benefits and drawbacks of the method in the light of the little scientific evidence available.

In terms of the quality of the food produced, the main advantages claimed for the method are that it causes the full retention of all food juices, flavours and aromas within the bag or pouch during the cooking process. Thus, it has been claimed that the water soluble vitamins

(Chapter 2), which would normally be lost by leaching, are retained within the food, especially if it is processed at low temperatures (in meat, typically, 60–70°C for immediate service only), or in juices trapped within the bag.

It is claimed that, in *sous-vide* cooking, natural flavours are retained to a higher degree than in conventional cooking. The need for flavour enhancers and additives—including salt, vinegar, wine, spices, herbs—is, therefore, reduced. Indeed, it has been suggested that in some recipes such items can be eliminated completely. This effect has been noted by several experienced users of the method and has led to the resultant claim that recipe cost can be reduced in *sous-vide* cooking due to the lower amounts of condiments and flavourings used.

In conventional meat cooking shrinkage and weight loss may be as high as 25–40% in meat depending on the heat input used. In *sous-vide* cooking this weight loss may be greatly reduced (i.e. to only 5–10% for some meats). This reduction in weight loss is a direct result of the use of a range of cooking temperatures well below 100°C which leads to the retention of much of the original food juices. This claim also impacts on the financial aspects of catering as well as those of quality. A saving of 10–20% in meat weight on cooking means 10-20% more meat remaining to be served at the end of the cooking process therefore leading to higher profits. Furthermore, if juice can be retained in meat and fish then succulence and textural enhancement should result.

The advantages mentioned above arise whether the *sous-vide* method is being used for immediate service of food products or as a cook–chill variant. Another advantage of the method, therefore, is simply the fact that it can readily be adapted for use as a cook–chill system. Indeed, exponents claim that the benefits in juice retention and quality are retained even when higher cooking temperatures are used to ensure complete pasteurisation before chilling. Thus, the method can be used to rationalise food production; using proven recipes, unskilled operatives to prepare raw food and pasteurise it under the supervision of a small number of professionally qualified and skilled technicians. The other advantages of cook–chill also accrue to cook–chill/*sous-vide,* i.e. dishes can be prepared in advance of need, dated to ensure stock rotation, re-heated (regenerated) at the point of sale by trained staff, and only small regeneration kitchens are needed with minimal equipment and less skilled manpower than a conventional restaurant.

In the hotel and restaurant sectors in particular a large menu selection is on offer and there is a variable demand for any dish during each service period. Raw materials are purchased well in advance and are usually stored prior to preparation, followed by another period of storage before eventual use or wastage occurs. Throughout the storage period the quality of foods inevitably diminishes and unless some method is adopted to retard this process spoilage costs will be high. The cook–chill/*sous-vide* system, therefore, offers some practical protection over traditional short-term storage methods whilst maximising the retention of the food's natural qualities.

Finally, the *sous-vide* technique could provide a useful means of ensuring high quality convenience foods for hospital patients requiring special dietetic meals. Satisfying meals could easily be developed which exploit the flavour retention properties of *sous-vide* foods to their maximum. With regard to special dietary requirements, this would enable the production of flavoursome meals without the need for potentially harmful ingredients such as salt, sugar, some fats. Indead, *sous-vide* products could be used to take full advantage of natural and fresh ingredients (e.g. fresh herbs, natural citrus juices, wine, spices) to enhance flavour.

The Hilton International hotel chain in Europe has been an established user of the method for some years. Their experiences in developing the method for their own needs led to the production of an internal document on *sous-vide* cooking in 1985. The advantages, described in this document, of *sous-vide* cooking over conventional cooking are summarised below:

(1) The cellular structure of the food does not change during cooking.
(2) As a properly controlled cook–chill method it is extremely hygienic, with no post-cooking contamination by air, handling or refrigeration odours.
(3) Excellent food cost control is gained.
(4) Up-grading of menu items is made possible by preparing food in advance.
(5) When used properly, the system can result in manpower savings.

In summary then, based on the claims made in the foregoing section, the *sous-vide* method might seem to be an extraordinary new method. It does, however, have a number of shortcomings, some of which might be seen by some as enough reason to prohibit its widespread use or even to ban the method altogether.

First, it has limited applications. On its own, *sous-vide*/cook–chill, like classic cook–chill, cannot provide an entire menu range because many items do not suit this unusual production method. For example, bakery items, many complex dishes and soufflés cannot be prepared successfully by the *sous-vide* method. It has been suggested that, contrary to popular claims as outlined above, it is best suited to operations where demand can be easily predicted.

In contradiction to the claims made for the method in terms of cost savings, it has been suggested that it leads to increased costs. These are said to be mainly due to the need to use only very high quality raw materials, the need for extra equipment (e.g. vacuum packaging devices and, possibly, specially designed ovens), and the need to package each item thus incurring the additional costs of the packaging material (inevitably to be borne by the consumer). Finally, as the method is somewhat labour-intensive, involving extra steps not normally encountered in standard cook–chill operations (pre-cooking procedures, packaging and vacuumising), there must be a knock-on cost in production.

Critics of the technique have suggested that its introduction into an existing business could lead to industrial relations problems. Caterers themselves may not accept the method at first, especially with its implications for extra hygiene control and an increased technological input. This point is based on research findings which show that technological change almost invariably leads to feelings of insecurity amongst employees and a fear of the unknown. This could lead to some measure of staff resistance to the introduction of the new technology as the change is seen as a threat to jobs and job quality. However, these criticisms may be levelled at any new technology and not specifically *sous-vide*. Indeed, they have been raised by research carried out on the introduction and implementation of classic cook–chill operations in the UK.

Finally, but most importantly, the method may pose a considerable threat to public safety if not properly used. The technique must be used within a controlled and hygienic environment, using high quality raw materials, proper packaging materials and very careful control of cooking (pasteurisation), chilling, chilled storage conditions and re-heating. A lapse in the monitoring or control of any of these parameters could lead to the growth of potentially lethal micro-organisms within the bag (see Chapter 3 on the growth of anaerobic organisms).

The *sous-vide* bag or pouch provides the ideal environment, given improper pasteurisation, for the growth of anaerobic organisms capable of growth at chill temperatures (e.g. *Listeria monocytogenes*—see Chapter 3). Even under conditions of complete pasteurisation, if the product is not held strictly below 3°C during chilled storage there is also the possibility of growth of spore-forming anaerobic organisms such as the clostridia.

What must be made clear is that all of the above stated benefits and disadvantages derive from assumptions made about the method which are based on the application of well-known theoretical considerations or from the writings or suggestions of chefs using the method. Few have yet been proven by experimental means. In the next section we will examine these claims drawing on results from the little research work that has been carried out to date.

EXPERIMENTAL RESULTS WITH *SOUS-VIDE* COOKING

Few serious studies of the *sous-vide* method have been carried out to date. Two such studies will be discussed here in the light of the comments made above. In the first study several products were investigated. These were chicken à la king, courgette provencale and white fish stuffed with shellfish. The products were cooked and chilled under strictly controlled conditions of temperature and time and stored at 1–3°C for up to 21 days. Microbiological and sensory analyses of the products were carried out. The sensory analyses consisted of both trained and consumer panels to assess changes in product sensory attributes as well as consumer preferences.

The microbiological analysis showed that after 21 days storage at 1–3°C, there was negligible microbial growth in the chicken and vegetable samples (Table 8.1). This data was corroborated in subsequent studies of other *sous-vide* products. One product, however, proved the exception to this general trend; this was the rolled, stuffed fish. The authors noted that this was cooked under less rigorous conditions (< 70°C) and under inadequate vacuum as a full vacuum would have crushed the product. In this case, counts of greater than 6×10^6 were recorded after 21 days' chilled storage, although at 14 days the product was still acceptable on a total count basis. The authors, therefore, stressed that success in inhibiting microbial growth in the *sous-vide* products they examined relied on adequate pasteurisation of products during the initial prime cooking stage, rapid chilling followed by adequate temperature control during storage. Soft products which caterers are reluctant to fully vacuumise pose special problems in the *sous-vide* process because inadequate vacuums within packs will inevitably lead to a higher risk of the growth of psycrophilic aerobes.

In this study the consumer acceptability trial showed that the chicken à la king rated very highly with consumers even after 14 days' storage (Table 8.2). At this time only 30·5% of consumers preferred the fresh sample whilst 69·5% either preferred the stored sample or had no preference. The courgette dish, however, did not show the same retention of

TABLE 8.1
Microbiological Analysis of *Sous-Vide* Food Products (from Light *et al.*, 1988)

Storage time 1–3°C (days)	*Total viable count (per gram of sample)*					
	Chicken[a]		*Courgette*[b]		*Fish*[c]	
	A	*AN*	*A*	*AN*	*A*	*AN*
0	70	30	> 20	> 20	102	167
7	60	> 10	> 20	> 20	260	170
14	80	> 10	20	> 20	48 000	1 600
21	95	20	> 20	> 20	$6{\cdot}6\times10^6$	69 500

[a] Chicken à la king.
[b] Courgette provencale.
[c] Rolled stuffed white fish.
A = total bacterial counts of aerobic micro-organisms.
AN = total bacterial counts of anaerobic micro-organisms.

TABLE 8.2
Consumer Preferences for Fresh and Stored *Sous-Vide* Products (from, Light *et al.*, 1988)

Samples tested with consumer panel[a]	*Chicken à la king*[b]			*Courgette provencale*[b]		
	No pref.	*Fresh*	*Stored*	*No pref.*	*Fresh*	*Stored*
Fresh compared with fresh	44	25	31	33	33	33
Fresh compared with 7-day-old sample	9	45	46	17	61	22
Fresh compared with 14-day-old sample	37	29	34	18	68	14
Fresh compared with 21-day-old sample	nd	nd	nd	25	68	7

[a] Each consumer panel consisted of more than seventy volunteers.
[b] Each panellist was given two randomly coded samples and asked 'which do you prefer or have you no preference?'
nd = not done.

acceptability during storage. Sixty-eight per cent of consumers in the trial preferred the fresh sample compared to the 21-day-old specimen, a result which suggests that the method of *sous-vide*/cook–chill may not suit all products in terms of long shelf-life. Indeed, the point made by this data is that the shelf-life of *sous-vide* foods is likely to be product-specific.

The results of the trained taste panel with both products showed that characteristics of colour, flavour and texture were all affected by storage (Figs 8.3 and 8.4). The chicken à la king showed changes in all of the major attributes identified by the trained panel as being important in defining its overall quality. The colour of the peas, the appearance of the meat, the consistency of sauce and the definition of individual flavours all showed small losses in quality characteristics at day 7 although all these attributes appeared to have recovered to their original levels by day 14. The strength of flavour in the dish showed an increasing trend which peaked at 14 days of storage. In general, therefore, it seemed that this product was reasonably stable over 14 days' chilled storage at 1–3°C.

In the courgette dish, however, pungency of smell, crispness, crunchiness and appearance were all detrimentally affected by storage as assessed by the trained taste panel and this was reflected in the lower acceptability of the stored product with the consumer panel. The 21-

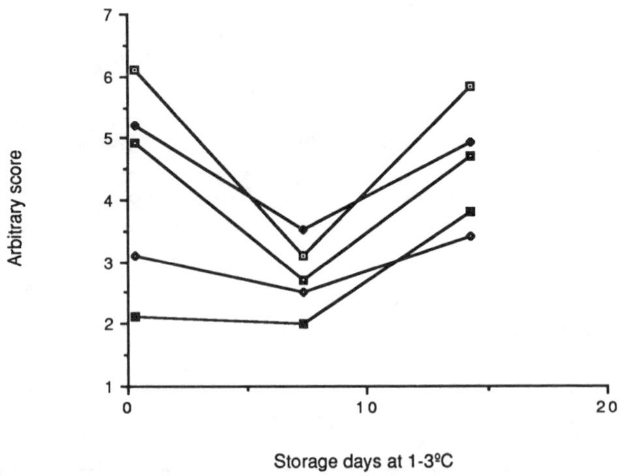

Fig. 8.3. The results of trained taste panel assessment of cooked, chilled chicken à la king prepared by the *sous-vide* method and stored for various times up to 14 days. (□), Pea colour; (◆), sauce consistency; (■), flavour discrimination; (■), flavour strength; (◆), appearance of meat.

day-old product was described by the trained taste panel as 'soggy', poorly coloured and unpleasant.

In this small study of three food products prepared using the *sous-vide*/cook–chill method several key points appear to stand out. These are:

(a) The method is not universally applicable to menu items used by caterers. Some will have far better shelf-life potential than others. However, those with a good sensory shelf-life appear to be extremely stable in terms of sensory quality.
(b) Good extension of shelf-life in terms of microbiological quality is possible using the method, provided, and this is the crucial point, very high standards of raw materials, preparation, cooking, chilling and storage are set and adhered to.
(c) In order to ensure full protection of products from aerobes during extended storage, packs should be properly evacuated. The use of soft products and inadequate vacuums to protect these dishes could pose high risks of unchecked microbial growth. In a system which claims to give high assurances of microbiological safety (and which could, thereby, lead to complacency on the part of users) this drawback is potentially a major hazard.

The authors of this study note that the 21 day shelf-life used in the work may not necessarily be desired or needed by caterers. However, they

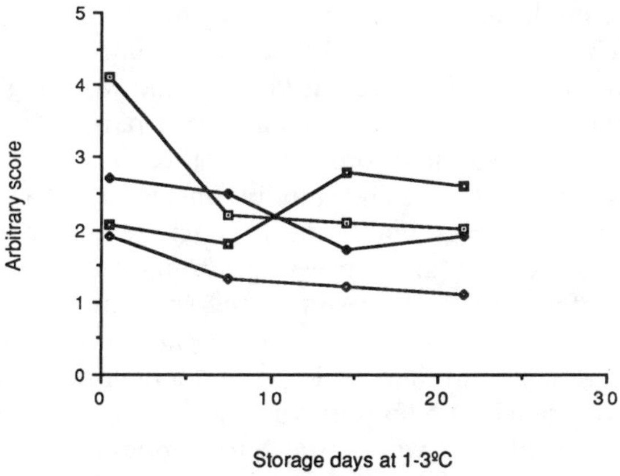

Fig. 8.4. The results of trained taste panel assessment of cooked, chilled courgette provencale prepared by the *sous-vide* method and stored for various times up to 21 days. (■), Pungency of smell; (□), crispness; (◆), crunchiness; (◆), pigment of flesh.

point out that the ability to obtain 21 days or more of safe chilled storage can be seen as a method of obtaining a margin for error which allows confidence in product stability over a shorter but more practical storage time (e.g. 8–10 days).

A later study of *sous-vide*/cook–chill involved the use of a single composite product, namely chicken ballotine, to investigate the efficiency and reliability of standard cook–chill techniques in the production of the dish. This study involved the measurement of temperature profiles in packaged products during cooking and chilling, as well as the temperature profiles of cooking and chilling equipment (i.e. a combination oven (see Chapter 5), an iced water chiller and air-blast chiller). The microbiological and sensory parameters of the fresh and stored product were also measured.

The prepared and packaged food products were heat processed in a standard convection wet steamer (with adjustable heat and steam control settings and a temperature range 45–100°C) at a pre-set temperature of 85°C. The chosen pasteurisation temperature for controlled thermalisation was 80°C. Two types of chiller were investigated; the standard electro-mechanical blast air type chiller and an iced water bath chiller. The oven and chiller were tested with two separate loadings, i.e. with the oven loaded to its full capacity with 72 vacuum packed chicken portions on shelves or with only 18 individual portions distributed evenly on the shelves of the oven.

In the first part of this study, i.e. the investigation of the potential problems arising from the use of standard catering methods in the preparation of a cooked, chilled '*sous-vide*' food product, the data from batch cooking in the specialised '*sous-vide*' oven used suggested worrying inconsistencies. It was shown that, whilst products in one part of the oven might receive excess heat treatment, products in other parts of the oven were under-processed (Table 8.3). The authors concluded that a serious implication of this finding was that some products might be inadequately pasteurised at the end of heat processing. From what we have discussed in earlier chapters this could result in the survival of psycrophiles in inadequately pasteurised products. If such organisms included anaerobic pathogens then such a product could present a high risk of food poisoning. Even if no psychrophiles survived the inadequate pasteurisation, such products, if abused with respect to temperature during chilled storage, could provide an environment for the growth of any other anaerobic spoilage or poisoning organisms. These results obviously have important implications for the design of effective ovens for the production of pasteurised *sous-vide* foods intended for long chilled shelf-life use

TABLE 8.3
Data from Three Individual Runs Made with the Combination Oven in the Cooking of Large (70 Portions) and Small (18 Portions) Batches of Chicken Ballotine by the *Sous-Vide* Method (adapted with permission from Schafheitle, 1988)

Oven loading (portions)	*Max. oven temperature (°C)*	*Time to reach pre-set oven temperature (min)*	*Time to reach 80°C in food on*		
			Top shelf (min)	*Middle shelf (min)*	*Bottom shelf (min)*
70	89	6	33	39	24
18	87	9	15	nd	24
18	86	2	26	nd	33

nd = not done.

and, thereby, for strict temperature monitoring and control during pasteurisation.

However, the data presented in this study on rapid chilling are equally worrying. In these experiments, a maximum chilling time of 90 min (to reduce from cooking temperature to < 3°C) was set as a requirement based on the recommendations of the UK Department of Health and Social Security guidelines for the production of pre-cooked chilled foods (DHSS, 1980). It was shown that the electro-mechanical blast air type chiller did not perform to the required specification in some instances. The authors pointed to previous research which showed that most caterers using the cook–chill system in the UK use this type of chiller. They suggested that their results have important implications for the use of such chillers for *sous-vide*/cook–chill production because they may not allow the chilling of foods to the highest possible specifications. However, they also showed that, even when loaded to maximum capacity, the iced water bath was able to chill cooked products to below 3°C well within the required time (Fig. 8.5).

Clearly, the control of pasteurisation and chilling in *sous-vide*/cook–chill products intended for long shelf-life use is dependent on the ability and efficiency of the equipment used to provide that control. In the food manufacturing industry such equipment is available but caterers use a wide range of cooking equipment, much of which can be adapted to *sous-vide* applications. The use of vacuum cooking in large scale methods has, in some cases, been accompanied by the design of simple yet controllable cooking and chilling equipment to carry out their specialised job (see

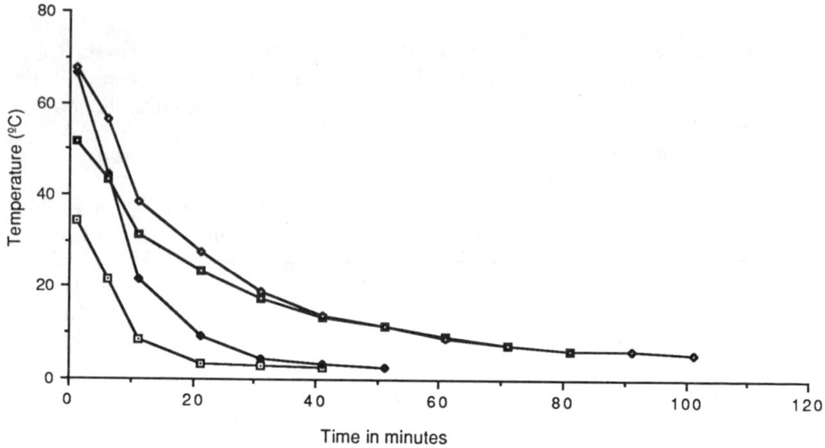

Fig. 8.5. Comparison of iced water bath and air-blast chilling of cooked chicken ballotine prepared by the *sous-vide* method. Air-blast chilling of 18-portion batches (■, ◆); iced water bath chilling of 18-portion batch (□); iced water bath chilling of 72-portion batch (◆).

Chapter 7). The same approach could be taken with *sous-vide*/cook–chill production.

The microbiological and sensory section of this study, in which products were made under highly controlled conditions of pasteurisation and chilled storage, corroborated the data of the earlier study. Little or no bacterial growth occurred even up to 21 days' storage at 1–3°C (Table 8.4) and its sensory profile suggested that this particular *sous-vide* product could withstand extended refrigerated storage under ideal conditions for

TABLE 8.4

Means (n = 10) of Bacterial TVC and Spore Counts for Chicken Ballotine Produced by the *Sous-Vide* Method and Stored Chilled (1–3°C) for up to 21 Days (adapted with permission from Schafheitle, 1988)

Storage time (days)	*Total viable count*		*Total spore count*	
	Aerobic/g	*Anaerobic/g*	*Aerobic/g*	*Anaerobic/g*
1	$4{\cdot}3\times10^1$	$2{\cdot}6\times10^1$	$1{\cdot}7\times10^1$	$< 10^1$
7	$9{\cdot}5\times10^1$	$7{\cdot}4\times10^1$	$3{\cdot}3\times10^1$	$1{\cdot}0\times10^1$
14	$2{\cdot}6\times10^2$	$6{\cdot}3\times10^1$	$5{\cdot}1\times10^1$	$8{\cdot}0\times10^1$
21	$1{\cdot}8\times10^2$	$1{\cdot}2\times10^2$	$2{\cdot}5\times10^1$	$1{\cdot}6\times10^1$

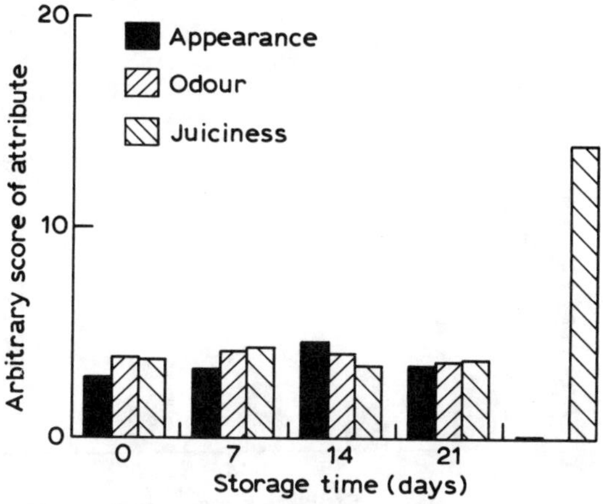

Fig. 8.6. Mean trained taste panel scores for appearance, odour and juiciness of cooked, chilled chicken ballotine prepared by the *sous-vide* method. The two columns on the extreme right of the histogram show the lowest and highest possible scores that can be given to attributes.

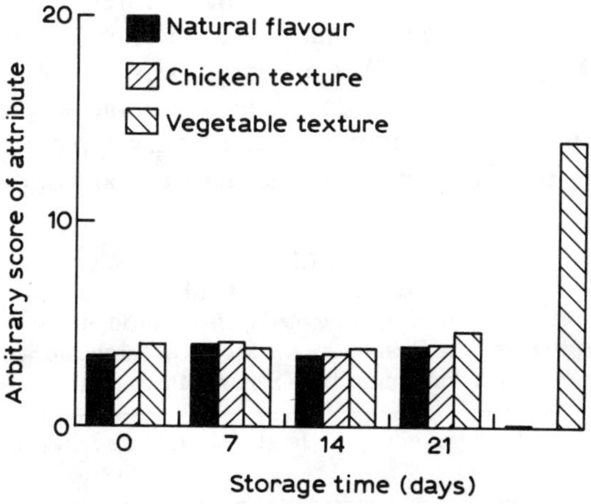

Fig 8.7. Mean trained taste panel scores for natural flavour chicken texture and vegetable texture of cooked, chilled chicken ballotine prepared by the *sous-vide* method. The two columns on the extreme right of the histogram show the lowest and highest possible scores that can be given to attributes.

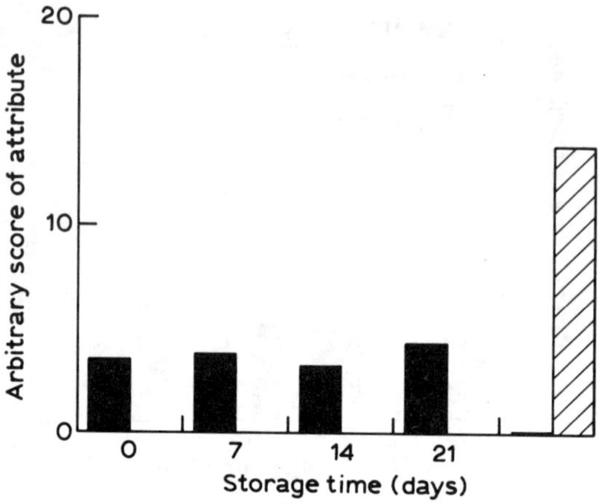

Fig. 8.8. Mean trained taste panel scores for overall acceptability of cooked, chilled chicken ballotine prepared by the *sous-vide* method. The two columns on the extreme right of the histogram show the lowest and highest possible scores that can be given to attributes.

up to 21 days and remain organoleptically acceptable. When stored for 7, 14 or 21 days the product could be identified by the trained panel in Triangle Tests, but it was found to be little different from the fresh product on quantitative sensory analysis even after 21 days' storage at 1–3°C (Table 8.5; Figs 8.6–8.8). Appearance was the only characteristic which showed any major difference due to storage so that difference between the 14-day stored sample and the fresh sample was statistically significant ($P = 0{\cdot}033$). There was little change in any of the other main

TABLE 8.5
Accumulated Hedonic and Sensory Scores (±standard deviations) for Attributes of Chicken Ballotine Prepared by the *Sous-Vide* Method and Stored Chilled at 1–3°C for up to 21 Days. The Results are the Means of Replicate Scores Obtained from Six Taste Panel Sessions (Schafheitle & Light, 1989)

Day	*Appearance*	*Odour*	*Juiciness*	*Flavour*	*Chicken texture*	*Vegetable texture*	*Acceptability*
0	2·8±1·5	3·8±2·3	3·7±1·8	3·4±1·8	3·5±2·1	4·0±2·5	3·5±2·5
7	3·2±1·7	4·1±1·8	4·2±2·2	4·0±1·9	4·1±2·0	3·8±1·8	3·8±2·3
14	4·5±2·3	4·0±2·2	3·4±1·7	3·4±1·8	3·5±2·6	3·8±1·9	3·2±1·7
21	3·4±1·6	3·6±2·4	3·7±1·8	3·9±2·2	4·0±1·7	4·6±2·7	4·3±2·3

attributes listed by the trained taste panel, i.e. odour, juiciness, natural flavour, texture of the chicken and texture of the vegetable component (Figs 8.6–8.8).

At the time of writing, these data are the sum total of the research findings to date on the viability of the *sous-vide*/cook–chill method and they point, simultaneously, to the exciting prospect of long shelf-life, high quality cooked, chilled foods. More importantly, they highlight the possibility of high risks in terms of food poisoning if the method is not used properly under rigorously controlled conditions. The authors of the second study (Schafheitle & Light (1989). *Int. J. Food Sci. Tech.*, **24**, 199–205) summarise their findings thus:

> In the light of the present results it would appear that, after using the appropriate techniques and complying with stringent working practices, the *sous-vide* cooking method in combination with controlled chilled storage can be used to successfully preserve food products for up to 21 days. However, on the basis of this study there is not sufficient evidence to provide conclusive proof of the system's commercial viability . . . (in catering applications) . . ., particularly with regard to potential problems arising from the abuse of products in terms of temperature and time of heat processing and also chilling and chilled storage. Evidently, further, more specific research is needed to establish time and temperature tolerances, pasteurisation levels, possible toxicological transfer of substances from the packing material during the slow pasteurisation process and, especially, microbiological hazards as outlined above.

FURTHER READING

Anon. (1988) The *sous-vide* processing method barred to US restaurants. *Food Engineering*, April, 19–21.

Demetriadi, J. (1986) Method *Sous-vide. Hospitality*, July/Aug., 34.

LeMaire, W. H. (1988) Too many cooks. *Food Engineering*, April, 70.

Lewis, A. & Light, N. (1988) A survey of cook–chill catering in the UK and the need for the extension of product shelf-life. *Food Science & Technology Today*, **2**, 214–17.

Light, N., Hudson, P., Williams, R., Barrett, J. & Schafheitle, J. M. (1988) A pilot study on the use of vacuum cooking as a production system for high quality foods in catering and retail. *Int. J. Hospitality Management*, **7**, 21–7.

Pring, A. (1986) *Sous-Vide*; Boon or threat? *Caterer and Hotelkeeper*, 24th July, 19.

Raffael, M. (1984) Revolution in the kitchen. *Caterer and Hotelkeeper,* 16th Aug.
Raffael, M. (1985) Cooking in a vacuum—perhaps the best thing since sliced brioche. *Guardian,* 27th Sept.
Schafheitle, J. M. & Light, N. D. (1986) La Cuisine *Sous-Vide*: An introductory Handbook. The Food and Hospitality Research Centre, Dorset Institute, Poole, Dorset, UK.
Schafheitle, J. M. & Light, N. D. (1989) *Sous-vide* preparation and chilled storage of chicken ballotine. *Int. J. Food Sci. Tech.,* **24,** 199–205.
Schafheitle, J. M. & Light, N. D. (1989) *Sous-vide* cooking and its application to cook–chill—what does the future hold? *J. Contemporary Hospitality Management,* **1,** 5–10.
Sheppard, J. (1987) The big chill. A report on the implications of cook–chill for public service. London Food Commission. Report no. 15.
Sheppard, J. (1988) The *Sous-Vide* Handbook, London, Convotherm Ltd.

CHAPTER 9

Quality Assurance—a Criti

INTRODUCTION

In any food preservation system the most imporant aspect of control and organisation is the quality of the final product. It may be possible to invent any number of technologically sound methods for preventing the organoleptic deterioration of foods but if food poisoning bacteria can grow in the products then the method is nothing more than useless. Similarly, it might be possible to prevent all forms of micro-organisms from growing in a food but if the eating quality is destroyed then, again, the technique will have no application. Thus, it is necessary to design a sound system of food preservation which results in safe, high quality food products. However, because of the risks of microbiological contamination and the fact that all such methods must rely on a number of steps being carried out successfully (see Chapter 4), without careful and judicious monitoring and control of all of these steps, then, once more, the method will be redundant.

The key to the prevention of breakdown in a method of food preservation is *good quality assurance*. Quality assurance can be defined as those activities and functions concerned with the attainment and maintenance of quality. In the past, the methods used in such procedures were termed quality control but this approach is now considered by many to be inadequate. In quality control the overview of the process is often retrospective, with checks being made on products or precursors of products after the event. In quality assurance, the process itself is continuously monitored and set up in such a way as to avoid contaminations or breakdown in quality. In other words, the process is carefully considered in all its multiple aspects, risks or hazard points are identified and graded and, at all stages, are strictly controlled. In such a system there is little need for retrospective checks because quality is built into the product as it is made. In other words, quality is assured.

There has been a consistent rise in the number of reported outbreaks of food poisoning in the UK and world-wide over the last 50 years. The most common causes of food poisoning in the UK are salmonellosis and campylobacter enteritis, whilst outbreaks associated with other organisms such as *Clostridium perfringens, Staphylococcus aureus* and *Bacillus cereus* are more rare and have shown little increase. Evidence suggests that the majority (up to 70%) of the cases are caused by the consumption of foods produced by the catering or food service industry. The catering industry is large and diverse covering a range of commercial and public outlets such as restaurants, fast food outlets, public houses, hotels, outside caterers, industrial caterers, hospital and other institutional caterers, forces caterers, holiday camp and leisure industry caterers and others. These outlets employ a wide range of methods of food production, cook–chill being just one of many. The staff employed in the industry are equally diverse in background and education and show a high level of turnover. Hygiene management under conditions extant in the industry to date, therefore, has been haphazard and difficult.

The increasing use of systems catering approaches like cook–chill and the reliance on centralised production has highlighted the need and the opportunity for the introduction of a rationalised approach to quality assurance (and, thereby, hygiene management). Systematised methods such as risk analysis and hazard analysis critical control points (HACCP) have a key role to play in the future of food manufacture and systems such as cook–chill catering. Various authors have suggested the use of this latter system in catering and food service but, to date, little use has been made of these techniques in the catering industry world-wide.

This chapter deals with some of the short-comings in quality assurance in the cook–chill catering industry today and reviews methods which will help to radically improve this situation and provide a means of ensuring good management control in cook–chill operations. In order to discuss these issues it is necessary to begin with a review of food poisoning as a phenomenon.

FOOD POISONING OUTBREAKS IN THE UK IN RELATION TO THE CATERING INDUSTRY

Since the control, over the last forty years, of water or food-borne diseases such as typhoid fever, paratyphoid fever and brucellosis, *Salmonella* and *Campylobacter* have become the main organisms responsible for food poisoning outbreaks in the UK but the numbers of reported cases of food poisoning due to these and other micro-organisms continues to rise. In the UK, in 1984, over 20 000 cases of campylobacter enteritis were reported as well as more than 13 000 cases of salmonellosis. Between 1978 and 1985 there were over 5000 cases of campylobacter enteritis associated with the consumption of raw or improperly pasteurised milk whilst in the remaining cases, over 100 000, the cause of infection was not established. The dramatic rise in notified cases of campylobacter enteritis has been largely attributed to the successful introduction of improved laboratory methods for identification of the organism.

Other organisms associated with food poisoning were *Clostridium perfringens*, which increased from 30–40 notified incidents between 1970 and 1974 (average 10 incidents per year) to 68 incidents (1716 cases) in 1984 alone; *Staphylococcus aureus* (associated with poor personal hygiene in food handlers), which has shown little increase in the last ten years; and *Bacillus cereus* (usually associated with rice dishes). *Listeria monocytogenes* has been associated with rare outbreaks of food poisoning in recent years although it should be noted that incidents associated with this organism have been increasing. This may be due to the fact that only recently have reliable tests been available for the identification and quantification of *Listeria monocytogenes* and new awareness of the organism has highlighted the need to study its epidemiology. *Yersinia enterolitica* too is an organism which has given cause for some concern with respect to cooked, chilled and chill stored foods because it also grows at low refrigeration temperatures (Chapter 3).

A review of food poisoning outbreaks in the UK for 1970–9 showed that nearly 70% of incidents were directly attributable to foods prepared in the catering or food service industry. In this study, the catering outlets included restaurants, hotels, clubs and holiday camps (responsible for 17·1% of incidents), hospitals (11·0% of incidents), banquets, dinners, receptions and parties (12·2% of incidents), institutions (9·3% of incidents), schools (8·8% of incidents), canteens and meals-on-wheels (8·5% of incidents) and ships and aeroplanes (0·4% of incidents). In this study it was evident that salmonellosis was the main cause of food poisoning in the commerical catering sector (restaurants, hotels, clubs and holiday camps, banquets, dinners, receptions and parties, and ships and aeroplanes) whilst *C. perfringens* was identified as the main cause in the 'non-profit' sector (hospitals, institutions, schools, canteens and meals-on-wheels)—see Table 9.1.

TABLE 9.1
Relationship of the Place of Food Consumption to the Number of General and Family Outbreaks of Food Poisoning in England and Wales 1970–9 (adapted from Roberts, 1982)

Place or function	*No. of cases*			*% of total incidents*
	Salmonella *spp.*	C. perfringens	*Total*	
Restaurants, hotels, clubs, holiday camps	83	48	179	17·1
Hospitals	28	80	115	11·0
Banquets, dinners, receptions, parties	90	18	127	12·2
Institutions	9	73	97	9·3
Schools	13	62	92	8·8
Canteens, meals-on-wheels	7	76	89	8·5
Ships, aeroplanes	1	—	4	0·4
Non-catering	165	30	341	32·7

It was also apparent that more than 60% of incidents were casued by the need to prepare food in advance, 40% were caused by food being stored at ambient temperature, 32% through food being inadequately cooled, 29% through inadequate re-heating and 19% through the use of contaminated processed food. Only a relatively small proportion of cases

TABLE 9.2
Factors Contributing to Outbreaks of Food Poisoning in England and Wales 1970–82 (adapted from Sheard, 1986)

Contributing factor	*% outbreaks in which factor recorded*[a]
Food prepared too far in advance	57
Food stored at ambient temperature	38
Food cooled too slowly prior to refrigeration	30
Food inadequately re-heated	26
Use of contaminated cooked food	17
Undercooking meat/poultry	15
Inadequate/uncontrolled thawing of frozen food	7
Cross-contamination raw to cooked food	6
Hot-holding below 63°C	5
Infected food handlers	4

[a] Total more than 100% because more than one factor was often associated with the outbreak.

were due to infected food handlers and in these cases most were carriers of *S. aureus*. Similar findings have been reported for US catering establishments and a later British study (Table 9.2). It should be noted, however, that, to date, there has not been a recorded incident of food poisoning associated with cook–chill catering or food products produced by cook–chill caterers.

Much is now known about the growth characteristics of the organisms named above (and discussed in Chapter 3), which have been implicated in the majority of food poisoning cases reported in the UK over the last ten years. Some are relatively easily destroyed at pasteurisation temperatures and do not form spores, others are spore-formers, some are able to grow at low temperatures and so on (see Chapter 3). From our knowledge of these organisms and the recipe requirements of the full range of menu items served through catering outlets in the UK it is possible to predict the best combination of processing and storage conditions for every food needed to ensure safety of the food. Much research has been directed at

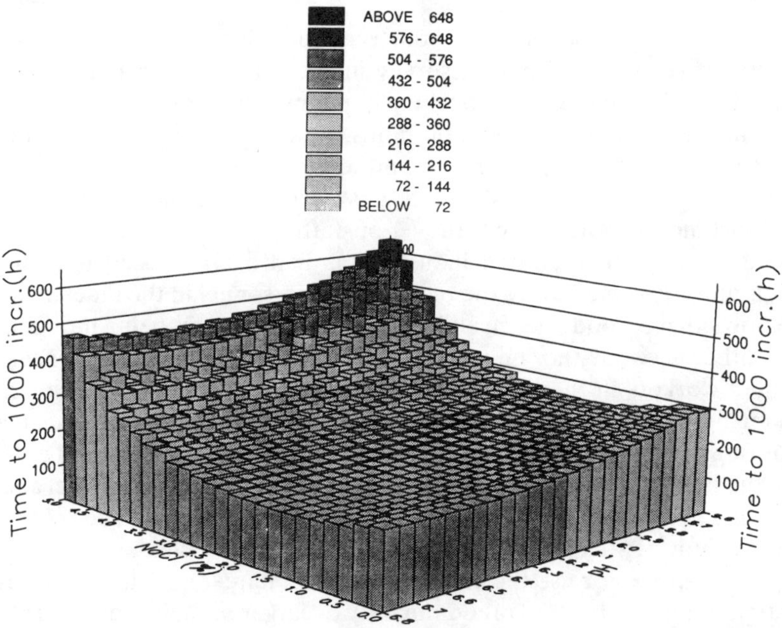

Fig. 9.1. Example of data from modelling of the growth of salmonellae showing the effects of pH and salt concentration on growth rate at constant temperature (10°C) (permission to reprint this figure was kindly given by P. J. McClure, M. Truman & T. A. Roberts).

accumulating and collating databases on the growth of food spoilage and food poisoning organisms in foods under specified conditions with the objective of providing food manufacturers and caterers with mathematical models from which shelf-life and stability could be predicted (Fig. 9.1).

This broad understanding of pathogenic organisms and their occurrence in food-stuffs may be seen as the first step in the introduction of systematic procedures for quality assurance into the cook–chill catering industry. It follows that if models of microbial growth are available then it should be possible to construct systems which minimise or totally inhibit micro-organism growth in specified foods under specified conditions. The second step in setting up such a system must include a recognition of the nature, requirements and problems of cook–chill catering.

PROBLEMS OF QUALITY ASSURANCE IN THE COOK–CHILL CATERING INDUSTRY

The catering and food service industry world-wide is large, diverse and proliferative. For example, a survey in the UK revealed that approximately 10% of the total UK working population in 1983 was employed in the catering and accommodation industry, a large proportion of whom were involved in food preparation and service. Similarly, this population of workers is largely semi-skilled or unskilled and transient so that most establishments show a high turnover of staff.

Historically, the catering industry has had a craft-based influence, chefs and cooks were once the only trained personnel in the industry and a knowledge of food science and microbiology was rare, being thought of as neither necessary nor important. The proportion of untrained or semi-skilled workers in the industry was concomitantly high. This picture is changing with the introduction over the last few decades of food science, catering and management courses for school-leavers interested in entering the industry up to honours degree and even post-graduate level and a growing body of research literature to underpin and support such courses and the industry itself. In addition, many companies have their own training programmes or use the expertise of consultants, although such programmes tend to be used predominantly by larger organisations and vary widely in their type and usefulness.

The food outlets of the catering industry may be sub-divided into the commercial (e.g. hotels, restaurants, fast food outlets, public houses, holiday camps, travel catering, whole food vending, banquets, dinners, etc., i.e. 'outside' catering, contract catering, clubs) and the non-profit making (e.g. hospitals, schools, meals-on-wheels, social services catering, some industrial canteens, forces catering). Other sub-divisions may be made on the basis of quality and size, i.e. high, medium and low quality operations, large, medium and small scale operations. Cook–chill catering is used in each of these sectors of the industry.

In the food manufacturing industry, one production line is often geared to the manufacture of one product for a set time and the total product range of one company may be relatively limited, e.g. 10–20 items. In catering, the diversity of the industry itself is equally matched by the huge variety of menu items which might be part of the regular production cycle (see Appendix 1). To facilitate these extensive menus, a wide variety of raw products are used, ranging from fresh to processed, common to exotic.

Traditional catering operations are often haphazard and poorly controlled, managed by the 'artistic' chef who may have little or no interest in technology or systematic methods of control. In such situations it seems likely that methods of quality assurance such as HACCP would be difficult to apply as they would impinge on the flair and individuality of the chef-manager. However, a major benefit of the 'systems' approach to catering, as exemplified by cook–chill catering, is that there are a number of identifiable, reproducible steps in the operation which are used throughout the industry (i.e. in operations in different sectors and of different size). Because of the technological nature of these methods and the use of a time buffer, the awareness of the need for control is higher. The hours of work in such operations are more regular than in traditional catering and, therefore, staff turnover should be lower. These are very good reasons to assume that a systematic approach to quality assurance might well be applied to these methods.

In a survey (1986–7) of UK cook–chill catering units (see Part 2) it was shown that the vast majority of operations did not employ a food technologist (or equivalently qualified person) on site to assist in the management of food production and hygiene control nor did they use the services of consultants with similar qualifications. Many of these operations were producing 5 000–10 000 or more cook–chill meals per day with a potential chilled storage life of up to five days.

This survey broached the question of training for staff involved in cook–chill operations. Answers given by catering managers revealed that in operations which were graded as successful (on the criteria of financial parameters, food quality, temperature control and staff and consumer satisfaction) extensive training in all aspects of the cook–chill operation had been undertaken at all levels of staffing. In operations graded as unsuccessful (using the same criteria) little or no training had been given (Table 9.3). The implications of these differences in approach were later made clear when the question of temperature control was broached. It was found that personnel in the successful operations understood and used the time/temperature monitoring devices supplied with cook–chill equipment (see Chapter 5) whilst those in the unsuccessful operations did not. In the latter, records were rarely studied and were generally disposed of at the end of each day. These findings highlight the disturbing differences in approach to systems and hygiene management in the cook–chill catering industry. It is clear that there is a need for the dissemination and uptake of knowledge about food microbiology and quality assurance systems.

Finally, the equipment used by cook–chill caterers is also diverse (see Chapter 5) and can be ill-designed from the point of view of hygiene control. Much modern catering equipment is based on traditional designs and takes no account of batch preparation methods, bulk pasteurisation and cleaning and hygiene needs. This has been partly attributed to the fact that most equipment is purchased by senior management who never use it themselves and who take no advice from semi- or unskilled workers

TABLE 9.3
Differences in the Extent and Type of Training Given to Personnel at Successful and Unsuccessful Cook–Chill Operations in the UK (adapted from Walker & Light, 1988)

Type of training	*Successful (% staff)*	*Unsuccessful (% staff)*
Pre-cook–chill		
Visits elsewhere	56	15
Manufacturers' training	63	31
Formal induction	19	0
Use of literature	56	8
Past experience	19	15
College courses	13	8
Total pre-cook–chill training	94	46
Post-cook–chill		
On-the-job	69	15
Observation only	31	39
No training	0	15

who have to use, clean and maintain it. A recognition of these problems has begun and equipment manufacturers seem to be trying to change their somewhat traditional approach in favour of 'designer' systems which are aimed at specialist applications. Indeed, in 1988 in the UK an expert committee, which included representatives of the equipment manufacturers, was set up to produce a booklet on hygiene management and training.

THE AIMS OF FOOD QUALITY ASSURANCE

The objective of quality assurance is to ensure that an operation is under control. The traditional means of assuring the microbiological quality and safety of meat and other food products is based on inspection and testing (quality control), e.g. the inspection of a meat plant for compliance with the EEC Hygiene Regulations on meat. The problem with many food regulations is that they do not distinguish between those requirements essential for safety and quality and those that are merely cosmetic and not needed for the assurance of food safety.

Traditional methods of assuring the microbiological safety of foods are not very effective and this is well illustrated by the continually increasing food poisoning statistics. Any preventative strategy designed to combat food poisoning at its root, i.e. at the stages of production and preparation for service, must recognise that it is the latter that presents the major challenge for control. More than 95% of the food-borne incidents associated with meat, for example, are not as a result of poor control during processing but the failure to follow proper hygienic practices during preparation for consumption. The preventative approach to quality assurance, already being widely taken up by the food manufacturing industry, can be applied to all sectors of the food industry.

It is well established that three types of quality are associated with food; its nutritive value, its microbiological status and its sensory (organoleptic) quality. Poor quality assurance in a food production and preservation system results, therefore, either in food of very low eating quality or food which is potentially poisonous or both. In the first case, the operation will probably go out of business or, in other cases, its low standards may have to be borne by consumers who have little choice (e.g. hospitals, prisons, social services catering). Should a food poisoning incident occur in an operation, its business may be seriously affected but, much worse, people will suffer the unpleasant effects of food poisoning and possibly even die as a result.

The retention of nutritional value, an often forgotten value in modern catering, was fully discussed in Chapter 2 and will not be further discussed here.

The assurance of microbiological quality should be considered as the main requirement for any food production system, including cook–chill catering operations. Through the choice of high quality raw materials, careful menu and recipe selection and controlled cooking procedures the organoleptic quality of the final product can be assured, in most instances.

The next section considers the concept of good manufacturing practice (GMP) and discusses a system of quality management which covers all aspects of quality (HACCP).

GOOD MANUFACTURING PRACTICE (GMP)

The concept of good manufacturing practice (GMP) in the food manufacturing industry is not new, yet this is a principle that has been slow in finding a use in the catering industry. The basic principles involved in GMP are concerned with setting out a total description of every routine and action involved in a process so that standards can be set and monitored for each. Systems can then be set in place to assure quality in the final product. The Institute of Food Science and Technology (UK) define Good Manufacturing Practice as being:

> . . . considered as that part of a food and drink control operation aimed at ensuring that products are consistently manufactured to a quality appropriate to their intended use . . . GMP has two complementary and interacting components, the effective implementation of well-designed manufacturing operations and the effective implementation of a well-designed quality control/quality assurance system.

In a recent study, the elements of the GMP system recommended for use in the food manufacturing industry were analysed and critically assessed in terms of their application to catering operations. Table 9.4 shows some of the ways in which the characteristics of catering operations differ from those of the food manufacturing industry. In this limited study 20 cook–serve, cook–freeze and cook–chill operations were surveyed and scored on the basis of the use of principles of GMP (or good catering practice—GCP—as it was described in this study). The maximum achievable

TABLE 9.4
Some Examples of Differences between Catering Operations and Food Manufacturing Operations (adapted from Collison *et al.*, 1987)[a]

Characteristics of the food manufacturing industry	*Characteristics of the catering industry*
Limited product range manufactured on a large scale	Wide product range manufactured on a relatively small scale
Production	Production and service
Even level of production	Production levels in peaks and troughs
Product and ingredient specifications well defined	Product and ingredient specifications often ill-defined or non-existent
Formal written food control procedures	Food control often ill-defined and unwritten
Science/technology-based operation	Craft-based operation
Capital intensive	Labour-intensive

[a] Note that many of the characteristics typical of food manufacturing could be used to describe a well-run cook–chill unit but that most of the characteristics typical of catering practices in general should not apply to a well-run cook–chill unit.

score was 18 and the scores for cook–chill units ranged from 5–17 (28–94% of total possible score). The lowest scoring units in the study (0 score) were all traditional cook–serve operations, with cook–freeze units showing a similar range to the cook–chill units, i.e. 3–17 (17–94% of total possible score). Interestingly, the units with the highest scores (15–17) were cook–serve, cook–freeze and cook–chill operations but the majority were in what was classed as the lowest market sector, i.e. hospitals and school meals and one fast food outlet. The low scoring operations, on the other hand, were all at the upper end of the market sector (Table 9.5), e.g. restaurants and hotels. The high scoring group kept written records of standardised recipes, measured and recorded temperatures, did not recycle hot food and had written food control procedures. However, only 2 of the 20 units had a Food Control manager or equivalent (a cook–freeze unit and the fast food outlet).

These results illustrate the benefits of using systematic written procedures and implementing strict rules and regulations concerning the cleanliness and hygiene of premises, equipment, personnel, food-stuffs and procedures. It is common, in well-run cook–chill catering units, to

TABLE 9.5
Scores Allocated to Catering Operations in a Survey of the Use of the Principles of Good Catering Practice (adapted from Collison *et al.*, 1987)

Score range	*Total no. of units in range*	*Technology (no. of units)*		*Market sector (no. of units)*		
		Cook–freeze/ cook–chill	*Cook–serve*	*Upper*	*Middle*	*Lower*
0–2	5	0	5	3	1	1
3–5	4	2	2	4	0	0
6–8	1	0	1	0	1	0
9–11	2	2	0	0	1	1
12–14	2	1	1	0	1	1
15–17	6	3	3	0	0	6

find the practice of relying on reasonably strict rules of operation, experienced management and the taking of samples of raw, cooked and stored food on a random routine basis for microbiological testing. However, the use of retrospective microbiological testing of the final product (a method much favoured in the past) is no longer regarded as useful as, by the time the results of such tests are available, the food has normally been consumed.

The best and most comprehensive approach to quality assurance in a cook–chill catering operation is through the instigation of a systematic methodology. However, this does not negate the need to follow certain basic rules for the setting up and running of the operation along controlled and hygienic lines. For example, a list of some of these can be made as follows:

(a) always choose highest quality raw materials and specify and check the quality of incoming supplies;
(b) keep food handling to a minimum;
(c) ensure proper hygiene in premises, plant and operatives;
(d) operatives should use only clean, well-kept kitchen wear;
(e) for vacuum packaged samples always properly vacuumise the product (air pockets could promote faster growth of micro-organisms)—similar care should be taken to ensure proper packaging when using MAP;

(f) heat to pasteurisation temperature at the centre of the food for the required time;
(g) chill rapidly (e.g. to 0–3°C within 90 min) and store properly at the same temperature;
(h) re-heat to at least 70°C at the centre of the food;
(i) have food samples, kitchen swabs and raw materials occasionally randomly checked for microbiological contamination;
(j) always monitor air temperatures in blast chillers, freezers (if they are used), chilled and frozen storage rooms and in any transport used;
(k) monitor temperatures in food at those points in production where certain targets must be met e.g. above 70°C during prime cooking; below 3°C at the end of chilling; below 3°C on arrival at the end kitchen before storage or regeneration; above 70°C at the end of cooking and regeneration.

These basic principles underlie a broader, more systematic approach to quality control. One of the newer and more widely written about methods is hazard analysis critical control points (HACCP).

HAZARD ANALYSIS CRITICAL CONTROL POINTS (HACCP)

Hazard analysis critical control points (HACCP) is a method for total quality assurance which is being taken up widely by the food manufacturing industry. It provides a system of preventative hygiene management in which the hazards associated with foods or the ingredients of food, equipment used to transport, prepare, store and serve food and the relevant potential human contaminating factors may be analysed and understood and the various points in the food chain, from harvesting to consumption, may be categorised to specify critical points at which control needs to be exerted to prevent unacceptable food safety risks. By means of this rational, systematic and easily documented procedure, the preparation of food for consumption can be monitored and effectively managed in terms of hygiene and food safety. Although, as we have noted, the method has been widely applied since its introduction in the food manufacturing industry it has seen little use to date in the catering and food service industry, particularly in the UK.

The use of microbiological analysis of food samples from catering and food service outlets has long been acknowledged to be a relatively futile procedure in terms of the *prevention* of food poisoning. Whilst the food manufacturing industry may be able to hold food samples long enough to wait for microbiological analysis to clear them for use this is generally not the case in the catering industry. Even most cook–chill units (which have an average food storage life of 2–3 days) do not keep food products long enough for microbiological analysis to have any preventative role. There is a real need, therefore, for a system of management which allows the identification of hazards in such operations and the critical points at which control may be exerted to prevent food contamination and subsequent food poisoning incidents occurring.

The idea of applying the HACCP method to catering and food service systems is not new. As early as 1977, Bobeng & David produced models for the application of HACCP methods for the quality control of food service systems. Since then a number of studies have been published identifying the uses and applications of the method to catering and food service systems. These studies recognised that most food service systems can be categorised and described using a flow chart system. It can be seen that this approach is similar to that of GMP (mentioned earlier) in which each step of the process used is documented for control purposes. Having described the process in detail the next steps are to identify and rank the hazards involved in the process and then identify, from this, the control points and, most importantly, the critical control points.

Identification of Hazards

Initially, in any HACCP analysis it is possible to identify certain groups of hazards which may be expected in a process by considering factors which are associated with the production of food products. This type of analysis is illustrated by the groups of hazard listed in Table 9.6.

Microbiological hazard analysis requires an appreciation of the types of micro-organisms that can cause spoilage of a food product or harm to the consumer. It also requires a detailed understanding of how each of these hazards could arise in a process (and, therefore, by definition, a profound understanding of the process itself) and a ranking of the probability of their occurrence at each step of the operation. The identification of hazards, therefore, requires a combination of microbiological knowledge and process information. This knowledge base may include epidemiological or commercial information when a product or product type has

TABLE 9.6
Identification of Hazard Groups by Consideration of Critical Food Product Factors

Food product factors	*Parameters affecting hazard*
Storage	Ambient—wet/dry; chill; frozen
Formulation/preservation	A_W; pH and effect of acidulant; chemical preservatives, including gas packing; competitive microflora; structure/fat, freeze/thaw stability; redox and oxygen exclusion
Packaging	Permeability; integrity; protection—tampering/sabotage
Consumer practices	Misuse; cooking—traditional, microwave, slow; re-heating (catering establishments); thawing; cross contamination; structure—freeze/thaw stability; preparation/reconstitution, time–temperature; 'topping up'; recipe enhancement
Target groups	General catering; institutional catering; infants; geriatric; immunocompromised; consumers–normal babies

A_W = Water activity.

been in a market place for some time or in the case of new product, only technological information. It is important to consider the types of micro-organisms of concern likely to be present in raw materials or introduced during handling or processing and the effect of each processing or handling stage on their ability to grow, survive or die. For example, the cooking of frozen chicken may be scored highly due to the known loading of chickens with salmonellae and *Listeria monocytogenes*, whereas the service of the final re-heated cook–chill food may score much less.

Critical Control Points

A control point is any location, practice, processing step or procedure in a process where control must be exercised to assure quality. A critical control point (CCP) is a location, practice, processing step or procedure where control must and can be exercised in order to prevent one or more of the identified hazards from occurring. The degree of concern at each CCP depends on the severity of the hazard that needs to be controlled, e.g. with respect to cook–chill *C. botulinum* group E has a minimum recorded growth temperature of 3·3°C and is, therefore, a more severe hazard than, for instance, *C. perfringens* whose minimum growth temperature is above 10°C. The degree of concern at the CCPs also depends on the likely frequency of the hazard occurring i.e. the risk. Therefore, it is important to rank CCPs in terms of overall concern in relation to severity of a hazard needed to be controlled and the probability of the occurrence of the hazard. If more than one hazard needs to be controlled then control would normally be directed to the most important of these.
thick cooked products, thin cooked products, sauces and brews, fruit,

From this type of analysis it is possible to identify those CCPs where control must be most effective and secure and thus to concentrate efforts on the control of these particular steps. In some instances contamination may be identified where control cannot be effectively exercised. It is of course possible to eliminate or reduce the importance of CCPs by making fundamental changes to a procedure or step, e.g. changing the formulation of a menu item to eliminate raw products with a high microbiological risk or introducing a biocidal washing procedure for a particular preparation area.

Application of HACCP in Cook–Chill Catering Operations

The use of HACCP in the food industry is made more straightforward by the fact that, in most instances, only one product or a limited product range is being produced by any one system. However, cook–chill catering systems are, by need and definition, producing a range of menu items usually based on a 21-day menu cycle (see Chapter 13). This means that several hundred different menu items may appear on the menu cycle involving a wide variety of food components. It is a daunting task to score each menu item in terms of its potential hazard but, if the system is to work, this approach must be taken.

The study of the system and its components, the allocation of hazard scores and the identification of critical control points is normally undertaken by a team of people. This team is composed of people expert in the various areas used in or affecting the process, i.e:

(a) a chairperson to organise the meetings, determine the objectives of the group and co-ordinate the paperwork and records of the study;
(b) a food microbiologist to advise on the microbiological hazards involved;
(c) an engineer or process engineer who understands the plant used in the process and can advise on its use and problems;
(d) the process manager who understands the steps used and the techniques of food preparation and handling involved;
(e) other persons expert in the system, perhaps a chef, a purchasing officer and/or a person responsible for packaging and transport of the food-stuffs;
(f) a reliable secretary who can accurately record the discussions and findings of the group. There are computer programs available to help set up an HACCP system and so the use of an advanced word-processer linked to such a program is recommended.

The group, working together in well organised sessions, must examine the GMP and process control charts for the cook–chill system, allocate hazard scores and identify CCPs. Once this stage is complete, the system of HACCP can be implemented in such a way that control is exerted, with the necessary monitoring, to avoid the identified hazards. The process of setting up HACCP is complex and time consuming but provides in-depth, reliable control of processes in a way unprecedented in catering operations.

Several good examples of HACCP procedures for catering applications now exist in the literature, such as flow charts for the production and service of turkey (Fig. 9.2) and a suggested general HACCP check list for catering systems (Table 9.7).

Four major critical control points for HACCP models of food service systems like cook–chill catering have been identified in research studies. These are:

(a) ingredient control and storage;
(b) equipment sanitation;

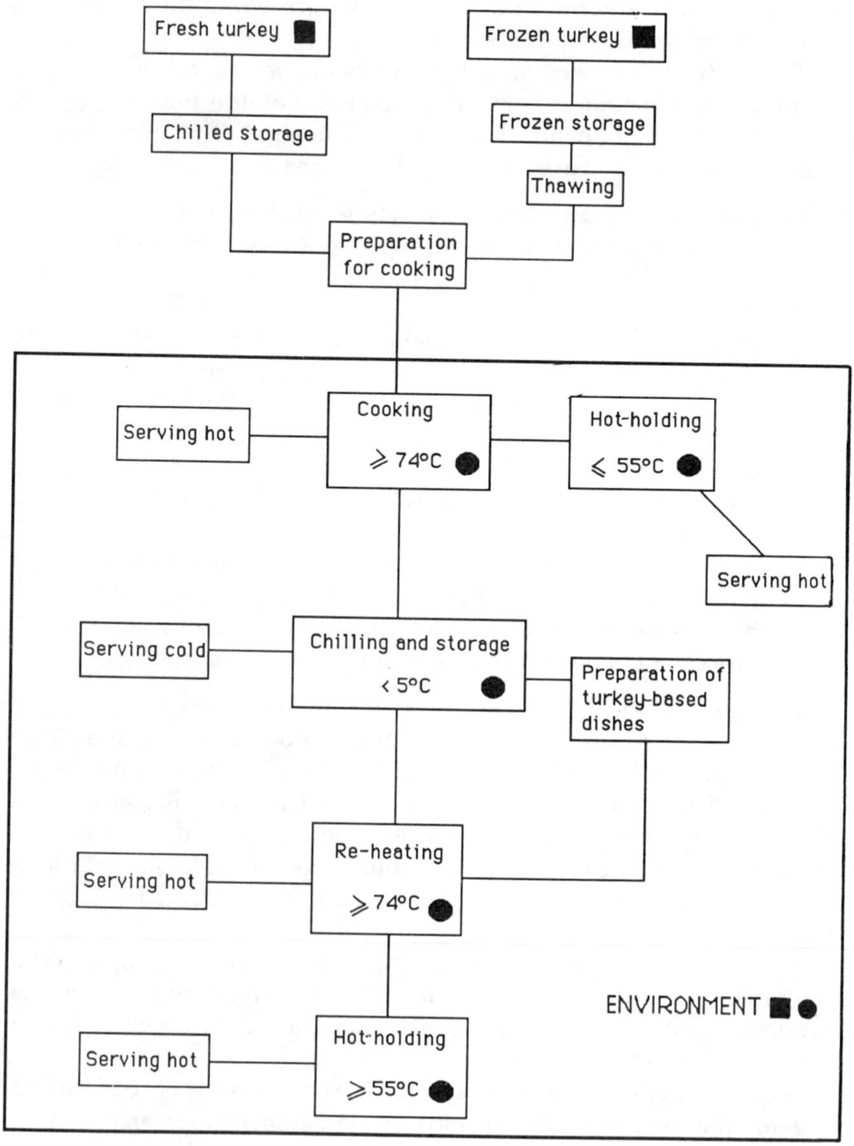

Fig. 9.2. Flow diagram showing the production procedure for turkey. ■, Raw materials contamination; ●, critical control point (CCP).

(c) personal sanitation, and
(d) time/temperature relationships.

According to the literature there are six classes of hazard which should be revealed by HACCP in catering and food service. These are:

(1) Potentially hazardous foods.
(2) Pathogenic food-borne organisms.
(3) Employee practices which may cause or fail to prevent transfer of pathogens to foods.
(4) Time/temperature conditions that permit the survival of, promote the multiplication of or fail to destroy food-borne pathogens or fail to detoxify poisonous substances.
(5) Procedures that may contaminate food after heating.
(6) Environmental conditions that may permit the maintenance of food-borne pathogens and their transfer to food from food contact surfaces, air, water or vectors.

In addition, HACCP in food service or catering situations, should be carried out to:

(a) identify potentially hazardous foods or ingredients;
(b) identify likely sources and points of contamination at each step of the operation;
(c) determine the potential for micro-organisms to survive at each step of the operation;
(d) determine the potential for micro-organisms to multiply in food products by establishing time/temperature relationships.

Use of these parameters in combination with the type of flow chart for operations shown in Fig. 9.2 should allow efficient hygiene management in catering situations. Indeed, despite the variety of operations in the catering industry, any catering operation or part of an operation can be described as a system and, therefore, can be illustrated as a flow chart. In fact, seven recipe processes have been described for the production of thick cooked products, thin cooked products, sauces and brews, fruit, vegetables and starches, breads and batters, cold combinations (chilling) and hot combinations (hot-holding). Similarly, control points (a point in a process where a potential food hazard may exist) and critical control points (a point in a process where control eliminates or reduces hazard) have been summarised for conventional, cook–freeze and cook–chill processes (Table 9.7).

TABLE 9.7
Example of Critical Control Points in Terms of a Generalised HACCP Model for Catering Food Production and Food Service Systems (adapted from Bobeng & David, 1977)

Operation	*Critical control point*			
	Ingredient control and storage	*Equipment sanitation*	*Personnel sanitation*	*Time and temperature relationships*
Procurement	×			
Preparation		×	×	×
Heating				×
Hot-holding				×
Chilling and chilled storage		×		×
Freezing and frozen storage				×
Thawing		×		×
Portioning and assembly		×	×	×
Portioning, assembly and distribution		×	×	×
Chilled storage and distribution		×		×
Re-heating				×
Service		×	×	×

In theory it appears that the HACCP principles are applicable to the food service and catering industries. To date, as mentioned earlier, the uptake of the method by the UK catering industry has been negligible although there are disparate instances of catering managements attempting to enforce similar or related systems. Why then, has the system not become more widely used? There are two reasons for this; first, a lack of awareness and knowledge of the technique and a suitable understanding of the basic principles which allow its proper use and second, a lack of training of food handling personnel in the implementation of efficient food hygiene systems. Two major requirements in a cook–chill operation, after the establishment of the HACCP system, are proper training of all personnel in the use of all relevant operation and monitoring methods (see other chapters) and the implementation of monitoring systems, i.e. the physical tools of quality assurance.

MONITORING AND THE TOOLS FOR QUALITY ASSURANCE

Checking that a processing or handling procedure at an established CCP in a system is properly carried out and is under control involves the systematic observation, measurement and recording of significant factors for control. It is highly important that the monitoring procedure should be able to identify an out-of-control situation before it leads to a hazard. Monitoring procedures should be specified in documents together with the actions to be taken if there are deviations from the specified control requirements.

The proper implementation of HACCP in a cook–chill operation, therefore, depends on the use of specific tools for the monitoring of food materials, equipment, personnel and premises at control points. Apart from the visual inspection of food raw materials, personnel, plant and premises, perhaps the singular most important parameter in the operation is temperature. As we have previously discussed, temperature is critical in:

(a) the storage of raw materials prior to use,
(b) the environment for food handling,
(c) heat processing (pasteurisation),
(d) rapid chilling,
(e) portioning,
(f) packaging,
(g) chilled storage,
(h) distribution of the cooked, chilled food (i.e. by vehicle to satellite kitchens or within a single site, e.g. to the wards of a large hospital),
(i) the centre of the food after re-heating.

Many pieces of equipment designed for cook–chill catering are now fitted with automatic temperature reading and recording devices (Chapter 5). However, most equipment must be operated by people and few systems are foolproof. It is only through use of proper monitoring procedures that the benefits of the systems approach and, more importantly, proper quality assurance can be achieved. In a survey of cook–chill users in the UK in 1986–7, it was shown that a disturbingly high proportion of caterers were either using equipment that did not have adequate temperature monitoring systems (Fig. 9.3) or the monitoring devices were in place but were not being properly used (Table 9.8). This data surely

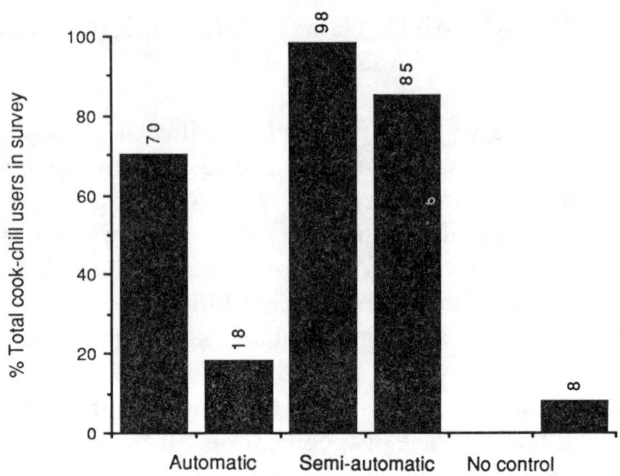

Fig. 9.3. Survey of the use of temperature monitoring devices by UK cook–chill catering operators (from Walker, 1988).

points to the fact that, although systems may be in place, they must be properly used. Only by the adoption of a systematic model for monitoring of every process step in food production can total quality be assured.

TABLE 9.8
Regularity of Temperature Monitoring of Chilling Devices (adapted from Walker, 1988)

Regularity	*Successful (% units)*	*Unsuccessful (% units)*
Blast chillers		
Every production run	61	14
Daily	39	27
Weekly	0	5
Less than weekly	0	41
No blast chiller	0	14
Chill store		
Regular intervals through day	87	32
Daily	13	14
Weekly	0	9
Less than weekly	0	32
Never	0	14

Much is now known about the growth characteristics of food spoilage and food poisoning micro-organisms and perhaps two of the key factors which influence their growth, and certainly the two that are easiest to control and monitor, are time and temperature (see Chapter 3). As we have seen throughout this chapter, microbial growth is the single most important issue in food production systems, including cook–chill catering. By judicious use of well-defined and well understood time and temperature monitoring systems, as in the HACCP model, many of the most serious hazards and risks of such a system can be avoided.

CONTROVERSY

Starting in 1986–7, comment began to be made about the risks of food poisoning associated with the consumption of re-heated cooked, chilled foods. With the reported rise in the number of cases of listeriosis (see Chapter 3), albeit low, this comment increased until, in 1988–9 it reached a peak of coverage in national and local newspapers, on radio and television in the UK. This publicity, dubbed by one catering magazine as 'Listeria hysteria', was due, largely, to the uncertainty about the growth and pasteurisation characteristics of *Listeria monocytogenes* and the well-described and very tragic effects of listeriosis (spontaneous abortion in pregnant women and death in sensitive people—see Chapter 3).

Much research work is still continuing into the characteristics of this organism but it is known that:

(a) it grows only very slowly at 1°C and even at 4°C its generation time is measured in days;
(b) it is relatively easily destroyed by heat and, therefore, can be eliminated by adequate pasteurisation procedures.

At this time, then, it seems safe to assume that this organism will pose no more of a threat to human safety, in terms of the consumption of cook–chill foods produced by caterers, than any other pasteurisable organism provided all the proper controls and monitoring procedures are used during the preparation, storage and regeneration of cook–chill products. The assurance that such controls are in place can only come from the use of a proper quality assurance system as we have discussed in the foregoing chapter.

FURTHER READING

Bauman, H. E. (1974) The HACCP concept and microbiological hazard categories. *Food Tech.*, **28** (9), 30–4, 74.

Bobeng, B. J. & David, B. D. (1977) HACCP models for quality control of entrée food service systems. *J. Food Prot.*, **40** (9), 632–8.

Bryan, F. L. (1974) Identification of food hazards in food service establishments. *J. Environ. Health*, **36,** 537–40.

Bryan, F. L. (1978) Factors that contribute to outbreaks of foodborne disease. *J. Food Prot.*, **41,** 816–27.

Bryan, F. L. (1981) HACCP approach to epidemiology rationale and application in food service operations. *J. Environ. Health*, **44,** 7–14.

Bryan, F. L. (1981) Health hazards of food service establishments. *Food Tech.*, **35,** 78–87.

Bryan, F. L. (1982) Foodborne disease risk assessment of food service establishments in a community. *J. Food Prot.*, **45** (1), 93–100.

Curtis, J. E. & Huskey, J. E. (1974) HACCP analysis in quality assurance. *Food Prod. Devel.*, **8** (3), 19, 24, 25.

Galbraith, N. S., Barrett, N. J. & Sockett, P. N. (1987) The changing pattern of foodborne disease in England and Wales. *Public Health*, **101,** 319–28.

Morris, N. (1987) Designing for hygiene? *Design 460,* April.

Munce, B. A. (1981) The role of the food service industry in the microbiological safety of food: a review. *Food Technol. Aus.*, **33** (7), 326–31.

Munce, B. A. (1984) Hazard analysis and critical control points and the food service industry. *Food Technol. Aus.*, **36** (5), 214–17, 222.

Peterson, A. C. & Gunnerson, R. E. (1974) Microbiological critical control points in frozen foods. *Food Tech.*, **28,** 37–44.

Roberts, D. (1982) Factors contributing to outbreaks of food poisoning in England and Wales 1970–79. *J. Hyg. Camb.*, **89,** 491–8.

Sheard, M. (1986) Working party on food poisoning in hospitals, institutions and residential establishments. *Environ. Health*, **93** (10), 263–6.

Snyder, O. P. (1986) Microbiological quality assurance in food service operations. *Food Tech.*, **40,** 122–30.

Taubert, C. A. (1982) Defining sanitation hazards and critical control points in food service operations. *J. Foodservice Sys.*, **2,** 171–5.

PART 2.

THE IMPLEMENTATION AND MANAGEMENT OF COOK–CHILL CATERING SYSTEMS

CHAPTER 10

Evidence from a Survey

INTRODUCTION

In this section of the book we will describe some of the practical implications of introducing cook–chill systems into catering organisations. In doing so we will draw on a number of different reports on actual cook–chill catering outlets. However, much of our evidence will be taken from the results of a nationwide cook–chill survey which was carried out in the UK by the authors during 1986. For ease of reference this survey will be referred to as the UK Cook–Chill Survey throughout the rest of this book.

In the UK Cook–Chill Survey, over 80 cook–chill catering operations were visited out of a total population of 240 units identified at the time. In each unit a very large amount of information was collected from a variety of respondents by means of three questionnaires. The complexity of the resulting database of information and subsequent detailed analyses was such that a brief explanation of the survey itself is necessary. Therefore, in this short chapter, we will examine the structure of the survey and introduce some of the concepts and methodologies used during the course of the survey.

THE UK COOK–CHILL SURVEY

The introduction of cook–chill is often mistakenly seen by caterers as a relatively straightforward transformation and this idea has been emphasised in some equipment manufacturer's sales packages. However, evidence from the UK Cook–Chill Survey showed repeatedly that, in reality, its introduction is far from simple, being complex and often problematic. One obvious effect of cook–chill catering is that it changes the whole concept of the work patterns within the innovating organisation. In other words, cook–chill catering has a central technological base and introduces a more systemised approach to food production.

It was clear from exploratory work carried out before the survey proper that the interaction of people within catering organisations would be a critical factor in determining whether the introduction of a cook–chill catering system was successful or not. Thus, the UK Cook–Chill Survey paid particular attention to the human element within the overall context of the process of change. With this in mind, it was important to observe the process of technological change (which was either complete or taking place at the time of interview) by examining the full range of activities which made up the process of that change, i.e. initiation, decision-making, implementation, evaluation and future use.

Thus, the UK Cook–Chill Survey was the first systematic, in-depth research undertaken which concentrated investigation on the introduction of cook–chill technology into a wide range of different catering organisations. The main hypothesis of the study was that the activities of successful cook–chill units would be expected to differ from the activities of unsuccessful cook–chill units at all phases of the change process and that comparison of successful and unsuccessful units could shed light on the key factors determining proper implementation and use of cook–chill catering systems.

Given the complexity of the change process described in brief above, its examination necessitated the acquisition of attitude, opinion and factual data from a variety of respondents. Information was, therefore, collected from three sets of informants. First, a key decision-maker in each unit (usually the catering manager or person of equivalent status), second, a selection of cook–chill operatives and third, actual consumers of the cook–chill food produced by the establishments involved in the survey.

Early exploratory work defined those areas which were considered important for a study of the transfer of cook–chill technology. These variables could be identified at different phases of the process of change (i.e. decision-making, implementation and evaluation) and at each phase, issues could be broadly defined as being either functional, technical or social. Figure 10.1 indicates the main areas for detailed investigation at each phase and at each level (i.e. functional, technical or social) of the change process.

Detailed information was obtained from the catering manager (or informant of equivalent status) on the activities leading up to, during and

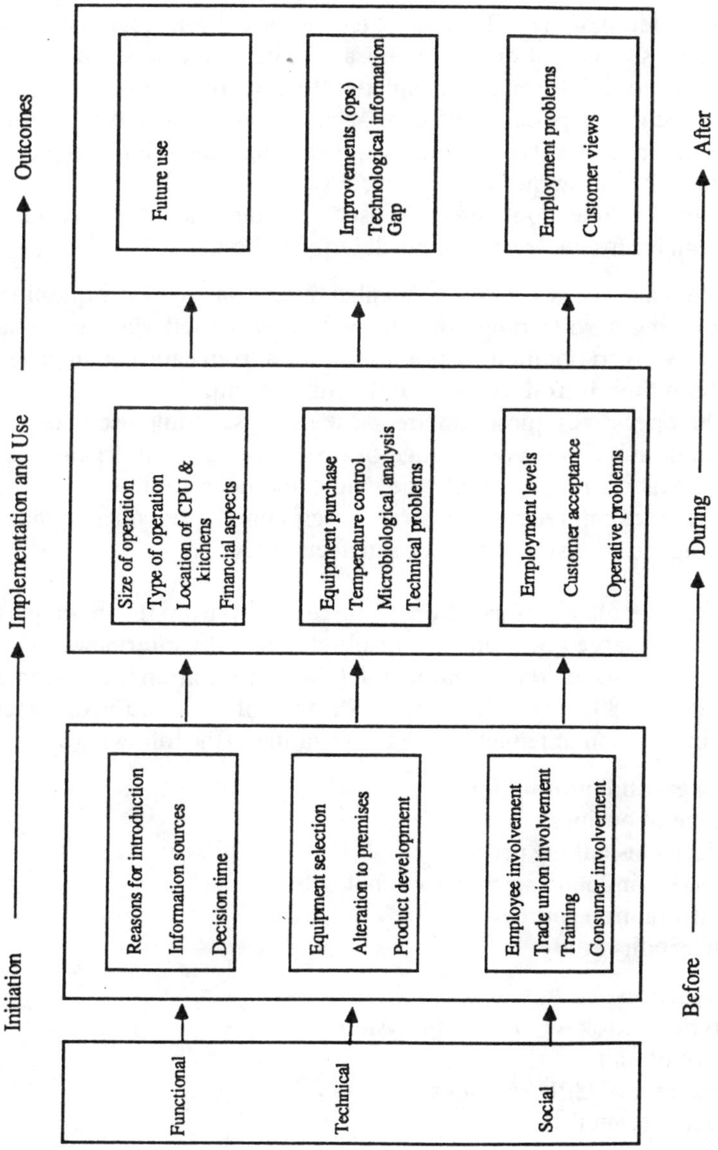

Fig. 10.1. Key factors in the transfer of cook–chill technology.

after the introduction of cook–chill. A key aspect of this research was that it attempted to obtain a complete picture of the process of change and thus provide detail of the full impact of technological change, not only on the financial aspects, but on the organisation, its employees and its customers. Therefore, information from operatives on the actual use of and, thereby, practical experience of cook–chill systems and the responses from consumers on levels of satisfaction regarding cook–chill food eaten in the establishment in question, were considered critical.

Three types of questionnaire were used to obtain the necessary information from each set of respondents described above.

(i) The main questionnaire—a detailed questionnaire of 102 questions covering a wide range of information was used via a personal interview to obtain in-depth information from catering managers about their introduction of cook–chill catering.
(ii) The operatives questionnaire—a skeleton schedule used via personal interviews with a range of operatives in each organisation combined with covert interviewing and observation.
(iii) The consumer questionnaire—a self-completion questionnaire designed to assess consumer satisfaction levels.

Obviously, the bulk of information was ascertained from the main questionnaire. This large questionnaire resulted in lengthy interviews, often lasting over two hours, but a wealth of detailed information was obtained from each of the 80 units in the sample. The bulk of information collected in the main questionnaire includes details relating to the following areas:

a. Background information
   type of business
   decision-making process
   factors influencing choice of cook–chill
   information sources
   feasibility study

b. Production
   type of cook–chill catering system
   size of unit
   number of CPU/satellites
   delivery method

c. Finance
   annual turnover

capital expenditure
profit/loss
payback policy

d. Purchase–regeneration
purchasing procedure
equipment
maintenance
product development
recipes
temperature control
containers
transportation
waste
service

e. Personnel
numbers employed
length of service
turnover
labour reduction
benefits
training

f. Problems
functional
technical
social

g. Customers
type
awareness of cook–chill
methods of informing
customer satisfaction

h. Evaluation
expenditure
problems
critical factors in the process

i. Future developments

Information from operatives was based around a broad set of concepts rather than individual questions. These related to the use and abuse of cook–chill, especially training, problems and any benefits and drawbacks as perceived by the operative staff.

The consumer questionnaire was short, concise and designed specifically to elicit levels of consumer satisfaction with cook–chill food. In addition, it collected information on the range of food being consumed through the system.

An analysis of the information obtained in the UK Cook–Chill Survey provided an invaluable insight into the process of changing to a cook–chill catering system, from initiation of the idea through the introduction process to implementation, use and assessment of outcomes from a variety of viewpoints, i.e. from managers, operatives and consumers.

The first question which needed to be addressed related to identifying the relative success of each organisation surveyed. This then enabled a detailed comparison of the activities undertaken by the most successful cook–chill units in the sample with those of the least successful units in the sample.

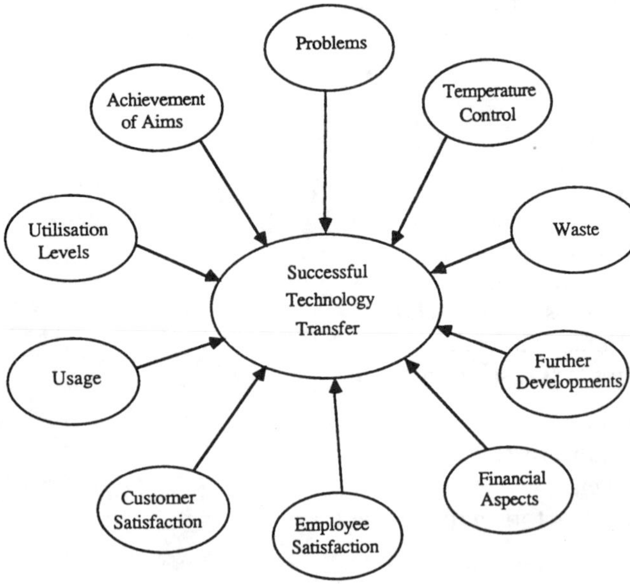

Fig. 10.2. Factors selected for assessment of success.

A multi-faceted definition of success was used to measure relative success. This definition was based on the understanding that the introduction of cook–chill is a complex, iterative process and as such any analysis of it should reflect this complexity.

Ten success factors were used to determine the relative success of each organisation, based on its multi-phased, multi-layered characteristics (as indicated in Fig. 10.1). These were based on levels of usage, effective utilisation, achievement of aims, temperature and waste control, financial efficiency, customer and employee satisfaction and the extent to which any problems encountered had been overcome. Each one of these success criteria was a carefully selected measure with both cook–chill-specific and general aspects regarding the technology taken into consideration. A brief review of these factors follows (see also the summary in Fig. 10.2). The factors used to determine success were categorised under the sub-headings of the three levels of the process of change which we identified earlier.

Functional Factors

The first success factor differentiated between users (at the time of interview) from non-users. The non-user category resulted from those catering establishments visited who had stopped using cook–chill catering as a system of food production and were hence automatically classed for whatever reason as unsuccessful cook–chill users.

Since much of the early cook–chill literature describes cook–chill as a total catering system, the second factor measured utilisation levels as indicative of success. It was realised that utilisation is a complex measurement and not just a function of percentage use. It was, therefore, measured as 'efficiency of utilisation', in other words a maximisation of the use of available capacity.

The next functional factor was a qualitative measure which took into account the number of problems encountered during introduction, implementation and use of a new cook–chill system and whether these problems still remained, had worsened, been solved or whether new problems had developed after start-up.

Finally, the inclusion of a factor which assessed the achievement of financial targets was considered essential. Almost always in technological change, the financial achievement of an organisation impacts on success, although it may not always be immediately evident or easy to isolate from other factors. Thus, in any measure of success its inclusion is imperative.

Technical Factors

Two technical factors were identified as being specific to the introduction of cook–chill in terms of the technological change involved. The first, temperature control, was more a measure of 'proper use' than success *per se*, however, its inclusion was justified because of the problems which could run counter to success if temperature control was inadequate or abused.

Second, a waste factor, indicated both the technical merits of portion control and the level of acceptance by consumers in terms of plate waste, although differentiation was not made between these two measures of waste.

Social Factors

Two groups of people were directly affected by the introduction of cook–chill, namely, the operatives and consumers. The behavioural factors chosen in the study simply measured the acceptance and satisfaction of both employees and consumers with regard to the introduction and use of cook–chill. The employee factor related to the operational aspects of the cook–chill system and the consumer factor to the 'end-product experience'.

Scores were allocated for all of the success factors in each cook–chill unit surveyed and the total scores resulted in a 'league table' of the cook–chill catering units with the most successful units at the top and the least successful units at the bottom. The upper and lower quartiles were selected as being the most successful group and the least successful group respectively. This provided the two groups needed for a comparison of the activities at all phases and at all levels of the cook–chill transfer process.

The following chapters describe some of the results of this survey in detail. Chapter 11 gives an overview of the key activities found to be significant in affecting the successful introduction of cook–chill in the UK Cook–Chill Survey. Chapters 12 to 14 cover in more detail, factors affecting success, with particular attention paid to financial aspects, menus, products, research and development and the effects on the different personnel involved. The last chapter summarises the impact of the transfer of cook–chill technology in the form of a 'blueprint' for success based predominantly on those factors found to affect success in the UK Cook–Chill Survey. Finally, it assesses the future use and development of cook–chill systems in catering.

FURTHER READING

Walker, A. E. (1988) The transfer of technology in the catering industry: a study of UK cook chill catering operations. PhD Thesis, CNAA/Dorset Institute of Higher Education, Poole, Dorset, UK.

Walker, A. E. (1989) *Success or Failure in Cook Chill Catering—A Study in Technology Transfer*, Bradford, Horton Publishing Ltd.

Walker, A. E. & Light, N. D. (1988) Cook–chill catering: How to make it work. *Catering,* **VIII** (3), 45–7.

Walker, A. E. & Light, N. D. (1988) Is cook–chill being put to best use? *Hospitality,* Feb., 15–17.

CHAPTER 11

Factors Affecting the Successful Introduction of a Cook–Chill Catering System

INTRODUCTION

There have been many words of wisdom aired regarding the requirements for setting up a cook–chill operation. Some of these are based upon sound practical experience, others are based on mere conjecture. This chapter aims to give an overview of factors affecting the successful introduction of cook–chill catering operations. It is not intended as a comprehensive guide to setting up a cook–chill operation. Rather it presents some of the major factors which have been shown, in practice, to affect the success or otherwise, of a cook–chill catering system. Whilst most of the evidence presented is taken from the UK Cook–Chill Survey, other research and anecdotal examples are given where appropriate.

A common debate on the issue of technological change is whether the decision to innovate is based on a technology 'pull' or 'push'. In other words, is the decision based on the lure of available technology (i.e. technology 'hype') or is there an inherent need to change the system and hence a requirement for new technology? Almost inevitably, any decision to change is financially motivated. Whilst ultimately all reasons for change in industry impact upon financial gain, some reasons for introducing change are less overtly financial than others. In addition, for reasons other than clear financial ones, the actual financial impact is more difficult to determine (see Chapter 12).

Any form of technological change has many component activities and has been shown to be a complex and iterative process. Figure 11.1 is a simplified diagram of some of the activities involved in the process of innovation and shows its progression in a number of clear stages (initiation,

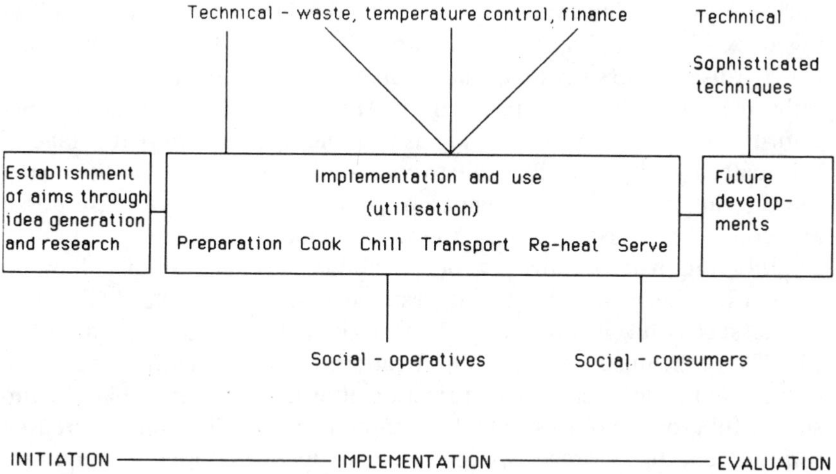

Fig. 11.1. Diagrammatic representation of the three main phases involved in setting up a new cook–chill catering operation, initiation, implementation and evaluation.

implementation, use and evaluation). Research has shown that, prior to technological change, many of the pre-implementation activities lay the foundations for success. For example, in the UK Cook–Chill Survey, the decision-making process, reasons for introduction, information sources, research and development, alteration to premises, employee participation and training were all found to impact on success (see later chapters). Thus, this chapter concentrates on activities which take place before the introduction of a cook–chill catering system. In particular, attention will be concentrated on those activities mentioned above.

THE DECISION-MAKING PROCESS

In the UK Cook–Chill Survey a sample of 80 cook–chill units were visited out of a total of 240 identified at the end of 1985. In the sample as a whole there was a tendency to assume that cook–chill catering was the most appropriate method to suit the needs of the innovating organisation. Little consideration was given to any possible alternative catering systems. Analysis of the successful and unsuccessful groups revealed little difference in the amount of consideration given to other catering systems. In fact, many operations in both groups dismissed the possibility of modifying an existing system in some way other than using cook–chill. A finding which causes some surprise because, on the whole, the units in the survey utilised cook–chill for a very small percentage of their total food production. For example, for the sample as a whole, the median percentage of total catering output attributable to cook–chill was only 29%—the rest of the catering output being provided by other systems such as cook–serve and cook–freeze. Nonetheless, there was evidence that more of the successful group had investigated the possibility of other systems, albeit on a superficial basis, than the unsuccessful group. For example, 14 (60·9%) of the successful group had looked at cook–freeze, compared with 7 (31·8%) of the unsuccessful group. In addition, the successful group were more able to show reasons for rejecting alternative systems than the unsuccessful group. For example, high capital costs; inflexibility; increased requirement for storage space; more recipe development required; higher energy consumption and a need for expensive additives and stabilisers were some of the reasons against the introduction of cook–freeze cited by the successful managers. The unsuccessful group, on the other hand, tended to dismiss cook–freeze without any further investigation than an awareness of the name.

However, most organisations surveyed regarded a feasibility study as a necessary part of the decision-making process. The content of these studies varied widely between the successful and unsuccessful groups. A wider range of topics were included in the feasibility studies completed by the members of the successful group than the unsuccessful group. For example, the successful group paid attention to the need for training, implications for employees, financial requirements, equipment needs, temperature control, research and development and microbiological testing. In contrast, the unsuccessful group were more likely to rest content with exploring potential cost savings in detail, whilst only giving passing attention to other issues.

The successful group used the feasibility study to decide on such factors as capacity requirements, the type of cook–chill system to introduce, the number of central production and satellite kitchens to have, the distribution methods required, the type and extent of training required for different employees and the need to inform consumers. Whilst the unsuccessful group addressed some of these issues, it tended to be on a more superficial basis and it was more likely that much of the detailed investigation was left out of the feasibility study and only considered as an afterthought after implementation.

DECISION TIME

The period between first awareness of cook–chill as a possibility and the actual introduction of a working cook–chill system (i.e. going 'live') is termed the decision time. There was some evidence in the UK Cook–Chill Survey which supported the notion of an optimum decision time before implementation takes place. Although there appeared to be little difference between the two groups with regard to the average length of decision time (see Table 11.1), the standard deviations for the two groups differed considerably with the unsuccessful group having a wider range of decision times than the successful group. These results leant support for an optimum decision time which appeared to be between seven and 24 months. However, having an optimum decision time alone does not guarantee success. There were members of the unsuccessful group who had made their decisions within the optimum range but had subsequently failed to introduce cook–chill successfully. Success, therefore, is not only dependent upon an optimum decision time, but also, critically dependent on that time being spent effectively.

REASONS FOR INTRODUCTION

In the UK Cook–Chill Survey, managers gave a wide range of reasons for introducing cook–chill. These reasons can be grouped into those which are directly financial and those which are indirectly financial. The former

TABLE 11.1
Length of Decision Time to Introduce Cook–Chill in the Successful and Unsuccessful Groups in the UK Cook–Chill Survey

Decision time (months)	*Successful group*		*Unsuccessful group*		*Total no. units*	*%*
	No. units	*%*	*No. units*[a]	*%*		
Up to 6	5	21·7	8	47·1	13	45·0
7–24	15	65·2	5	29·4	20	50·0
25+	3	13·0	4	23·5	7	17·5
Total	23	100·0	17	100·0	40	100·0

[a] Only 17 cook chill units were able to disclose this information out of a total of 22 in the unsuccessful group.

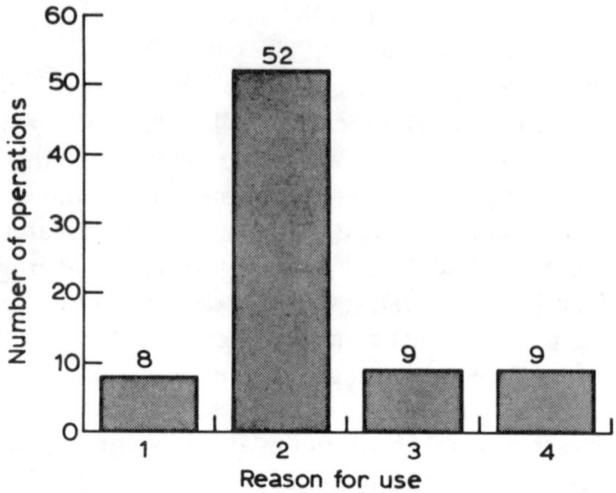

Fig. 11.2. Reasons given in the UK Cook–Chill Survey for setting up new cook–chill catering units. 1, To feed consumers remote from the production unit; 2, to reduce costs; 3, to improve productivity; 4, to improve quality.

group consists of a need to reduce costs, whilst the latter includes a desire to improve quality, improve 'productivity' and feed 'remote' consumers. A number of respondents indicated more than one reason for the introduction of cook–chill, but in such cases there was, almost without exception, one overriding reason which influenced this decision. Figure 11.2 shows the reasons cited by respondents as the *main* influence in their decision to introduce cook–chill. In a few cases there were one or two reasons cited which were not considered prime motivating factors with regard to decision making but rather as a bonus—namely, to improve hygiene and to improve safety. Neither of these were cited as main reasons. Fifty-two (65%) of the operations surveyed stated that cost reduction was their main reason for the introduction of cook–chill and of these, labour costs were seen as the most important area for reduction. In a few cases, other cost savings were expected in addition to those through labour reduction (such as energy and food purchasing), but were seen as minor savings in comparison with those anticipated from a reduced need for labour.

Improvement of end-product quality was cited as the main reason for the introduction of cook–chill by nine (11·25%) of the operations in the sample. For example, those in small rural schools in outlying areas, remote wards in a sprawling hospital complex without their own catering facilities and nightshift workers who were temporally, rather than physically remote from the main catering facility. Most of these units had experienced particular difficulties with their catering systems prior to cook–chill and tended to use methods which involved excessive warm-holding which resulted in dehydrated, unpalatable meals.

In these organisations, there were three different catering systems in use before the introduction of cook–chill. Five operations were operating a cook–warm-hold–serve catering system, two units had a system of hot transportation (which could also be classed as a cook–warm-hold–serve system) and two a cook–freeze system. In all of these cases food quality was thought to be sub-standard by the managers and consumers involved. In the first two groups, this was because of excessive warm-holding and in the latter, a result of detrimental effects on food quality caused by the freezing process.

Warm-holding has long been associated with many catering outlets. It has been described as a process unique to catering which often has deleterious effects on quality (see Chapter 2). One report on cook–chill states that one of the potential benefits from using cook–chill is that it removes the necessity for prolonged warm-holding as chilled food can be re-heated immediately before it is required instead of being kept warm for long periods after cooking and prior to service.

In the UK Cook–Chill Survey, it is this distinction which those managers who wanted to improve end-product quality, had seized upon. Cook–chill, by the elimination of warm-holding had the potential to improve end-product quality.

A further nine operations (11·25%) in the survey wanted to optimise work production. In other words, in a loosely defined way, they wanted to improve productivity. This group consisted of those respondents who said their primary aim was to optimise work production within the confines of their organisation rather than to reduce costs *per se*. In other words, the drive for increased productivity was for reasons other than direct financial gain; for example, a lack of physical space in main kitchens or to service peripheral units or to cope with a sudden increase in demand.

Finally, eight managers said that the decision to introduce cook–chill rested on the need to feed remote consumers such as nightshift workers, hospital patients and factory workers. In the latter case there was a special need for the work-force to remain in close proximity to their place of work during breaks. This resulted in a need to take the food to the work-place (i.e. factory floor) rather than the work-force 'going to' their meals in, say, a canteen or restaurant located elsewhere.

Other writers have given similar reasons for the introduction of cook–chill, but tend to refer to all the potential benefits globally. In doing so, they have a tendency to reflect an ideal situation, based on theoretical rather than empirical evidence. For example, for the UK National Health Service, the system has been described as being able to:

a. create viable business units (prompted by competitive tendering);
b. divert revenue and create more economic ways of serving food;
c. create effective capital equipping costs with longer payback periods;
d. optimise use of skilled staff;
e. make food transportation easier.

Of course all of these reasons have lured caterers towards the cook–chill option but the UK Cook–Chill Survey served to illustrate that in many cases the final decision is influenced by a single overriding factor, such as the ability to reduce costs. Furthermore, there were other notable trends between the successful and unsuccessful groups. Some of these were sector-specific and therefore the following discussion will deal with each sector in turn.

Within the UK National Health Service, catering establishments face their own particular problems. The recent UK government policy of placing services ancillary to essential health care (for example, cleaning, catering, and laundry) out to competitive tender (see earlier) has resulted in a determined effort by catering managers to seek cost effective ways of retaining their catering facility in-house. It is also worthy of note that in the health service much of the decision-making takes place at regional level, thus, individual hospitals, with their own unique idiosyncrasies, may have limited input in the decision-making process. The threat of privatisation was frequently mentioned by National Health Service catering managers in the survey sample as a whole. Clearly, the cost factor was expected to figure highly in the competitive tendering process.

However, of the eight NHS units which fell in the successful group (see Chapter 10), their decisions were not solely based on cook–chill's cost-cutting potential. Rather, other factors were stressed as being either more important or at least as important as the cost factor. For example, the ability, through systemised procedures to improve hygiene; improve the temperature of food to the patient; and improve the organoleptic quality of food to the patient. All these three were implicated as being unacceptably poor in the Platt, Eddy and Pellet (1963) report on hospital catering and although improvements have been made since that report was published there is still a pressing need to further improve and maintain standards.

In short, managers in the successful group were viewing cook–chill as a total catering process rather than as a direct means of cost-cutting. One of these managers described their motives quite succinctly:

> We were under pressure from the government to privatise and so were looking for cost-cutting methods but we saw other advantages such as improving hygiene in the kitchen and at ward level, increasing variety and most important, an ability to feed patients on demand at ward level.

There were only three cook–chill units from the hospital sector in the unsuccessful group. Without exception these units saw the introduction of cook–chill purely as a cost-cutting process. The decision-makers in these units gave no consideration to other possible consequences of introducing the system. This had ramifications on the eventual success of the cook–chill operation in question and was symptomatic of a tendency for these organisations to cut corners (and hence reduce immediate capital costs) throughout the change process (see Chapter 12). This observation was apparent throughout the unsuccessful group and was not limited to NHS cook–chill users.

A similar trend to that in the health sector emerged in the education and welfare sector in that successful cook–chill operations (4 units, 40% of education sector) indicated a wider variety of reasons for introducing cook–chill than the unsuccessful ones who were almost exclusively financially motivated. In particular, the effects of recent UK government announcements to reduce expenditure on school meals has meant that a greater emphasis has been placed on cost-reducing activities. In some extreme cases, provision of school meals has consequently ceased altogether. Three units in the education and welfare sector (30% of the sector as a whole) fell into the unsuccessful group. Two of these units had their own particular problems. Neither system realised its objective of producing good quality food products for their meals-on-wheels service and both found inherent problems subsequent to introduction insurmountable. More precisely, temperature was difficult to control during transportation, storage and regeneration. The situation was equally poor whether it was at chill temperature or regeneration temperatures. The reasons for this were inadequate transport facilities and poor end unit equipment, especially where the end unit was a voluntary day centre or, more usually, a recipient's home. Domestic equipment is rarely accurate and lacks the controls required to ensure cooked, chilled foods remain within safe parameters (i.e. at chill temperatures below 3°C and above 70°C during regeneration). In short, the systems were uncontrolled, resulted in an inferior end-product and were potentially dangerous.

In the hotel and leisure sector, 8 operations (57·1% of all hotel and leisure sector) fell into the successful group and 3 (21·4%) into the unsuccessful group. In the majority of cases, cook–chill was introduced to cope with a banqueting service, but in a few cases other uses were apparent, for example, to feed hotel workers, in à la carte menus and for room service. On the surface, there was little difference between successful and unsuccessful units with regard to their reasons for introduction. There were differences, however, in attitude. The successful managers paid greater attention to the needs and concerns of both the work-force and the consumer, as well as the financial attraction of introducing cook–chill systems. In contrast, respondents from the unsuccessful group tended to have a preoccupation with cost reduction.

Finally, in the industrial sector, there was further evidence of more informed reasons for the introduction of cook–chill in the successful group than the unsuccessful group. More specifically, the successful group based their choice on well-thought out policies which fitted their own particular organisational needs and did not restrict their vision to economic gain. The unsuccessful group, on the other hand, tended to fall victim to sales literature 'hype' and placed much emphasis on the potential for cutting costs through the introduction of cook–chill catering systems. Organisational needs and the suitability of cook–chill catering for their requirements were not considered. A common response being 'our only aim was to reduce labour costs'.

In summary, across all sectors, the successful group applied a wide ranging approach to the question of cook–chill and had more depth of knowledge with regard to reasons for its introduction. In comparison, the unsuccessful group rested content with the notion that this was the system to reduce costs.

SOURCES OF INFORMATION

Early cook–chill users were at a disadvantage as far as background information was concerned. Very little research had been carried out and there were few other caterers using cook–chill. Those that did exist were often not in comparable situations and experiences of one establishment were often difficult to parallel elsewhere. In addition, although what liter-

TABLE 11.2
Breakdown of the Main Sources of Information Used by the Successful and Unsuccessful Groups in the UK Cook–Chill Survey

Information source	*Successful group*		*Unsuccessful group*	
	No. units	%	*No. units*	%
Equipment manufacturer (one only)	4	17·4	19	86·4
Equipment manufacturers (more than one)	14	60·9	3	13·6
Trade magazines	7	30·4	6	27·3
Consultants	12	52·2	15	72·7
Other cook–chill users (one only)	3	13·0	5	22·7
Other cook–chill users (more than one)	14	60·9	7	31·8
Seminars/conferences	12	52·2	14	63·6
DHSS guidelines	21	91·3	5	22·7
Own knowledge	10	43·5	3	13·6

ature did exist detailed some of the essential microbiological, temperature, nutritional and sensory implications, it was not concerned with the practical implications of introducing such a system.

However, more recently, the situation has improved and a range of information sources on most aspects of the cook–chill system are now widely available. This is an important factor as the results of the UK Cook–Chill Survey showed that source of information was a critical early factor affecting the successful introducton of cook–chill. Table 11.2 shows a breakdown of the information sources used by members of the successful and unsuccessful groups respectively. It is clear that success cannot be guaranteed by the reliance on a single or limited range of informants. Overwhelmingly, success was linked with amassing information on all aspects of cook–chill from a variety of different viewpoints. For example, Table 11.2 shows that unsuccessful groups often rested content with information gained from one of three sources, a single equipment manufacturer, seminars/conferences and consultants (including the Electricity Council). In contrast, the majority of the successful group gleaned information from a number of different information sources, in particular, a range of equipment manufacturers, other cook–chill users, consultants, their own prior experience and the DHSS guidelines (see Chapter 4). Perhaps a major additional factor was the tendency of the unsuccessful group to rely solely on the advice obtained from their limited sources, whereas the depth of information collected by most of the successful group resulted in a much more balanced knowledge base.

A number of specific differences between the two groups were apparent. For example, the advice of other cook–chill users was rarely sought by the unsuccessful group and where it was, it was limited to only one other user. Moreover, in some cases, that other user was often running an operation non-comparable with their own (e.g. a small industrial unit gaining information from a large school meals unit).

Concern over temperature control in cook–chill catering operations has been recently highlighted and, thus, information regarding temperature (and, consequently, microbiological control) is of the utmost importance. In the UK, the DHSS guidelines for cooked, chilled food are the most used reference with regard to safe levels of temperature and microbiological control but even these have been criticised for their inadequacy and were reviewed and updated during 1988/1989. It is surprising therefore that a high percentage of the unsuccessful group manifested a total disregard for any such information with very few consulting the DHSS guidelines at all. Over 70% of the unsuccessful group had not used the guidelines and over 20% were unaware of their existence altogether. The succcessful group, on the other hand, put great emphasis on the adherence to the DHSS guidelines. However, because these guidelines are not mandatory, there is no vehicle which requires cook–chill users in the UK to operate within strict temperature and microbiological guidelines. The situation in France, however, is somewhat different, there, legislation regarding temperature and microbiological control in cook–chill food production is in force and all cook–chill users are required, by law, to comply with this strict code of practice (see Chapter 1). Any cook–chill user not adhering to the rules laid down in the legislation is liable to prosecution.

In the UK Cook–Chill Survey, prior knowledge of cook–chill did seem to have an effect on increasing awareness of the necessary requirements for the introduction of cook–chill. Ten (43·5%) of the successful group had some previous knowledge or experience of cook–chill from actual use, college courses, seminars and literature. Only 3 (13·6%) of the unsuccessful group had such knowledge.

The use of consultants in an advisory capacity appeared to be a hit and miss affair. There was little difference between the two groups with regard to the use of consultants as an information source. However, there was a difference between the two groups with regard to the extent to which they relied on consultant's advice and recommendations. The successful group tended to rate the information given by consultants poorly, whereas the unsuccessful group rated consultant's involvement highly.

In summary, the key factor, with regard to information sources, appears to be intrinsically linked with thorough and wide-ranging background research by individual caterers. Undoubtedly, much information can be transferred from one situation to the other, but each catering unit has its own peculiarities which must be considered. The experiences of both the successful units and the unsuccessful units serve to illustrate the value of 'getting it right' from the beginning and this involves thorough development work and gathering a wide range of relevant information from a number of different sources.

RESEARCH AND DEVELOPMENT

There has been considerable concern recently with regard to cook–chill systems within the UK. Much of the concern stems largely from the dearth of research and development into cook–chill. It is of paramount importance that research and development is carried out before the introduction of cook–chill especially with regard to microbiological quality, temperature control and product development. Although these topics are all dealt with in detail in other chapters, they will briefly be considered here with respect to the results of the UK Cook–Chill Survey which provides information on the extent to which research and development was considered important and used within the cook–chill units in the sample.

Recent evidence with regard to the microbiological quality of cook–chill food has raised important questions with regard to the safety of food produced by this method (see Chapters 3 and 8). In the light of this information, it is of utmost importance that microbiological testing is carried out on swabs from environmental sources (e.g. kitchen work tops, equipment, walls, etc.) and random samples from incoming and unprocessed stored products. Each individual cook–chill operation has its own idiosyncrasies and, as previously stated, the lessons from one unit will not necessarily apply to another, which makes individual testing essential. In addition, the current thinking of many microbiologists, researchers and

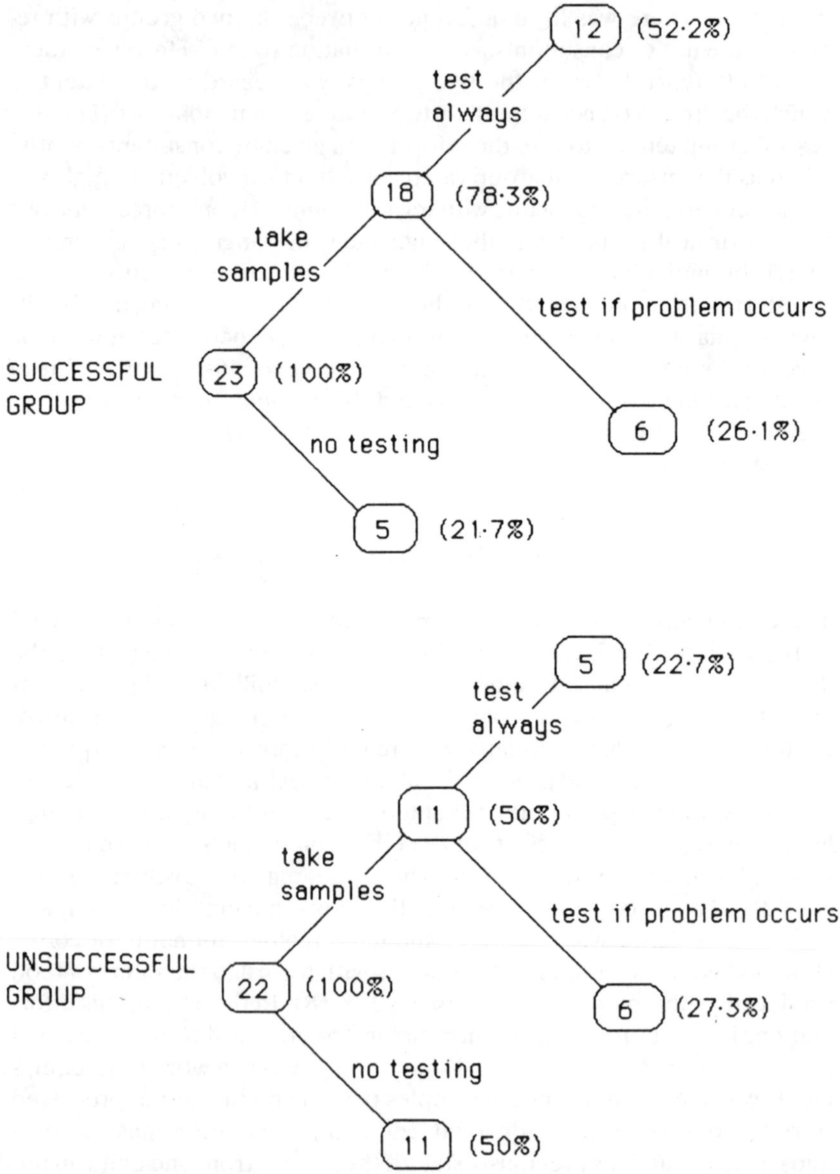

Fig. 11.3. Diagram summarising the use by successful and unsuccessful cook–chill catering units of microbiological testing as a method of checking food quality and safety.

environmental health officers, places emphasis on ensuring the microbiological quality of incoming food from suppliers rather than testing samples of the food once it has been through the cook–chill process (the reasons for this are explained fully in Chapter 9). However, there is a need to establish that the working practices of a cook–chill operation are such that microbiological contamination does not occur during production and one step in this process is to carry out thorough development work prior to introducing cook–chill. This includes full testing of all products microbiologically, organoleptically and for temperature control throughout the system. The latter is important for establishing weak areas of temperature control within the system.

The UK Cook–Chill Survey identified the levels of microbiological testing amongst the units surveyed. Considerable differences were noted between the two groups. Figure 11.3 illustrates these differences with regard to the incidence of microbiological testing between the successful and unsuccessful groups. A greater percentage of units in the successful group (78·3%) were conducting microbiological analysis than the unsuccessful group (50%). Most of the testing referred to that which was done after preparation i.e. on samples taken during chilled storage prior to regeneration. In the light of recent evidence regarding the validity of microbiological testing, this may not be the most effective practice. However, it is indicative of the successful group's greater concern over control and shows that the resources were in place to introduce microbiological testing at source. In fact, some successful units were indeed already carrying out this practice.

One of the outstanding features throughout the UK survey was the apparent trend of successful managers to show a proactive approach to introducing cook–chill whereas the unsuccessful managers tended to be reactive. This phenomenon manifested itself in a number of ways and research and development was no exception. The successful group placed more importance on the advantage to be gained by fulfilling in-depth research and development prior to introducing cook–chill.

Other studies have also stressed the importance of effective research and development strategies showing that catering is not the only industry where research and development forms a key aspect in effecting successful 'technological changes'. For example, a study of chemical processing and scientific instrument industries showed that in all cases the extent to which effective research and development had been carried out beforehand had a distinct influence on eventual success or otherwise. In other words, research and development correlated well with success.

ALTERATION TO PREMISES

Undoubtedly, whatever the situation, the introduction of cook–chill will have a profound effect on the total organisation. This will affect not only the people within the organisation but the 'structure' of the organisation itself. For example, certain questions need to be addressed such as; what capacity needs to be produced, what type of cook–chill system is needed, how many central production kitchens will be required and hence how many satellite units, what sort of distribution methods are required (which in turn influences the type of transport system used). The outcome of these and many more issues which need to be addressed during the decision-making process will affect the need for any alterations to existing premises or whether there is a need for a completely new building.

The results of the UK Cook–Chill Survey showed that there was little difference between the successful and unsuccessful units with regard to the incidence of structural alteration. Approximately 50% of each group reported the need for major structural work. However, there was a difference between the two groups in that the successful group carried out their alterations in anticipation of cook–chill, whereas the unsuccessful group left it until the system was up and running, only to find that structural alterations were essential. This further reinforces the tendency for the successful group to advocate proactive styles of management as opposed to the more reactive styles of the unsuccessful group.

EMPLOYEE INVOLVEMENT

There have been a number of examples where proposed cook–chill units have fallen by the wayside because of industrial relations problems. There are two cases which demonstrate this, in particular. First, the case of Jersey health authority which first initiated a cook–chill proposal in the early 1980s, purchased the equipment and were set up and ready to go when industrial action forced its postponement. Inadequate attention had been paid to the wider implications of the technology transfer (in particular, its effect on the work-force in terms of a change in skills, working hours and remuneration). Second, Wakefield health authority suffered severe disruption to its cook–chill programme facing not only employee resistance through the unions involved but also from the local community. This was not only due to inadequate preparation of the work-force and trade unions involved, but also because of an inappropriate choice of regeneration system (see Chapter 1).

According to many of the catering managers interviewed in the UK Cook–Chill Survey, many of the industrial relations problems which had occurred were due to employees receiving only a limited training and knowledge of cook–chill. Questions which concerned employees but were not addressed by management included, what effect was the introduction of cook–chill going to have on the work-force in terms of their job stability, job content, skill requirements, training and remuneration? The threat of pending job and financial insecurity as a result of the introduction of a cook–chill catering system was often cited by managers interviewed as a major source of discontent amongst employees. In addition, confusion was often exacerbated because the catering managers themselves had inadequate knowledge of cook–chill as a system of catering and the full implications of its introduction, especially on the work-force.

In the UK Cook–Chill Survey there were wide-ranging differences between successful and unsuccessful units, not only in the numbers of units informing their employees but also whether employees were continually updated on the progress of cook–chill, received training and were compensated financially for any loss of overtime and weekend work. These issues will be covered in more detail in Chapter 14. In short, however, success was linked with effective and thorough communication between management and work-force regarding the introduction of cook–chill and the practice of seeking and acting upon the views of employees on the practical aspects of the introduction of cook–chill.

CONSUMER INVOLVEMENT

There has been a plethora of popular newspaper and magazine articles recently discussing various aspects of the cook–chill system. Some of these have been based on sound evidence whilst others have suffered from sensationalism and media 'hype'. However, they have alerted members of the general public to the implications of introducing cook–chill systems into catering establishments and retail outlets. Furthermore, it prompts the question of whether or not there is a need to inform consumers of any treatment the food being eaten may have undergone. In doing this, however, there is a danger of creating a less favourable situation by imparting inadequate or distorted information on the consumer. Any information given to consumers must be clear and simple—not an easy thing to do with a complex system such as cook–chill. A straightforward explanation may be, quite innocently, misinterpreted resulting in unnecessary anxiety regarding the food being eaten.

Of the 80 cook–chill managers interviewed in the UK Cook–Chill Survey, opinion was divided as to the relative merits of informing consumers of the introduction of cook–chill. There were two distinct factions, those who felt that the customer has a right to know 'what's happening to the food being served' and those who thought that 'what they don't know, won't hurt them'. It was clear from the results that there was a sector influence determining which view was held rather than any link to success. Those in sectors which commanded a captive consumer audience tended to hold the views of the former group whereas those in sectors which had a casual consumer base were more likely to hold with the latter view.

The main difference between the members of the successful and unsuccessful groups who informed consumers was in the method used to impart that information to consumers. The most notable difference being the extent of formal and informal methods of transferring information to potential consumers of cook–chill food. The successful group made more use of formal methods, such as seminars and conferences, information leaflets and specific cook–chill open days. Whereas, the unsuccessful group relied on informal methods such as 'the grapevine' (see Table 11.3). These findings were reinforced by some of the comments received by managers in both groups. First, those in the successful group:

TABLE 11.3
Methods Used to Inform Consumers by Success Group in the UK Cook–Chill Survey[a]

Method	*Successful group*			*Unsuccessful group*		
	No. units	*%*	*Overall %*	*No. units*	*%*	*Overall %*
Formal						
Seminars	9	64·3	50·0	2	33·3	10·5
Literature	5	35·7	27·8	3	50·0	15·8
Special meal	3	21·4	16·7	1	16·7	5·3
Sub-total	14	100·0	77·8	6	100·0	31·6
Informal						
Discussion	12	85·7	66·7	13	100·0	68·4
Other	8	57·1	44·4	5	38·5	26·3
Sub-total	14	100·0	77·8	13	100·0	68·4
Total	18	—	100·0	19	—	100·0

[a]Some units used more than one method.

> We instructed customers on the changes through a video and a display. . . . Cook–chill was introduced and explained at a works canteen sub-committee lunch. . . . We had a product display and demonstration day. . . . We held a film show, visits and coffee meetings.

In contrast to these extracts, which indicate a general level of organised information exchange, were responses from some of the unsuccessful group managers:

> . . . they're aware but its from the grapevine. . . . They don't really understand it, it wasn't done very well because we tried to keep it quiet. . . . Everyone got to know because of the catering workers dispute, it got in the papers, on the local TV and even went on the radio.

These latter excerpts from interviews indicate that althouth consumers had been informed it was often done in an *ad hoc* fashion. In sum, the actual decision of whether or not to inform customers had little effect on success, however, if the decision to inform customers was taken, the methods used to do so did have an impact on success—more formal methods being more successful than relying on casual conversation and rumour.

CONCLUSIONS

Numerous commentators have stressed the need for careful planning, background research and a thorough knowledge of cook–chill before its introduction. However, despite such warnings and recommendations examples of ill-thought out, potentially unsafe cook–chill operations are apparent. Evidence from the UK Cook–Chill Survey showed that it was often activities during the planning stage which sowed the seeds of eventual success or failure.

A number of notable differences have been described which have been shown to affect the successful introduction of cook–chill catering systems. In particular, the UK Cook–Chill Survey revealed numerous differences between successful and unsuccessful cook–chill users. The major ones are summarised below:

- Decision-makers in the successful group were more proactive, whereas those in the unsuccessful group were more reactive.
- There was greater employment stability amongst managers and decision-makers in the successful group than the unsuccessful group.

- The introduction of cook–chill was more 'needs related' in the successful group than the unsuccessful group.
- The successful group carried out more thorough and efficient development work than the unsuccessful group although the total decision time was not necessarily shorter.
- The successful group were more orientated towards research and development than the unsuccessful group.
- The successful group had a higher capital expenditure per unit of capacity than the unsuccessful group.
- The successful group acquired more harmonisation of equipment into the total catering system than the unsuccessful group.
- Effective communication was a salient feature of successful cook–chill units whereas in unsuccessful units communication was scant where it existed at all.
- Employees in the successful group were given a wider range of, and more in-depth training for, cook–chill than employees in the unsuccessful group.

In short, there were factors which differentiated between successful and unsuccessful cook–chill units throughout the technology transfer process and we will elaborate on these in the remaining chapters.

FURTHER READING

Anon. (1988) Cook–chill: getting it right. *Hospital Caterer*, Jan./Feb., 7–8.

Department of Health and Social Security (1980) *Guidelines on Pre-cooked, Chilled Foods*. London, HMSO.

Light, N. D. (ed.) (1986) *The Future of Cook Chill Technology in Catering*. The Food and Hospital Research Centre, Dorset Institute of Higher Education, Poole, Dorset, UK.

Platt, B. S., Eddy, T. P. Pellett, P. L. (1963) *Food in Hospitals*. Oxford University press.

Walker, A. E. (1988) The transfer of technology in the catering industry: a study of UK cook–chill catering operations. PhD thesis, CNAA/Dorset Institute of Higher Education, Poole, Dorset, UK.

Walker, A. E. (1989) *Success or Failure in Cook Chill Catering: A Study in Technology Transfer*, Bradford, Horton Publishing Ltd.

Walker, A. E. & Light, N. D. (1988) Cook–chill catering: How to make it work. *Catering,* March, 44–6.

CHAPTER 12

Financial Aspects Associated with the Introduction of Cook–Chill Catering Units

INTRODUCTION

In the initial stages of the decision-making process the main advantage of introducing cook–chill catering (as opposed to any other catering system), is its potential for greater cost effectiveness. Advantages, other than purely financial ones, have been claimed by some cook–chill catering protagonists, but are less clear cut and, to a certain extent, depend on the type of cook–chill system being installed coupled with the type of catering system it is replacing. Also, the installation of a cook–chill catering system necessitates considerable capital investment and the need to forecast pay-back periods and expected cost savings or benefits. Thus, it is not surprising that the financial aspects are seen by many to dominate the decision-making process.

COST SAVINGS ATTRIBUTED TO THE INTRODUCTION OF COOK–CHILL CATERING

The development of cook–chill in the catering industry was encouraged by frequent claims of massive financial benefits which could be gained through its introduction. Articles which appeared in the early 1980s suggested that cook–chill catering systems make savings in labour and food wastage, drastically reduce or reverse previous operating losses, tap new areas of profit, make savings in purchasing and control and provide economies of scale in the use of skilled staff.

More recently, some observers and users have questioned whether or not savings are actually realised. Undoubtedly, many operations have made substantial savings on labour costs alone by the introduction of cook–chill, but there are also many examples of cook–chill units which have been unable to retrieve their costs as anticipated.

In the early 1980s, many catering equipment manufacturers advertised the monetary benefits to be gained through the introduction of cook–chill. For example, the following quotes are taken directly from early 1980s sales literature describing cook–chill systems:

> A system which cuts the costs not the quality . . . improved financial, purchasing and portioning controls . . . reduced energy consumption . . . savings . . . savings . . . savings.

> When you purchase, lease or rent the [cook–chill] system you invest in a planned programme of economies. Savings can repay your investment within the first 12 months. Our comprehensive feasibility study for your operation will detail the costs—and the benefits . . . Reduced overheads and lower staff costs mean more value on the plate—plus big savings as well.

> Cook–chill is a remarkable system which gives customers the best of both worlds, the opportunity to produce consistently high quality meals in bulk utilising existing equipment, staff and energy resources to better effect than ever before—ensuring that it remains extremely cost effective.

> Overall staffing costs are reduced as bulk production techniques allow for a reduction in staffing levels. Significant savings on overtime result from full utilisation of the normal working day . . . cook–chill can produce a saving on food costs due to the possibility of bulk purchase.

> The ability to produce larger quantities of food reduces costs . . . staff are used more effectively—the more skilled in the preparation of dishes in central kitchens; the less skilled in regeneration and serving at the end kitchen . . . less equipment is generally required . . . it is used more intensively and is therefore more cost effective . . . Management has greater control over food costs, individual costs, quality and portion supervision, menu planning and demand forecasting . . . considerable energy savings are achieved by programmed use of the latest equipment and methods.

These excerpts suggest advantages which, for caterers struggling to keep within ever-tightening budgets, must have been difficult to resist investigating further. They also place a major emphasis on the financial benefits of installing the system.

However, most of the reports of cost savings have had little concrete evidence in support of these claims. Accurate costings which are readily available for comparisons are relatively difficult to obtain. Once again, this has been a poorly researched topic. In addition, there may be some reluctance for organisations to admit that the introduction of cook–chill was not cost effective. Yet, many cost appraisals have been and are carried out, but detailed, full costings are rare. Generally speaking, any costings carried out by individual organisations tend to concentrate on the costs associated with tangible rather than intangible (i.e. problem solving) items. The latter often hold the key to the successful introduction of a cook–chill catering system, but are in themselves difficult to isolate and measure.

The lack of accurate financial statements which fully describe the financial implications associated with the introduction of cook–chill, may be because a difficulty exists in separating actual costs attributable to cook–chill, from the rest of the financial concerns of complex organisations. Even financial data from the UK Cook–Chill Survey was unable to reveal any detailed financial statements for general comparisons across the board. This was largely because many of the organisations involved in the survey were either unable or unwilling to disclose detailed figures with regard to the cost of introducing their cook–chill system. Some costs (or cost savings) associated with the introduction of cook–chill were 'lost' in the general running of the organisation and, thus, became almost impossible to identify as being a direct result of the introduction of a cook–chill catering system. Nevertheless, certain cost factors were easily identifiable as directly attributable to its introduction. For example, the cost of specialist capital equipment, building alterations, consultants' fees, packaging, containers, labels, and transport costs. However, there were many other examples of hidden costs which were, more often than not, ignored as a cost attributable to the introduction of cook–chill systems. For example, training costs, maintenance, storage (both food and equipment such as trolleys, insulated boxes and trays), wastage (plate, kitchen and over production) and increased administration. Perhaps most importantly, the cost of problems caused during the learning curve for change were rarely, if ever, considered.

Thus, it is difficult to establish the real costs of introducing cook–chill catering systems. In the following sections we will identify some of the major cost factors which need to be taken into account during the introduction of cook–chill technology and we will also describe some of the different methods which have been used previously to measure them. In addition, we will present some of the results of an analysis of the financial data gleaned from organisations in the UK Cook–Chill Survey.

PREVIOUS STUDIES OF COSTS

One fairly comprehensive study of the cost to an organisation of introducing centralised catering systems was carried out in the late 1970s by the Department of Education and Science (DES) (Cowell, 1980). The study concentrated on cook–freeze catering but similar principles can equally be applied to cook–chill catering.

The DES study advocated that the main aim in costing a system should be a feasibility study to assess capital and recurring costs, manpower implications and the minimum economic viability of introducing a centralised system. It was suggested that this is probably best achieved by estimating potential meal numbers and demand patterns, equipment requirements (including storage), staffing levels, planning and 'costs' (see later).

Assessment of meal numbers is often difficult to achieve and, this study claimed, is best done over a period of time in order to more accurately assess demand. Ideally, such a study needs to include an assessment of the relationship between demand, production and storage which, hence, take into account the storage life of products and the maximum amounts of food product that can be stored at any one time. With regard to cook–chill catering, this means that specifications must be developed which assess chilling and chilled storage costs and calculate re-heat capacity for a meal period where demand for food is greatest. This, in turn, enables a more accurate assessment of new equipment needed for the safe and efficient running of the system. For example, the number and size of chiller equipment and chilled storage areas at the CPU and chilled storage and regeneration equipment at the satellite units.

The DES study also stressed the importance of defining total storage requirements. Obviously, this includes all raw materials stores (dry goods, refrigerators, freezers, etc.), separate chilled stores for the finished products and full equipment stores, especially for trolleys and other transportation equipment. Centralised systems, like cook–chill catering, require provision for bulky equipment for transportation of food to satellite units which, unavoidably, occupies valuable floor space.

This is a factor which must be adequately planned for in order to avoid the dangerous practice of blocking thoroughfares in the CPU with casually-stored equipment (a common sight during the UK Cook–Chill Survey.

Whilst the above mentioned constitute the areas where the most capital outlay occurs, cost savings are to be made through reduction in staffing levels and minimisation of food and energy waste. Cook–chill systems are more capital-intensive than conventional systems in that they have a more concentrated technological base. This leads to a lower manning requirement, thus, cost savings usually result from a reduced need for skilled labour at the end kitchens (see Chapter 14). However, cost savings are particularly difficult to evaluate and with this in mind the DES study suggested that, ideally, a full scale work study be carried out to assess all of the financial implications. This exercise ensures that any assessment of the minimum size of team required to cover production in the central unit is as accurate as possible. In reality very few organisations indulge resources in such studies prior to introducing cook–chill.

Finally, the report considered evaluation of costs. It recommended that investment in a new food service system should be evaluated by the measurement of net additional capital costs and comparisons with changes in recurrent costs *for the full life of the system*—an important undertaking which builds-in the necessity to update and expand or contract systems as changing circumstances dictate.

Another of the early UK documents which attempted to detail the financial aspects of cook–chill catering was the Electricity Council's *Planning for Cook–Chill,* which advocated, in its section on finance, 'a simple feasibility study to determine labour cost savings in particular when measured against a capital cost for a project' and that 'careful consideration must be given to the equipment requirement related to the range of meals and meal items to be produced'.

This latter point is important and helps, from the outset, to gear the whole project towards the individual requirements of the implementing organisation. The report recommends design of an 'economically satisfactory system' a jargonistic term which presumably refers to a stringent assessment of menu needs related to the capability of equipment incorporated into a realistic layout design.

To estimate the cost of such systems, this particular document recommends the division of overall cost into fixed and variable costs and, in addition, to relate these costs to the price of the products produced from the system.

The formulation of a planning group made up of a variety of 'experts'

TABLE 12.1
Summary of Project Ownership and Operating Costs

A. Ownership costs			
Initial fixed capital cost of project.			
(a) Equipment cost			
(b) Ancillary controls etc.			
(c) Electrical installation			
(d) Water, drainage, waste			
(e) Building cost, increase to house the project	+		
(f) Any decrease in cost because of the project	–		
(g) Transport purchase			
(h) Other installation costs			
Initial fixed capital cost			
B. Estimation of fixed costs, excluding wages			
(a) Amortization period (years to recover initial fixed capital cost)			
(b) Interest rate on loan capital to cover initial fixed capital cost			
(c) Capital recovery factor			
(d) Equivalent annual uniform cost [from (a) (b) and (c) above]			
(e) Interest at____% on £____working capital			
(f) Rates			
(g) Rentals			
(h) Insurance			
(i) Indirect overheads, e.g.	(i) Water supply, electricity distribution		
	(ii) Administration services etc.		
(j) Maintenance allowance			
Total annual fixed cost			
C. Wages			
(a) Operators			
(b) Supervision			
(c) Management			
(d) Distribution			
(e) Indirect overhead, i.e.	(i) NHI on wages		
	(ii) Superannuation		
Total annual wages bill			

TABLE 12.1—*contd.*

D. Estimation of variable costs (operating costs)	
(a) Annual energy and fuel costs	______
(b) Packaging costs	______
(c) Distribution costs/transport costs	______
Total annual variable cost	______
E. Working capital	
(a) Raw material costs	______
(b) Finished product cost	______
Total working capital	______

is advised. Such a group would comprise experts in the following fields, building, finance, work study, refrigeration and catering. The main objectives of the group are to realistically assess the financial, engineering and design requirements of a cook–chill system specific to the establishment in question. It is argued in this document that the feasibility study be one of comparison between cook–chill and the cost of an alternative conventional system. It is interesting to note here, that in the UK Cook–Chill Survey, despite the fact that a large proportion of managers utilised the services of the Electricity Council, very few actually investigated alternative catering systems to cook–chill in any depth. Detailed costings recommended in the Electricity Council's document are summarised in Table 12.1.

The recommendations as laid out by this document omit several pertinent issues critical in the organisational changeover process in the implementation and use of cook–chill catering. Foremost in these omissions, is one of the most important aspects of the whole cook–chill technology transfer process, namely, *training*. The need to cost and incorporate effective training schedules (from senior management through to unskilled operatives) into any implementation plan needs heavy emphasis. These is no doubt that operative and management training is an absolutely essential component of any cook–chill system and must not be ignored, particularly in view of the requirement in such systems for good food, personal and plant hygiene control (see Chapter 9). Provision for training costs however are conspicuous by their absence in the Electricity Council's document guide. Similar lack of such provision was apparent in a large number of cook–chill establishments interviewed in the UK Cook–Chill Survey (see Chapter 14).

Clearly, both the DES report and *Planning for Cook–Chill,* point the way to essential components of a financial feasibility study. Both reports emphasise that substantial savings are made through reductions in labour costs, an issue which we will now consider in more detail.

COST SAVINGS THROUGH REDUCTION IN LABOUR

As we saw earlier in this chapter, cost savings through the reduction of labour may be made by shedding unwanted labour hours or reducing job skill requirements and, thus, the need for more expensive kitchen employees. It is not always possible to achieve projected labour savings (see Chapter 14) or if they are achieved, to do so without incurring any extra costs to the organisation. In other words, it is not always possible to shed labour through 'natural wastage' methods. More often than not, there are substantial costs to an organisation involved with labour reduction processes. On some occasions, these costs take the form of redundancy payments or, on other occasions, they can be less 'direct' through, for example, dissatisfied employees taking industrial action. The cost of losing labour must be built into the feasibility study and realistically assessed for its worth against the total cost implications associated with the implementation of a cook–chill catering system.

A major feasibility study, for one of the largest UK cook–chill catering programmes to date, claimed substantial energy savings, in addition to labour savings, on the basis of comparing the energy costs of an existing conventional catering operation and the potential energy costs with a new cook–chill system. However, most of the savings cited were attributable to more efficient equipment usage and as such are invalid for comparative purposes. That is, the study assumed the conventional system to be energy efficient—an assumption almost certain to be wrong. Had a fully efficient conventional system been compared with the new cook–chill catering system the energy savings claimed may not have been so substantial. However, it could be claimed that a greater potential for process control in cook–chill catering systems allows the caterer a means of incorporating more energy efficient methods into the production system.

In large cook–chill operations it is theoretically feasible that energy savings can be made. For example, take a situation where a number of kitchens operate within an establishment, say a group of university halls of residence, where a full complement of kitchen staff are employed in each kitchen. The introduction of cook–chill catering could mean that all but one of these kitchens need function as a primary food production kitchen, the others taking on the role of satellite units, for chilled storage and regeneration purposes only. Energy usage in these new satellite units would thus be reduced to the requirements of chilled storage equipment and a short period during regeneration. In such situations, a net reduction in energy usage is plausible. It is more doubtful whether energy savings are possible in smaller applications where similar economies of scale are not encountered. On the other hand, energy costs in some large situations could substantially increase following the introduction of a cook–chill catering system. Take, for example, a hospital complex where, prior to the introduction of cook–chill, food was prepared in a main kitchen and transported to ward level in heated trolleys. The introduction of cook–chill could result in a net increase in equipment through the installation of chilled storage units and regeneration equipment at ward level. A net increase in energy costs is almost inevitable in such situations.

Rationalisation of a catering system through the introduction of cook–chill catering may also lead to savings in food waste. Food waste can be divided into production waste and plate waste. It has been claimed on several occasions that cook–chill catering can help to reduce both of these forms of waste. The former by bulk ordering food (and particularly fresh produce) to exact specifications in terms of size, quality and degree of pre-preparation required. This becomes more feasible in a centralised system such as cook–chill because the amount of a particular food item being ordered at any one time is often substantial enough for a supplier to ensure that the products provided meet the specification. Plate waste, on the other hand, can be lessened as the use of cook–chill catering allows for the amount of food regenerated to be regulated. In other words, only enough food to cater for the actual number of customers served need be re-heated. This is especially true of units incorporating microwave regeneration into their system.

It has also been claimed that savings can be made in food purchasing itself through the introduction of cook–chill. The basis for these claims lies in a centralised catering system being able to submit bulk orders and, thus, having more bargaining power with suppliers than smaller, individual units. In some larger situations this may indeed be possible. However, evidence from the UK Cook–Chill Survey suggests that savings in food purchasing are not widespread and tend to be associated only with organisations able to place very large food orders. There was some indication though that centralised purchasing did increase the threshold levels enough for caterers to demand higher specifications than they were able to prior to cook–chill, and in this sense were probably getting 'better value for money' (see Chapter 13).

In summary then, equipment manufacturers, the Electricity Council and the DES have all investigated the potential theoretical savings or changes in costs to an organisation with regard to the introduction of cook–chill, but little has been gleaned, to date, on what cost savings are actually achieved in practice. Are potential savings always realised? If not, why not? and if they are, are they as substantial or higher than first estimated? Some of these questions were addressed in the UK Cook–Chill Survey which we will describe in the following sections.

EVIDENCE FROM THE UK COOK–CHILL SURVEY—GENERAL FINANCIAL ASPECTS

This section describes some of the general financial characteristics of organisations involved in the UK Cook–Chill Survey. The main emphasis of the results presented here concentrates on the level of total capital expenditure analysed with respect to a number of different factors. For example, expenditure, by sector of industry; by type of cook–chill; by percentage of the total catering attributable to cook–chill; and as a percentage of the total annual turnover of the organisations concerned. Subsequent sections address other issues such as size, capital expenditure per unit of capacity and a summary of what capital was actually spent on.

Total Capital Expenditure by Industry Sector

All sectors of the catering industry using a cook–chill catering system were included in the sample chosen for interview in the UK Cook–Chill Survey, with the exception of fast food outlets. The actual number of units in each sector that were analysed is shown in Table 12.2 and is proportional to the sectoral representation in the total population (i.e. all the catering establishments using a cook–chill system at the time of the survey). There was little difference between the total expenditure of private as opposed to public sector organisations, however, public sector cook–chill units tended to be larger and took advantage of special discount systems afforded to them. Table 12.3 shows a breakdown of total capital expenditure by industry sector. Although the actual figures are not very meaningful when considered in isolation from factors such as maximum capacity (see later), Table 12.3 does indicate that spending in

TABLE 12.2
Breakdown of Units Visited by Sector

Sector	*No. units visited*	*No. units in total population*
Industrial	38	114
Hotels and leisure	16	48
Hospitals	16	48
Education and welfare	10	30
Total	80	240

TABLE 12.3
Total Capital Expenditure by Sector

Expenditure (£000)	*Sector*				
	Industrial (units)	*Hotels (units)*	*Hospitals (units)*	*Education (units)*	*Total (units)*
0–10	1	0	1	0	2
11–50	6	0	4	2	12
51–100	3	1	4	0	8
101–1000	6	5	1	5	17
1001+	5	1	0	0	6
Total	21	7	10	7	45

Note: Private and public sector organisations were distributed amongst all four sectors, i.e. some private sector organisations were in the education and hospital sectors as well as the hotels and industrial sectors.

different sectors of the industry varies. For example, despite the fact that the largest cook–chill units were found in the hospital and education sectors, the largest capital investments were in the industrial sector. Of course, this provides no information on whether expenditure was directed in the right areas, nor does it attempt to make any comment on this aspect which is dealt with in more detail later in this chapter. It merely suggests that the amount of total capital expended on cook–chill by different organisations is affected by a number of criteria unique to each organisation. However, useful comparisons can still be made (see later).

Total Capital Expenditure by Type of Cook–Chill

Four main types of cook–chill were identified as being used by catering establishments in the UK survey. These were defined as:

Type A Electro-mechanical chilling with mass regeneration.
Type B Cryogenic chilling with predominantly mass regeneration.
Type C Electro-mechanical chilling, vending with individual regeneration.
Type D Bought-in chilled food with either mass or individual regeneration.

Of the 80 establishments in the sample, 58 units (72·5%) utilised a version of Type A; four units (5%) Type B; four units (5%) Type C; and 13 units (16·25%) Type D. Type A cook–chill accounted for 86·7%, 87·5% and 90% in the hotel, hospital and education sectors respectively, whereas in the industrial sector the proportion using Type A cook–chill was significantly lower (56·4%). Consequently, the majority of units using Type D cook–chill (92%) were in the industrial sector. This fact alone may account for the spread of expenditure in the industrial sector described in Table 12.3. This implies that expenditure for Type D cook–chill establishments should be lower than in other types of cook–chill system. This assumption is supported by the results shown in Table 12.4. The lowest capital expenditure occurred in organisations opting for Type D cook-chill. By far the greatest amount of capital outlay was experienced by organisations introducing Type B cook–chill systems. Mean expenditure in this category outstripped the more common Type A cook–chill by over 300%.

TABLE 12.4
Total Capital Expenditure by Cook–Chill Type

Total expenditure	*Cook–chill type*			
	Type A	*Type B*	*Type C*	*Type D*
Mean (£)	297 000	990 000	51 500	30 400
Median (£)	144 100	873 500	51 500	30 000
Number of units	42	3	1	5

Total Capital Expenditure by Percentage Cook–Chill

Percentage cook–chill refers to the percentage of the total catering output attributable to cook–chill in the organisations in the survey sample. In the UK survey, there was an overall trend towards a low percentage use of cook–chill (median output attributable to cook–chill = 29%). However, percentage cook–chill should not be confused with size of unit which is measured by capacity levels (see later). Some of the units in the sample utilised cook–chill for less than 25% of their total catering output but had a higher maximum capacity than other organisations who utilised cook–chill for a much greater percentage of their total output. Nonetheless, as a general rule of thumb, total capital expenditure increased in the sample as a whole, as percentage use of cook–chill also increased (see Table 12.5).

Total Capital Expenditure as a Percentage of Annual Turnover

Table 12.6 shows total capital expenditure expressed as a percentage of the annual turnover of organisations in the UK Cook–Chill Survey. Almost half the units invested less than 10% of their annual turnover in cook–chill and, the majority of these had an annual turnover between £1m and £5m. There were a few units which invested more than 30% of their annual turnover but, in such cases, the transfer to cook–chill technology incorporated a massive refurbishment programme which included building a totally new central production kitchen. From these

TABLE 12.5
Total Capital Expenditure by Percentage Cook–Chill

Total expenditure	*% catering attributable to cook–chill*			
	1–25	*26–50*	*51–75*	*76–100*
Mean (£)	181 977	191 562	409 917	582 320
Median (£)	46 500	111 500	300 000	571 500
Number of units	20	10	4	10

TABLE 12.6
Total Capital Expenditure Expressed as Percentage of Annual Turnover

Turnover (£000)	*Expenditure as % of turnover*						
	0–10	*11–20*	*21–30*	*31–40* (*number of units*)	*41–50*	*50+*	*Total*
51–100	1	1	0	0	0	1	3
101–500	2	3	1	1	0	1	8
501–1 000	4	1	0	0	1	0	6
1 001–5 000	13	2	5	1	0	2	23
5 001+	2	0	3	0	0	0	5
Total	22	7	9	2	1	4	45

results, establishments embarking on a cook–chill project can expect to invest anything up to 30% of their annual turnover into the system.

HOW IMPORTANT ARE COST SAVINGS?

The importance of the cost reduction factor to operations introducing cook–chill was overwhelmingly apparent in interviews with catering managers in the UK Cook–Chill Survey. Every single manager interviewed stated that cost savings played a highly significant part in influencing their decision to embark on the road to cook–chill, especially in the initial stages. However, despite the fact that cook–chill was often perceived as a 'great money saver' the results of the cook–chill survey indicated that such savings were not always achieved in practice. It was cook–chill users in the successful group who were more likely to approximate to their anticipated savings than those in the unsuccessful group.

The majority of units in the successful group were significantly larger than those in the unsuccessful group, in terms of numbers of meals produced per day but there were examples of successful small units and unsuccessful larger units. Whether the potential to achieve costs savings was directly attributable to size is not clear from the results of the survey. Undoubtedly, larger cook–chill operations were more able to maximise the benefits possible through economies of scale and, hence, found it easier to realise their projected cost savings more easily than smaller units.

However, achievement of savings was not exclusively restricted to the larger organisations, nor was a failure to recoup costs limited to small operations.

The findings of the survey show that the total capital expenditure associated with the introduction of cook–chill ranged from £7000 to £1·85m and maximum capacity ranged from 20 meals per day to 30 000 meals per day. However, the system which invested the greatest capital was not the largest nor the system with the lowest initial investment, the smallest. Nevertheless, overall there was a strong correlation between maximum capacity and total capital expenditure (see Fig. 12.1). In general, the larger the organisation, the higher the initial capital investment.

Table 12.7 shows a summary of the total capital expenditure of organisations of different sizes in the UK Cook–Chill Survey measured against the maximum capacity of their respective systems. The average total capital expenditure for cook–chill units with a maximum capacity up to 500 portions per day was £33 500 (median £34 000) ranging from £7000 for a 150 portion per day maximum capacity to £64 000 for a 400 portion per day maximum capacity. Whereas, for units with a maximum capacity between 5000 and 10 000 portions per day, the mean total capital expenditure was £510 000 (median £394 000) ranging from £250 000 for a 6000 portion per day maximum capacity to £873 500 for a 7500 portion per day

TABLE 12.7
Capital Expenditure by Capacity Levels

Maximum capacity (portions per day)	*Total capital expenditure*			
	No. units	*Mean (£000)*	*Median (£000)*	*Range (£000)*
1–500	13	33·5	34	7–64
501–1 000	6	297·6	57	41·5–1 290
1 001–5 000	20	300	144	7–1 250
5 001–10 000	6	510	394	250–873
10 001+	5	625·5	300	200–1 850

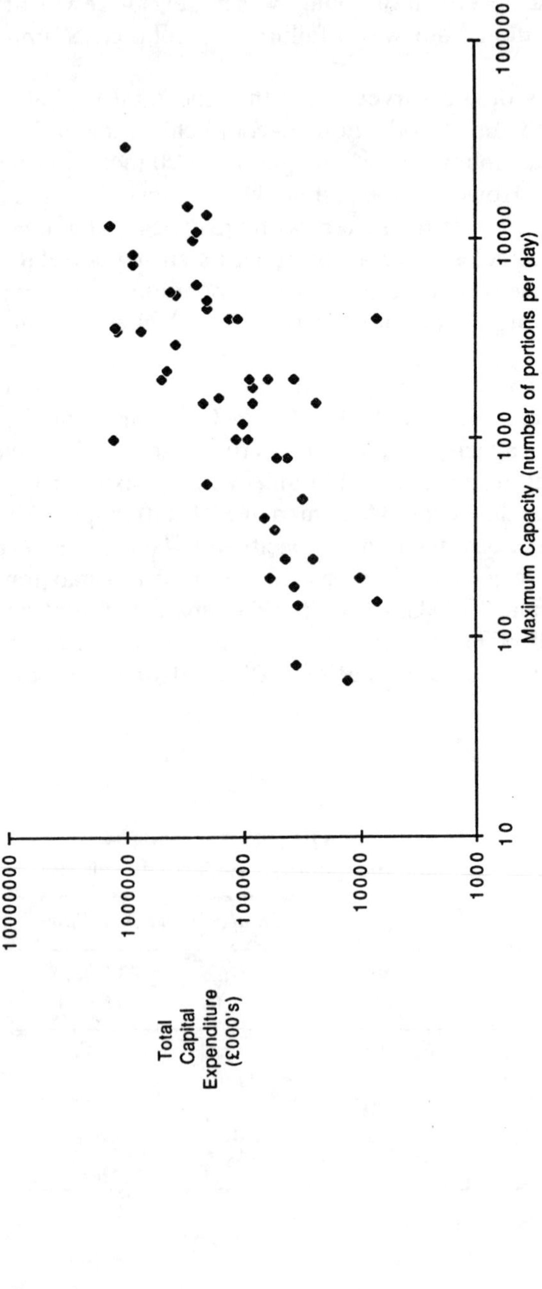

Fig. 12.1. Scatter plot of maximum capacity and total capital expenditure.

TABLE 12.8
Total Capital Expenditure Per Unit of Capacity by Size of Cook–Chill System

Maximum capacity (portions per day)	*Expenditure per unit of capacity*			
	No. units	*Mean (£)*	*Median (£)*	*Range (£)*
1–500	13	170	160	47–486
501–1 000	6	324	102	52–1 290
1 001–5 000	20	108	50	2–347
5 001–10 000	6	72	73	27–117
10 001+	5	41	23	15–114

maximum capacity. Table 12.8 gives a clearer indication of the difference size has on costs. For example, in this survey, the average cost *per unit of capacity* ranged from £170 in the smaller units (up to 500 portions per day) to £41·10 in the larger units (10 000+ portions per day). These results emphasise the advantages possible through economies of scale.

Further examination of the results of this calculation showed that, for similar sized organisations, units in the successful group spent slightly more money per unit of capacity than those in the unsuccessful group. This provides us with the first piece of direct evidence which supports the notion that cook–chill catering, in common with most new technologies, cannot be introduced successfully 'on the cheap'.

In addition, it must be remembered, that any savings claimed as a result of the introduction of cook–chill are irrelevant if a system fails to comply with, as a minimal measure, the safety criteria demanded by the DHSS guidelines.

PRICE BARGAINING

Despite the fact that in general there was a good correlation between maximum capacity and total capital expenditure, there were a number of discrepancies where organisations of similar capacity invested vastly different amounts in their respective systems. For example, one organisation invested nearly £1·3 million in a system which was producing (at the time of interview) 1000 portions per day, whereas another organisation, also producing 1000 portions per day, invested a comparatively small

£90 000. Given that both purchased the necessary chillers, storage facilities and regeneration equipment—why did their expenditure vary so greatly? Clearly, other forces were instrumental in this process, not least, perhaps, the different discounts favoured by equipment manufacturers to respective organisations, including deals made in anticipation of future expansion and developments. In addition, the ability for some organisations to utilise existing equipment to maximum effect has a substantial influence on total expenditure.

It is possible that large catering organisations operating a cook–chill system, may get a 'better deal' on equipment purchased than smaller organisations irrespective of the percentage of total catering output attributable to cook–chill. This is simply because of the stronger financial bargaining position of the larger organisation. However, despite this and given that price bargaining is an accepted practice between manufacturers and purchasers alike and the fact that maximum capacity and total capital expenditure correlated well together (significant at $P < 0{\cdot}0001$ level), the financial data from the UK Cook–Chill Survey certainly provides us with useful information on likely expenditure for cook–chill systems at different capacity levels.

WHAT WAS THIS CAPITAL SPENT ON?

When the data from the UK Cook–Chill Survey was analysed to show how the capital had been spent in setting up new cook–chill catering operations, it became apparent that there were several items which were priority purchases for the majority of establishments. These can be grouped roughly as expenditure on background research, equipment, building alterations, redundancy payments and consultancy fees. Table 12.9 compares the percentage distribution of the total capital expenditure for the successful and unsuccessful groups. The differences are striking. Whilst the successful group spread its expenditure relatively evenly between background research, new equipment and building alterations, the unsuccessful group concentrated its expenditure almost exclusively on a combined package of new equipment and consultancy fees. Furthermore, this consultancy was received predominantly from the equipment manufacturers themselves. In other words, the unsuccessful group were led by the equipment manufacturers.

The appearance of such establishments in the unsuccessful group is hardly surprising considering their average investment in background

TABLE 12.9
Breakdown of Capital Spent Per Item of Expenditure by Success Group

<table>
<tr><th rowspan="2">Item of expenditure</th><th colspan="3">Successful group Total capital</th><th colspan="3">Unsuccessful group Total capital</th></tr>
<tr><th>Mean</th><th>Median</th><th>%</th><th>Mean</th><th>Median</th><th>%</th></tr>
<tr><td>Background work (£000)</td><td>102</td><td>61</td><td>34</td><td>2.8</td><td>7</td><td>2</td></tr>
<tr><td>Equipment (£000)</td><td>116</td><td>103</td><td>38</td><td rowspan="2">114</td><td rowspan="2">28</td><td rowspan="2">80</td></tr>
<tr><td>Consultancy (£000)</td><td>0·5</td><td>0</td><td>neg</td></tr>
<tr><td>Redundancy (£000)</td><td>1</td><td>0</td><td>neg</td><td>23</td><td>15</td><td>16</td></tr>
<tr><td>Building (£000)</td><td>83</td><td>2</td><td>28</td><td>2·4</td><td>0</td><td>2</td></tr>
<tr><td>Training (£)</td><td>68</td><td>0</td><td>neg</td><td>59</td><td>0</td><td>neg</td></tr>
<tr><td>Total capital (£000)</td><td>300</td><td>179</td><td>100</td><td>141</td><td>35</td><td>100</td></tr>
</table>

research amounted to only 2% of the total capital expenditure. This is an exceptionally low figure, particularly when it is considered that on average members of the unsuccessful group only spent £141 000 on cook–chill anyway, which means that £2820 was allocated to conducting background research. In contrast, the average spending on background research in the successful group amounted to 34% or an average £102 000. It seems fair to assume a strong correlation between this lack of resourcing for proper background research and failure. It also emphasises that cook–chill catering cannot be implemented successfully by cutting financial corners. Chapter 11 provides evidence that the decision-making period (a large proportion of which is taken up by background research) is the most crucial period during a system change process (such as introducing cook–chill catering) and often holds the key to eventual success.

TRAINING COSTS

There was a lack of financial resources set aside specifically for specialist training in a number of units in the UK Cook–Chill Survey. Although the data in Table 12.9 show that there was little difference between the successful and unsuccessful group in the amount of money set aside for training *per se* (£68 000 and £59 000 respectively), further investigation revealed that the two groups differed quite substantially in the way the resources for training were allocated—the successful group paying more attention to ensuring that training was relevant to their particular organisation and the unsuccessful group investing in training of a much more general nature. This subject is dealt with in more detail in Chapter 14.

OTHER IMPLICATIONS OF COST SAVINGS

The measure of the success of a cook–chill catering system include evidence of a strong and effective food safety policy. That is, massive savings are meaningless if the system introduced is not operating to accredited safety guidelines. Thus, only systems which operate within such guidelines (such as the UK DHSS guidelines for pre-cooked, chilled foods and the IFST cooked, chilled food guidelines) and produce quality goods can be compared fairly in terms of financial parameters and cost savings.

It was disturbing, therefore, to observe in a few units in the UK Cook–Chill Survey, a lack of suitable equipment for both cooking and chilling and other practices in CPUs which quite clearly contravened the guidelines recommended by the DHSS. For example, queues of trolleys, at ambient temperature, waiting for room in the chiller were worrying but, sadly, a relatively common sight. It is axiomatic that any claimed savings in situations such as these become null and void. In addition, there were other examples of food being microbiologically safe, but with customers noticing a deterioration in terms of organoleptic quality of the end product. It is important, therefore, to place any projected or realised cost savings into the context of the system they are supposed to be representing.

THE INCIDENCE OF COST SAVINGS ACTUALLY ACHIEVED IN PRACTICE

In its examination of results from the UK Cook–Chill Survey, this chapter has concentrated on the costs incurred through the introduction of cook–chill catering by an innovating organisation. It has been repeatedly cited, however, that the main attraction to catering organisations, particularly administrators in control of finance within the organisation, is the potential for saving money through the introduction of cook–chill catering. In particular, the ability to drastically reduce the wages bill. Little work has been carried out to date which investigates the extent to which these savings are actually realised in practice. In one recent case, highlighted in the UK trade press, the manager of a large hospital cook–chill unit stated that the savings expected through the introduction of cook-chill catering had, at that point, not materialised—even though the projected payback period had passed. There have been few well documented cases which support the ability of cook-chill catering to provide some of the ambitious savings previously advertised. On the other hand there were examples of cases where the introduction had been, financially speaking, a complete disaster, with an outright failure to recoup what was, in many cases, a massive initial capital investment.

Examination of financial data from the UK Cook–Chill Survey relating to payback periods indicates a certain amount of difficulty in achieving savings through the introduction of cook–chill catering. Failure to recoup costs was often caused by circumstances unforeseen in the planning stages, for example, industrial action, power failures, equipment inadequacies and consumer rejection of the system. In many cases, this was a direct result of bad planning and poor assessment of the total impact such a technological change meant to the organisation as a whole. It is hardly surprising, therefore, to find that of the 52 organisations who stated that their primary aim for introducing cook–chill was to reduce overall costs, nearly one-third failed to achieve this aim. However, there were two-thirds of the sample (35 units) who set out to reduce costs and succeeded in doing so. Moreover, a large proportion of these, at the same time, adhered as closely as practicably possible to criteria laid down in the DHSS guidelines. Not surprisingly, these units occupied places in the successful group. Part of the reason why these units were successful stems from their more modest expectations of cook–chill catering and, as a direct result of detailed planning in the early stages, a clear strategy for recouping costs over a realistic period. Consequently, the average payback period in the successful units ranged between three and five years. At the time of interview, the majority of units, later identified as successful, had achieved or were on target to achieve their projected savings.

SUMMARY

This chapter has discussed the financial implications of introducing cook–chill and presented findings from the UK Cook–Chill Survey on capital expenditure and associated cost savings. It is quite clear that there can be no magic formula for costing cook–chill catering which can be applied universally. Nevertheless, there are a series of categories which, quite clearly must be included in any financial study aimed at assessing the feasibility of cook–chill catering (or, for that matter, any other catering system). These are summarised early in the chapter and draw on research done by the DES, Electricity Council and the UK Cook–Chill Survey itself. In short, these categories include; equipment, building alterations, labour, maintenance, cleaning, background research and most important of all the need for training to be fully costed from the outset.

Evidence from the UK Cook–Chill Survey, presented in this chapter, provides useful examples of typical expenditure on cook–chill systems for different size units. It also shows any differences in capital expenditure per unit of capacity for organisations of a similar size. Thus, whilst larger units, as expected, invested more total capital in their systems than smaller units, their expenditure per unit of capacity was much less than in the smaller units. Clearly, this showed the possibility for economies of scale to be quite substantial. In addition, successful units (comparing like for like in terms of maximum capacity levels) spent more capital per unit of capacity than unsuccessful units.

Finally, the chapter draws attention to the differences between the successful and unsuccessful groups with regard to where their capital expenditure was directed. Perhaps the most important finding here was the almost complete disregard by members of the unsuccessful group to include allowance for spending on thorough background research; on average a mere 2% of the total capital expenditure being invested for this purpose. In the last analysis, the outcome following the introduction of cook–chill, in each of the establishments surveyed, expresses far more clearly than is possible in this book the importance of prudent expenditure when transferring catering production to a cook–chill system. Moreover, in the light of findings from the UK Cook–Chill Survey, the most important ingredients of this costing process must be, surely, provisions for detailed background research, effective training schedules for both management and operative level employees and a clear strategy for re-couping capital costs based on a thorough planning phase.

FURTHER READING

Anon. (1982) Developing new cook–chill operations. *Chilled Foods,* Apr. 36–8.

Armstrong, R. (1986) Problems and Benefits of Cook Chill Systems. Paper presented at The Future of Technology in Catering Workshop, Dorset Institute of Higher Education, Poole, Dorset, UK, April.

Cowell, F. (1980) Measuring the cost effectiveness of changing to a new catering method. In: *Advances in Catering Technology,* (ed. Glew, G.), London, Applied Science Publishers, pp. 345–57.

Electricity Council (1983) Cook–chill and cook–freeze. Caterelectric. EC document no. 4421/11/83.

Fosters (UK) *Cook–Chill in Action,* A Foster Refrigeration Publication.

Herrmann, H. (1987) The big chill—fact or fiction. Paper given at a Conference on Cook–Chill, Aberdeen.

Hobart UK Ltd. Cook Chill Regeneration Systems: The Cost Effective Approach to Catering, Hobart Manufacturing Company Ltd., London.

Oxford Regional Health Authority (1986) Cook–Chill Appraisal Report (Oxford). Prepared by R. M. Associates (consultants), July.

Oxford Regional Health Authority (1986) Cook–Chill Appraisal Report (East and West Berkshire). Prepared by R. M. Associates (consultants), July.

Regethermic UK Ltd (1982) *Regethermic Cook–Chill Systems,* Sales Brochure, London.

Scott, C. (1986) An interim report on costs and manpower requirements. British Telecom. Brighton, February.

Sheppard, J. (1987) The big chill. A report on the implications of cook–chill catering for the public services. London Food Commission, Report no. 15.

Stormont, G. J. & Harvey, J. F. (1971) An assessment of the Regethermic meals service system at the Royal London Homeopathic Hospital, July.

Walker, A. E. (1988) The transfer of technology in the catering industry: a study of UK cook chill operations. PhD thesis, CNAA/Dorset Institute of Higher Education, Poole, Dorset, UK.

Walker, A. E. & Light, N. D. (1988) Is cook–chill being put to its best use? *Hospitality,* February.

Walker, A. E. & Light, N. D. (1988) Cook–chill catering: How to make it work. *Catering,* March, 44–6.

West, P., Hughes, W., Witham, T. & Hull, S (1986) Report on energy costs for the implementation of cook–chill at Heatherwood Hospital, Ascot, file no. W1530, Dec.

Zanussi *Cook–Chill Sales Brochure.*

CHAPTER 13

Products, Menus and Research and Development

INTRODUCTION

During the early growth stages of cook–chill catering systems in the UK, proponents of the method suggested that an important advantage was the amount of menu flexibility it allowed, especially when compared to cook–freeze. The latter was notorious for its inability to handle successfully a wide range of products without a significant amount of recipe development, e.g. sauce-based products and egg dishes required special additives in order to survive the cook–freeze process. On the other hand, cook–chill catering, it was suggested, produced 'excellent, palatable results with all foods from poached eggs to haute cuisine dishes'. As we shall see in this chapter, evidence from actual users shows that, in fact, recipe development is a vital step in setting up and running a new cook–chill catering system.

The range of products found to be used in UK cook–chill operations (see Appendix 1), supports the view that cook–chill allows for a good degree of freedom in the menu products it can offer.

This chapter discusses the range of items used in cook–chill operations in the UK, gives sample menus, examines the extent to which research and development are both needed and used in practice and, draws on evidence from research on the range of recipes used, product development and food purchasing policy.

SUITABILITY OF MENU ITEMS FOR USE IN COOK–CHILL CATERING

Research work carried out recently in the UK has investigated the range of different products used in cook–chill catering systems in the UK across all sectors of the industry. A list of products which have been used in these cook–chill operations is long and varied (see Appendix 1). Such is the variety of meals produced that this survey alone identified over 100 different composite dishes, over 80 meat and fish dishes, over 55 different vegetable dishes, over 140 sweet dishes and a range of miscellaneous items. From this evidence, it appears that cook–chill catering enables a menu flexibility which was lacking in previous centralised systems such as cook–freeze catering. This view is, to some extent, supported by evidence from customers interviewed in 26 of the units visited in the UK Cook–Chill Survey. These customers consumed a total of 122 different products ranging from roast beef and Yorkshire pudding, sweet and sour pork and rice to apple crumble, bilberry tart and custard and a variety of vegetable accompaniments (see Appendix 2).

From this information, it would appear that cook–chill catering is not seriously limited by its menu capabilities. However, whilst recognising that new methods for processing 'problem' foods in cook–chill catering systems are constantly being developed, there is a small group of products which have consistently given poor quality end results and a larger group of menu items which, though acceptable, could still be greatly improved. In other words, contrary to some early claims for the method, cook–chill catering cannot produce top quality end-products from the complete range of conventional menu items. Difficult or unsuccessful menu items include battered and fried products, eggs and some bakery items.

CHANGES AND PROBLEMS OBSERVED IN COOKED, CHILLED FOODS

Perhaps the most serious changes in the quality of cooked, chilled foods are brought about by oxidation leading to detrimental effects on the food such as nutritional loss and noticeable organoleptic deterioration which, in some dishes, can bc apparent after only 3–4 days storage (see Chapter 2). One important research study which investigated the storage life of cooked, chilled food found that unpackaged cook–chill products consumed after their five day shelf-life period, even though safe microbiologically, were likely to be at their lowest limit of organoleptic acceptability.

In the UK Cook–Chill Survey, both managers and operatives were asked to comment on menu items which they had found to be unsuitable for use in their cook–chill operation. Many of them mentioned a deterioration in quality the longer a particular food item was stored and some referred to individual items specifically. For example, a number of users found that flavours in recipes incorporating the use of certain herbs were badly affected, bread products went stale and some vegetables exhibited substantial colour loss.

Overall, as Table 13·1 indicates, there was agreement between managers and operatives on which products were relatively unsuccessful when processed through a cook–chill system. In particular, fried products, chips, battered products and egg dishes caused problems in a wide range of different cook–chill catering establishments. One reason for this may be related to the assumption, made by many managers in the UK Cook–Chill Survey, that successful regeneration could be achieved by using a single piece of specialist regeneration equipment. The equipment utilised varied from unit to unit, but, infra-red, mass regeneration units were the single most common piece of equipment used for regeneration of food (51 units, 64%), followed by microwave ovens (39 units, 49%) and steam injection ovens (8 units, 10%). A few units utilised a combination of regeneration methods, and combination ovens (i.e. with steam injection) were found to be gaining popularity. Nevertheless, any one satellite unit tended to concentrate on one specific regeneration method.

It was clear, from interviewing many cook–chill catering users, that the use of a single, universal regeneration oven was a mistake and that, in reality, a number of different pieces of regeneration equipment were necessary to maximise end-product quality. On consideration this is not

TABLE 13.1
Foods Found to be Unsuccessful by Managers and Operatives

Unsuccessful foods	*No. of managers*	*No. of operatives*
Fried Products (not French fries)	53	57
French fries	57	57
Battered products	27	30
Eggs and egg products	25	27
Vegetables	11	20
Steak	11	11
Pasta	—	5
Others	46	39
Number interviewed	80	128

surprising. A steam oven used to regenerate a wet product to minimise drying out would not be able to re-heat crisp items without producing 'sogginess'. The problems identified by a significant number of users with regard to fried products and especially chipped potatoes (French fries) are prime examples. These types of product underwent extensive deleterious changes when processed through a cook–chill catering system, usually resulting in limpness or sogginess. A whole series of different techniques were employed to try and combat these effects including regeneration without lids, using additives such as salt and tomato purée in coatings and using pre-prepared products such as oven chips. However, all failed to improve the final product to an acceptable level of quality.

In the case of pre-plated meals, ingenious adaptations of containers were tried in an attempt to accommodate the varying needs of different products on the same plate (see Fig. 13.1). The adapted lid shown in Fig. 13.1 was an attempt to prevent, for example, a chicken leg in sauce and vegetables from excessive dehydration during regeneration, whilst at the same time retaining a crisp finish on the chipped potatoes. In some cases, this method succeeded, especially where meals were regenerated with a lot of space for hot air to circulate but, in others, it proved awkward to use and failed in its attempt to improve end-product quality. Some operatives found that in a fully loaded oven, condensation dripping from trays above caused products like chips to lose their crispness and resulted in an effect opposite to the one intended. In addition, during transportation, covers could become displaced such that they would be covering the meal items which should have been left exposed.

A wide variety of establishments using cook–chill catering are represented in Table 13.1. This includes, not only representatives from all four sectors of the catering industry covered in the survey, but also units differing in size and type of cook–chill method used. Nearly every manager interviewed and all the operatives approached, mentioned at least one product which they felt was inferior when produced through a cook–chill system to that produced through their previous catering system (in most cases cook–serve). Many operatives also commented that their opinions were based on the food as it left the central production unit and not on the final product after regeneration at the satellite units.

In sum, many products produced through a cook–chill system were seen as inferior, lacking crispness or being dry and overcooked, being burnt around the edges or containing cold spots. Pre-plated meals perhaps presented the most problems, simply because of the variety of individual regeneration requirements of component items with different

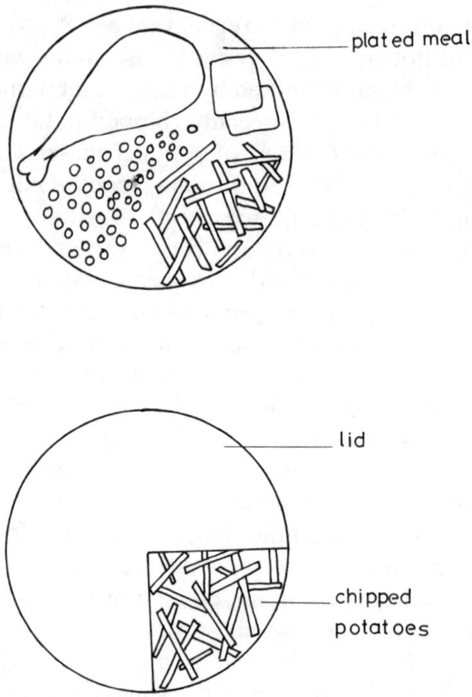

Fig. 13.1. Diagram showing the use of specially designed lids to selectively allow drying of cooked chilled foods. The top section shows a plate of food including soft, well hydrated products like a chicken drumstick in sauce, peas and other vegetables. The bottom section shows the same plate covered with a modified lid whose open portion is placed over the chips (French fries) to allow crisping of this component during reheating and to prevent or at least slow drying of the rest of the meal.

densities. Cold spots present in regenerated foods were, predominantly, the fault of the technology itself. In some cases, other problems could be overcome by an alteration of operative practices, which often resulted in a requirement for skilled operatives at the regeneration unit rather than unskilled employees as originally advocated (see Chapter 14).

FOOD PURCHASING

To some extent, variability in raw product quality can be overcome by setting strict food specifications on every ingredient included in the menu.

However, evidence from the UK Cook–Chill Survey showed that some small cook–chill users found difficulties in getting suppliers to adhere strictly to requested specifications simply because of the small amount of raw products involved. However, a good supplier should ensure that clear specifications should be met for both small and large users alike.

Complete development work also includes the production of detailed raw material specifications for suppliers. The quality of raw items is perhaps more critical in cook–chill catering systems than ordinary conventional cook–serve systems in that any changes to tried and tested recipes will affect the production of menu items of a consistent quality. Thus, in the UK Cook–Chill Survey, the majority of units (67, 94·4%) used established suppliers trusted to provide quality ingredients to their exact specifications (see Table 13.2). Centralised purchasing procedures were used by 16 (22·5%) of the units who also used established suppliers. Those using centralised buying were, inevitably, the larger units.

TABLE 13.2
Methods Used for Food Purchasing

Method	*No. of units*	%
Established suppliers	67	94·4
Centralised buying	16	22·5
No formal system	4	5·6
Total no. of units	71	100·0

Very few of the organisations interviewed had undergone a drastic change in their purchasing procedures. Certainly as far as the UK National Health Service and School Meals services were concerned, many managers were bound to National and Regional contracts which were difficult to alter. The main difference in all the organisations surveyed was that cook–chill had 'enforced' a more systematic method of ordering and purchasing even though their sources of supply had remained relatively unaltered.

PRIME COOKING AND REGENERATION

Whereas prime cooking adds a significant amount of value to the food in that it converts raw products into palatable, consumable menu items, the purpose of regeneration (or secondary cooking) is only to re-heat the food prior to serving. However, because products are re-heated to a minimum of 70°C as recommended by the UK DHSS guidelines for cooked, chilled foods, the food tends to cook for a second time. Some units in the survey attempted to counteract this phenomenon by undercooking food during prime cooking. This is a highly dangerous practice which carries a real risk of failing to inactivate fully any micro-organisms during the prime cooking phase. Thus, during storage, dangerous organisms not sensitive to low temperatures may multiply in the product before re-heating and thereby present a real risk of causing food poisoning.

A good example of this type is egg. One of the techniques used to combat detrimental effects of cooking, chilling, chilled storage and regeneration on the colour, texture and flavour of egg dishes in particular, is to undercook the egg during prime cooking. Eggs can be contaminated with *Salmonella* and, thus, many egg-based menu items may contain the potential to cause a food poisoning outbreak if prime cooking does not lead to effective pasteurisation. There is no easy solution to this problem, if items such as poached or fried eggs, omelettes and pancakes remain on the menu, other than perhaps introducing a call-order service at the end kitchen for items of this nature. It may then be argued that such activity renders the use of cook–chill redundant by increasing the labour requirement needed at the satellite units and reverting to a 'production kitchen' at the satellite unit rather than a finishing kitchen as intended by the system. Undoubtedly, proponents of the cook–chill catering system may argue the opposite, suggesting that cook–chill can be improved by the introduction of complementary systems (see Chapter 15). In fact, as mentioned earlier in this book, expecting one system to handle a full menu range may not, in the final analysis, be reasonable. Thus, a combination of cook–chill catering with a strategic amount of preparation using more conventional cook–serve methods, with perhaps some cook–freeze products could provide the most flexible approach.

The UK Cook–Chill Survey also showed that users were finding a need for a variety of different methods of regeneration at satellite units. For example, by far the most effective method of overcoming problems with fried products was to introduce a fryer at satellite units to regenerate chips, fried fish, fritters and other similar products. This did not mean that satellite units had to revert back to being the prime cooking units they were introduced to replace. Rather, they needed to be equipped with ovens or other equipment capable of regenerating a range of products to an acceptable level of quality. Such equipment could include, steam injection ovens, microwave ovens, fryers, a boiling table, plus, of course, chilled storage facilities.

As Table 13.1 suggests, inferior end-product quality was not limited to fried products alone, there were a number of other menu items which were of dubious quality after regeneration. Common to a large number of units were 'chewy' batter products such as Yorkshire pudding and pancakes; 'rubbery' omelettes and fried eggs; tough meat, particularly in the form of rump steak, chops and fillets served without accompanying sauces; burnt and dehydrated products, gravy dried up at the edges of plates; and secondary thickening of sauces during regeneration.

More encouraging evidence pertaining to the variety of products which can be prepared successfully using a cook–chill catering system has appeared in articles in the trade press and focus on individual cook–chill situations. This is supported to a degree by the wide range of menus offered by many operations (see Appendices 1 and 2).

However, there has been much controversy over the necessity to alter recipes for use in a cook–chill catering operation. Much early literature regarded recipe development as unnecessary and expensive. However, evidence from users interviewed in the UK Cook–Chill Survey dispelled this theory and showed that recipe development tailored to individual organisational needs can reap dividends. The next section reports findings concerned with the problems experienced by successful and unsuccessful cook–chill caterers with recipes whilst setting up and during the normal running of cook–chill catering operations.

THE ROLE OF RECIPE DEVELOPMENT IN PREVENTING PROBLEMS

Despite the fact that most of the successful group and a few of the unsuccessful group in the UK Cook–Chill Survey were involved in pre-installation recipe development, operations in both groups experienced problems with recipes at the implementation phase. There were a number of reasons for this. First, scaling-up recipes from the small quantities used in tests to full production often led to unforeseen problems. Second, inadequate or inappropriate recipe development was carried out prior to implementation and, third, some recipes or food items were only found to be incompatible with the cook–chill catering system through trial and error.

The majority of the successful group of cook–chill caterers experienced minimal problems with recipes after cook–chill. In contrast, the unsuccessful group experienced some serious product changes because of inadequate recipe development. As described above, the most common problems involved fried products and there was a statistically significant difference between the successful and unsuccessful groups with regard to the number of units experiencing difficulties with such products as battered fish, chipped potatoes and fried eggs. The successful group had solved most of their problems by substantial modification of recipes, methods and regeneration equipment used. Some of the solutions to problems have already been mentioned, for example, the installation of fryers at satellite kitchens. Other solutions included refreshing eggs in water just prior to service, re-heating Yorkshire pudding in a standard oven, and careful re-heating of bread products to reverse staleness.

It was apparent from the results of the UK Cook–Chill Survey, that many of the problems with cook–chill recipes are not insurmountable, but with careful research and development many difficult problems can be improved to such an extent as to ensure their acceptability by consumers.

Evidently, before constructing the menu a cook–chill caterer must be sure what foods cannot be processed through this method. The importance of the menu itself is the subject of the next section.

THE MENU

> For the man in the kitchen . . . the menu is the summary or digest of all his activities. The maitre chef in composing a menu is therefore not only producing information about a meal for the benefit of his clientele but is evolving a blue-print or basic plan on which the kitchen operation will develop (Fuller, 1966).

The significance of the statement '*evolving a blue-print or basic plan on which the kitchen operation will develop*' could not be more critical than in cook–chill catering situations. The menu is a key feature in any catering operation, but for cook–chill systems in particular, it is of paramount importance, for as Fuller goes on to say, 'The various parties receiving advance copies of menus will *plan their work according to its requirements.*' Thus, not only must each item on any cook–chill menu be compatible for use within that system, as we have already seen, but the menu also provides the main planning and scheduling strategy for the operation as a whole.

The menu cycle provides the central planning element for the cook–chill caterer and as such it should take the form of a rigid guide to production needs. This is because there has never before been a need for such strict accuracy and control in production, to attain an accepted level of quality, every time a particular menu item is produced, than in cook–chill catering. In short, menu planning is, possibly the most important factor in the structuring, scheduling and operation of a cook–chill production process.

MENU PATTERNS

Evidence from the UK Cook–Chill Survey indicates that cyclical menu patterns are the most commonly used in most sectors of the industry in the UK (see Table 13.3). However, there were examples of operations where other menu patterns were more appropriate. For instance, in many hotel and banqueting situations, set menus were more widely applicable and better suited to the catering services being offered. In those establishments operating cyclical menu patterns, longer cycles were more popular than shorter ones (see Table 13.3). In addition, the extent to which an individual establishment's clientele was 'captive' (i.e. restricted in choice of places to eat) was instrumental in the choice of cycle length. This was

TABLE 13.3
Menu Patterns of Cook–Chill Units in Different Sectors of the Catering Industry

Menu pattern	*Sector of industry*				
	Hotels	*Industrial*	*Hospital*	*Education*	*Total*
No pattern	3	3	0	0	6
2 week cycle	0	0	3	0	3
3 week cycle	0	0	7	1	8
4 week cycle	1	13	6	5	25
5 week cycle	0	7	1	2	10
6 week cycle	0	14	0	2	17
8 week cycle	0	1	0	0	1
Set menu	8	2	0	0	10
Total	14	40	16	10	80

noticeable in industrial and educational establishments in particular where a typical menu cycle was four weekly or longer. In hospitals, the length of cycle was often dependent on the type of hospital concerned. For example, 'long stay' and psychiatric hospitals tended to operate longer cycles than accident or 'short stay' ones, reflecting the more 'captive' nature of the former.

Figures 13.2–13.4 gives examples of different menus utilised in a variety of establishments using a range of menu patterns. Figure 13.2 shows

Appetisers

Melon

Halton skillet sautéed chicken livers,
bacon, mushrooms, Madeira wine and croutons

Half a pint of prawns with garlic mayonnaise

Cauliflower cheese soup

Poachers' broth 1642 game and vegetable soup made
to an original Lancashire recipe

Hors-d'oeuvre choose your own

Main Course Selection

St Georges Quay
smoked salmon, crab claw, brown shrimps,
prawns and oyster

Scampi Colonial
cooked in a light curry cream sauce
with banana, pineapple and pepper
served with braised rice

Duck and Cherry Pie
simmered in a Port wine sauce topped
with puff pastry and baked to order

Noisettes of Lamb Mascotte
tenderloins of lamb with artichoke, mushrooms,
Parisienne potatoes

Fillet steak

Escalope of Veal Suisse
cooked in butter with Swiss Cheese
and kirsch

Fig. 13.2. Example of a typical set menu in a hotel. (Not all items on the menu are produced through cook–chill.)

a set menu used in a high class hotel, Fig. 13.3, a six week cycle menu used in an industrial catering unit and Fig. 13.4, a four week cycle used in a social services catering unit. Clearly, from these examples, cook–chill catering can, and does in practice, have the capability and versatility to produce a whole spectrum of dishes from gourmet type meals shown in Fig. 13.2 through a variety of items shown in Fig. 13.3 to the more traditional British recipes shown in Fig. 13.4.

In this discussion, 'the menu' has been referred to as a key element in contributing to the success of a cook–chill catering system. What needs to be more clearly stressed, however, is that, in a cook–chill catering system, the menu must be thoroughly tested in order to ensure that the items represented on it are such that they can be relied upon to reproduce a desired quality every single time they are made. Inevitably, this requires extensive product development during the menu planning phase.

PRODUCT RESEARCH AND DEVELOPMENT

In an earlier section we discussed overall differences in approach to product development adopted by successful and unsuccessful cook–chill catering operations. In the present section we will expand this topic to investigate their actual procedures of research and development.

It has been stressed repeatedly, in writings on the catering industry, that, historically, the industry has been labour-intensive, craft-based, 'low tech' and, in general, slow to respond to change. In some situations, of course, this may still be true, but in general, this is becoming an increasingly out-dated viewpoint. In recent years, the catering industry has shown its ability to innovate, invest in 'high-tech' systems (such as computerised control systems as well as technologically-based systems such as cook–chill, cook–freeze and *sous-vide*) and has shown a marked movement towards accommodating 'healthy eating', ethnic cuisine and vegetarian foods. Within this more proactive environment, the role of research and development has naturally evolved as the use of more advanced methods of catering has increased.

With reference to cook–chill catering operations specifically, research and development takes on a number of different forms in practice. The first major difference in the use of research and development amongst cook–chill operations highlighted by the UK Cook–Chill Survey, was whether research and development was conducted in-house or through external organisations. In-house research and development was, by

WEEK 1			
Monday	Trout almadine with almonds and lemon	Grilled lamb cutlets with tomatoes and mushrooms	Sprouts Aubergine provencale Croquette potatoes
Tuesday	Liver provencale fried and coated in provencale sauce	Grilled poussins with grilled tomatoes	Peas Cabbage Baked potatoes
Wednesday	Blanquette of veal garnished with parsley	Barbequed rump steak marinaded in soy sauce, garlic and ginger	Sweetcorn Carrots New potatoes
Thursday	Casserole of duck with green peas	Roast loin of pork with apple sauce	Mixed vegetables Courgettes provencale Roast potatoes
Friday	Beef bourguignonne garnished with parsley	Fried fillet of haddock in batter	Broccoli Turnips Rice Baked potatoes
Saturday	Grilled ham with peaches coated in Madiera sauce	Canton fried chicken legs with lemon and tomatoes	Green beans Cauliflower Boulangere potatoes
Sunday	Roast leg of lamb	Chilli con carne	Peas Swede Roast potatoes
WEEK 2			
Monday	Pork in bean sauce	Lancashire hot pot garnished with parsley	Green beans Cabbage Baked potatoes

Fig. 13.3. Example of a six week cycle menu.

Tuesday	Grilled rump steak with grilled tomatoes	Broiled smoked haddock with Maitre d'hotel butter	Peas Cabbage New potatoes
Wednesday	Roast duck with black cherry sauce	Sausage lyonnaise with lyonnaise sauce	Sweetcorn Turnips Roast potatoes
Thursday	Grilled veal chop with mushroom sauce	Navarin of lamb garnished with parsley	Mixed veg. Carrots Boulangere potatoes
Friday	Fried plaice in breadcrumbs garnished with lemon and parsley	Kidneys with spring onions and cauliflower	Peas Courgettes provencale Baked potatoes
Saturday	Boiled leg of lamb with caper sauce	Pork chops charcuterie	Sprouts Swede Croquette potatoes
Sunday	Roast chicken fines herb with bread sauce	Stir fried liver	Broccoli Aubergine provencale Roast potatoes
WEEK 3			
Monday	Roast beef	Supreme of chicken Maryland with sweetcorn pancakes, tomato, bacon and banana	Sprouts Courgettes provencale Roast potatoes
Tuesday	Lamb curry with mango chutney	Pork chops flamande with Madiera sauce	Green beans Carrots Rice Baked potatoes
Wednesday	Veal chop Maitre d'hotel	Trout Grenobloise with lemon segments and capers	Peas Turnips New potatoes

Fig. 13.3—*contd.*

Thursday	Macoroni cheese garnished with tomatoes and parsley	Grilled rump steak with grilled tomatoes and mushrooms	Sweetcorn Aubergine provencale Boulangere potatoes
Friday	Turkey and mushroom pie	Fillet of haddock Florentine poached with spinach and covered in Mornay sauce	Broccoli Swede Croquette potatoes
Saturday	Grilled liver and bacon garnished with watercress	Shepherd's pie garnished with tomatoes and parsley	Mixed vegetables Cauliflower New potatoes
Sunday	Lamb cutlets persilee covered in breadcrumbs garlic and parsley	Vegetable curry with mango chutney	Peas Cabbage Rice Roast potatoes
WEEK 4			
Monday	Cod in batter garnished with parsley	Mixed grill garnished with watercress	Green beans Aubergine provencale New potatoes
Tuesday	Liver lyonnaise garnished with parsley	Grilled gammon steak with pineapple	Peas Carrots Baked potatoes
Wednesday	Braised chicken with leeks	Saute of kidneys chasseur	Mixed veg. Cauliflower New potatoes
Thursday	Roast loin of pork with apple sauce	Beef goulash garnished with parsley	Sprouts Swede Roast potatoes
Friday	Grilled rump steak with grilled tomatoes and mushrooms	Poached plaice in parsley sauce	Peas Turnips Croquette potatoes

Fig. 13.3—*contd.*

Saturday	Supreme of chicken Alexandra garnished with asparagus tips	Boiled silverside and dumplings	Broccoli Cabbage Boulangere potatoes
Sunday	Roast duck with apple sauce	Lasagne	Sweetcorn Courgettes provencale Roast potatoes

WEEK 5

Monday	Fillet of plaice garnished with banana, lemon and chutney	Roast leg of lamb garnished with watercress	Broccoli Carrots Roast potatoes
Tuesday	Devilled chicken legs garnished with grilled bacon and watercress	Chinese stewed beef garnished with spring onions	Sweetcorn Cauliflower Rice New potatoes
Wednesday	Grilled pork chops garnished with grilled tomatoes and mushrooms, apple sauce	Spaghetti bolognaise garnished with parmesan cheese	Peas Courgettes provencale Baked potatoes
Thursday	Spezzatino of veal garnished with parsley	Toad-in-the-hole garnished with tomatoes and parsley	Sprouts Swede Boulangere potatoes
Friday	Sweet and sour pork with bean shoots	Cod provencale with provencale sauce garnished with parsley	Mixed veg. Cabbage Rice Croquette potatoes
Saturday	Duck with almonds garnished with spring onions	Fried lamb's liver in breadcrumbs garnished with lemon quarters and parsley	Green beans Aubergine provencale Rice New potatoes

Fig. 13.3—*contd.*

Sunday	Grilled rump steak garnished with tomatoes and mushrooms	Roast chicken romarin roasted with rosemary, bread sauce and garnished with watercress	Peas Turnips Roast potatoes
WEEK 6			
Monday	Duck in orange sauce with sliced orange	Vegetable curry with mango chutney	Green beans Cauliflower Rice Baked potatoes
Tuesday	Carbonnade of liver cooked in beer and onions garnished with pickled parsley	Grilled rump steak garnished with mushrooms, tomatoes and watercress	Sweetcorn Peas Roast potatoes
Wednesday	Beef casserole	Haddock Monte Carlo garnished with points asparagus	Sprouts Turnips Boulangere potatoes
Thursday	Hamburger garnished with fried onion, tomatoes and mushrooms	Steak and kidney	Cabbage Courgettes provencale Croquette potatoes
Friday	Fillet of plaice *bonne femme*	Vegetable lasagne garnished with pickled parsley	Carrots Peas New potatoes
Saturday	Irish stew garnished with chopped parsley	Pork chop grilled garnished with tomatoes and mushrooms	Swede Broccoli Baked potatoes
Sunday	Roast beef	Chicken curry with mango chutney	Mixed veg. Aubergine provencale Rice Roast potatoes

Fig. 13.3—*contd.*

	Week 1	*Week 2*	*Week 3*	*Week 4*
MONDAY	Sliced pork Roast potatoes Butter beans Sprouts Apple pie & custard	Cornish pasty Croquette potatoes Sliced beans Raspberry mousse	Bangers & mash Sprouts Mixed veg. Bakewell tart	Beef casserole Boiled potatoes Carrots Cauliflower Cold rice & pear halves
TUESDAY	Steak & kidney pie Creamed potatoes Sliced beans Prunes & custard	Sliced beef Roast potatoes Sprouts Yorkshire pudding Apple frangipan	Sliced lamb Roast potatoes Sliced beans Peas Blackberry pie & custard	Faggots in rich sauce Jacket potatoes Whole beans Custard pie
WEDNESDAY	Cod Mornay Peas Croquette potatoes Bread & butter pudding	Chicken casserole Jacket potatoes Peas Carrots Mincemeat tart & custard	Cheese & onion quiche Baked beans Hash browns Fruit salad & jelly	Chicken à la king Mixed veg. Croquette potatoes Blackcurrant pie & custard
THURSDAY	Cottage pie Sprouts Carrots Canned pears & ice cream	Minced beef & onion pie Hash browns Cabbage Trifle	Ham & beans in barbeque sauce Croquette potatoes Strawberry cheesecake	Sausage & egg pie Baked beans True mash Rhubarb crumble & custard
FRIDAY	Battered haddock Peas Chips Rice pudding	Battered cod Sweetcorn Chips Gooseberry crumble & ice cream	Breaded plaice Peas Chips Apple pie & ice cream	Poached cod & parsley sauce Peas Boiled potatoes Baked chocolate sponge

Fig. 13.4. Example of a four week cycle menu.

definition, organisation-specific, whereas, external work could be split into two distinct categories; organisation-specific and organisation non-specific. In the former, an external research agency (educational establishments, research associations or even equipment/food suppliers) carried out research and development work on behalf of the organisation, unique to that organisation and usually instigated by that organisation. In contrast, the latter consisted of general research and development undertaken by outside agencies which could have been applied in more or less all cook–chill catering situations. This included academic (and other) independent research, and research and development which had been initially carried out in-house but which could equally have been transferred and used in organisations adopting similar cook–chill catering schemes.

The long lists of menu items described in Appendices 1 and 2 and evidence in sections above on the kind of problems encountered in the industry highlight the necessity for individual organisations to become involved in their own research and development work. In addition, the simple fact that no two cook–chill catering systems are alike, reinforces this need. Also, although two systems may appear the same, in that they are of a similar size and utilise similar recipe schedules, they may produce meals of very different quality. Some of the reasons for this lie in the existence of other variables which enter into the equation and have a significant influence on the end result e.g. different skill levels and innovating ability of chefs, differences in the capabilities or utilisation of equipment, unavoidable differences in initial product quality, differences in the skills and procedures at regeneration and differences in top management support.

TABLE 13.4
Facilities Available for Research and Development in UK Cook–Chill Catering Units

In-house R & D facility[a]	*No. of units*	%
Recipe development	33	44·6
Microbiological testing	13	17·6
Equipment testing	13	17·6
None	41	55·4
Total no. of units responding	74	100·0

[a] Some units had more than one in-house R & D facility.

Aspects unique to the cook–chill catering system dictate ways in which recipes are constructed and menu items are handled. For example, it is essential for successful chilling down to 0–3°C within 90 min of prime cooking and regeneration to a core temperature of at least 70°C to take place (DHSS, 1980). This has implications not only for the recipe used but also the physical parameters involved in producing a menu item. For example, research has shown that the requirements and limitations of the chilling medium, headspace in the container or package, geometrical shape of the product and depth of the product in the container as well as the physical properties of the food itself all have an effect on heat transfer. Indeed, the DHSS guidelines clearly state that the speed of chilling individual food items is affected by:

a. container size, shape, weight and construction material of the container;
b. food density and moisture content;
c. heat capacity of the food;
d. thermal conductivity of the food;
e. temperature of food entering rapid chill;
f. whether the container is provided with a cover, in which case chilling time is increased by 10% or more.

With these limiting criteria in mind, the guidelines also stress that the food distribution within its container, its depth and density are also critical factors. A maximum depth of 5 cm is recommended, the final depth depending on the density of the food. The updated (1989) UK DHSS guidelines for cooked, chilled foods go further and recommend maximum dimensions and sizes for products which cause particular problems, e.g. joints of meat and poultry.

In the absence of universally applicable recipe schedules and in the knowledge that establishments in various sectors of the catering industry have different menu requirements (as well as a growing competitiveness in certain sectors), individual organisations are, increasingly, finding a need to develop their own product and recipe specifications.

In the UK Cook–Chill Survey, a substantial number of units had the facilities needed to carry out their own in-house research and development (see Table 13.4). In addition, most managers interviewed agreed that certain aspects of cook–chill catering, such as specialist equipment, recipes, production techniques and storage time, required further research and development (see Table 13.5). If a follow-up study was conducted now, a further addition to the list shown in Table 13.5 must surely

TABLE 13.5
Managers' Responses in UK Cook–Chill Survey to Areas Requiring Further Research and Development

Area requiring R & D	*No. of units*[a]	%
Equipment	30	37·5
Recipes	28	35·0
Production techniques	28	35·0
Storage time	23	28·8
Visual Appearance[b]	13	16·2
DHSS guidelines	6	7·5
Other	37	48·9
None	9	11·3
Total no. of units	80	100·0

[a] Some managers thought more than one area required further research.
[b] Specifically relating to chilled products in a chilled vending machine.

be microbiological quality, particularly in the light of recent research in this area.

FACTORS AFFECTING SUCCESSFUL RECIPE AND METHOD DEVELOPMENT

The majority of organisations in the UK Cook–Chill Survey reported that at least some, if not all, of the recipes used in their particular organisation required modification in some way. Most commonly, the necessary alterations concerned proportions of ingredients (particularly liquid contents), methods and, in a few cases, the actual ingredients used themselves. Table 13.6 distinguishes between different modifications cited by respondents in the survey as requiring particular attention with regard to product formulation. For all types of modification, operations in the successful group had carried out more research and development than the unsuccessful group, and in all cases had a higher proportion than the sample as a whole.

However, finding a need to modify a product did not necessarily have a bearing on the eventual success of the cook–chill catering system. What

TABLE 13.6
Extent of Product Development in Cook–Chill Units in the UK Survey

Method used	*Successful group*		*Unsuccessful group*		*Total sample*	
	No. of units	*%*	*No. of units*	*%*	*No. of units*	*%*
All recipes	8	34·8	3	13·6	22	27·5
Sauces	17	73·9	10	45·5	53	66·3
Batters	11	47·8	5	22·7	34	42·5
Fried foods	11	47·8	7	31·8	35	43·8
Egg dishes	11	47·8	9	40·9	35	43·8
Methods	14	60·9	11	50·0	42	52·5
No development	2	8·7	5	22·7	11	13·8
Total no. of units	23	100·0	22	100·0	80	100·0

did appear to affect success was the *type* and *extent* of research work completed and *when* that work was carried out. Not only was there a tendency by the successful group to develop a wider range of products (Table 13.6) but the experimental techniques used in the development stage differed markedly (see Table 13.7).

A substantially higher proportion of the successful group (15 units, 65%) carried out their product development using a series of formal experiments conducted in-house, *prior* to the installation of cook–chill than

TABLE 13.7
Methods Used in Product Development in Cook–Chill Units in the UK Cook–Chill Survey

Method used	*Successful group*		*Unsuccessful group*		*Total sample*	
	No. of units	*%*	*No. of units*	*%*	*No. of units*	*%*
Formal experiments	15	65·2	7	31·8	45	56·3
Trial and error	12	57·2	14	63·6	39	48·8
Outside experiments	4	21·8	5	22·7	19	23·8
No development	2	8·7	5	22·7	11	13·8
Total no. of units	23	100·0	22	100·0	80	100·0

the unsuccessful group and the sample as a whole (see Table 13.7). Formal experiments helped to eradicate the majority of problems prior to implementation and use in all operations using them. The value of such experiments was reflected in comments from catering managers in the successful group:

> All recipes were made for cook–chill, we constructed new menus to fit into the system (medium sized, industrial unit). . . . The experiments were conducted by our chefs, nutritional analysis was completed and microbiological counts taken. This process is repeated for every new diet we introduce . . . and we have had only minimal recipe problems (large, hospital unit). . . . There were bound to be problems with liquid content due to the extra time in the oven (during regeneration) so experimentation was initially concerned with that. However, it's an ongoing development (medium sized, hospital unit). . . . It was methods and regeneration times that we experimented with initially, rather than recipes. As other products come into use, they are thoroughly tested (large, hospital unit).

In comparison, few of the unsuccessful group (7 units, 31·8%) had conducted formal experiments to develop acceptable end-products. Of those seven, only two elaborated on the extent of their experiments and the results obtained therein. In addition both of the managers who did elaborate, indicated that the process had been a difficult one:

> Trying to get chefs to experiment with a change of recipes was difficult (small industrial unit). . . . We're still looking to perfect even the simple recipes (medium sized, hospital unit).

TABLE 13.8
Methods Used By Cook–Chill Units in the UK Cook–Chill Survey to Hold Tested Recipes

Location of recipes[a]	*No. of units*	%
Computerised	18	22·5
Files	32	40·0
'Cook–chill book'	25	30·0
'Chefs experience'	27	33·8
Total no. of units	80	100·0

[a] Some units used more than one method.

A number of successful units had utilised trial and error as the basis for modifying recipes, but this was often completed alongside formal experimentation or the use of external research work. In contrast, the unsuccessful group tended to use trial and error as the sole method of recipe modification. Thus, product experimentation by members of the unsuccessful group was often unsystematic and ill thought-out. Problems associated with some recipes often only came to light after implementation had taken place. Many unsuccessful units were, at the time of interview, involved with a reassessment of their recipes and methods as a direct result of the poor results achieved at that time.

RECIPE SCHEDULES

Once experimentation with recipes achieves a satisfactory end-product which can be successfully reproduced to the same standard as each production cycle, the recipes need to be available in a form readily accessible to production unit staff. The UK Cook–Chill Survey revealed that caterers had identified a number of ways of achieving this (Table 13.8), varying from highly computerised systems, specialist cook–chill 'books' (produced by individual organisations), recipe files and a more traditional reliance on a chef's experience. The need for reproducibility in cook–chill catering systems renders the latter method dubious as to its reliability. Once a chef leaves employment, goes on holiday or is absent from work, then there is no record of the recipe used, no continuity and consequently the system is liable to break down. In any systemised process the need for reliable, agreed, written schedules is paramount. Production, hygiene and quality assurance methods and records are built around such documentation.

Well-developed recipe schedules should provide all the information required for the whole production team to process every single item on the menu successfully. This refers not only to accurate measurements for individual ingredients but also detailed methods, accurate time and temperature settings, variations for full and part loads and optimum storage life of a product. Some examples of recipe schedules used in actual cook–chill units are shown in Figs. 13.5–13.7. In addition to planning detailed recipes and methods, each production cycle requires careful advanced planning in order to ensure that recipe schedules can progress without hindrance. This process should circumvent problems such as a build-up of trolleys of cooked foods waiting to enter blast chillers and congestion in corridors

Recipe	Dauphinoise potatoes	*Recipe No.*	05.001
No. of Portions	100	*No. of Trays*	10

Ingredients	*Quantity*	*Instructions*
Potatoes	28 lb	Slice potatoes 1/4″ thick.
Salt	1/2 oz	Place in gastonorm trays and season
Pepper	1/2 oz	with salt and pepper.
Milk	12 pints	Pour on milk and grated nutmeg.
Nutmeg	1 oz	Cover with grated cheese and
Margarine	2 lb	margarine. Place in oven at 180°C for 1 h.
		Blast chill for 90 min at −10°C.
		Reheat time 20 min at 180°C.

Fig. 13.5. Example of a recipe schedule used in an industrial cook–chill unit.

and should ensure well-planned use of equipment and other production facilities. This is extremely important in the light of research evidence which has shown that excess vitamin loss occurs if there is a delay between cooking and chilling.

Another aspect worthy of consideration here is the differing shelf-lives of individual menu items. Evidence presented in Chapter 2 indicates that a number of products undergo a marked deterioration in quality over a five day shelf-life period. It is interesting to note here that the majority of units in the UK Cook–Chill Survey (60 units, 75%) stored chilled food for

Recipe	Steak and kidney pie	*Recipe No.*	01.001
No. of Portions	104	*No. of Trays*	13

Ingredients	*Quantity*	*Instructions*
Lard	1½ lb	Heat lard in pan. Add chopped onions and
Onions	3½ lb	sauté until onions become colourless.
Stewing beef	25 lb	Season diced stewing beef and kidneys and
Kidney	12½ lb	add to onions. Cook until meat is
Salt	1/2 oz	lightly coloured. Mix in the flour and
Pepper	1/2 oz	cook out. Mix in the tomato purée and
Flour	1 lb	mix in the stock. Add Worcester sauce
Tomato purée	1 lb	and adjust seasoning if necessary. Bring
Stock	1/4 pint	to the boil and simmer gently for 1½ h.
Worcester		Divide equally into trays and
sauce	1/4 pint	cover each with 1 lb puff pastry.
Puff pastry	13 lb	Blast chill for 90 min at − 10°C.
		Reheat time 40 min at 220°C.

Fig. 13.6. Example of a recipe schedule used in an industrial cook–chill unit.

Recipe Moussaka		*Recipe No.* 1.002
No. of Portions 96		*No. of Trays* 16
Ingredients	*Quantity*	*Instructions*
Margarine	7½ lb	Melt margarine and add chopped onions,
Onions	2¼ lb	crushed garlic and cook until onions are
Garlic	1 head	colourless. Add chopped mushrooms and
Mushrooms	5½ lb	cook for 2 min.
Minced meat	25 lb	Add the mince and brown lightly. Add
Tomato purée	15 oz	tomato purée. Add demiglacé and stew
Demiglace	5½ pints	gently for 1 h.
Aubergines	25 lb	Peel aubergines and cut into slices.
Flour	1 lb	Pass through flour and fry in oil until
Oil	3/4 pint	golden brown. Cut tomato into slices,
Tomatoes	10 lb	Place mince mixture into trays and cover
Breadcrumbs	1 lb	with alternate slices of aubergine and
Margarine	1½ lb	tomato. Season the surface with
Salt	1/2 oz	salt and pepper and sprinkle tops with
Pepper	1/2 oz	breadcrumbs and margarine.
Chopped		Blast chill for 90 min at −10°C.
parsley	2 oz	Reheat time 45 min at 180°C.

Fig. 13.7. Example of a recipe schedule used in an industrial cook–chill unit.

three days or less before regeneration (see Table 13.9). The results of later research work which investigated whether there was a need for extended shelf-life, correlated well with results from the UK Cook–Chill Survey, showing that mean storage time was around three days.

Storage life is an important factor to be taken into consideration during the research and development process. Individual recipes should be

TABLE 13.9
Different Storage Times Used in Units in the UK Cook–Chill Survey

Storage time (days)	*No. of units*	%	*Cumulative %*
1	29	36·3	36·3
2	10	12·5	48·8
3	21	26·2	75·0
4	2	2·5	77·5
5	18	22·5	100·0
Total	80	100·0	

thoroughly tested from start to finish including setting storage limits for individual products. With this in mind, it is absolutely imperative for cook–chill catering systems to introduce control systems to prevent the possibility of individual menu items being used after expiry of their storage life. The most common method utilised in practice—and possibly the most straightforward, is a colour-coded, date stamped labelling system such as those shown in Fig. 13.8.

DEVELOPMENT TIME

The results in Table 13.10 show that, typically, units interviewed in the UK Cook–Chill Survey, had a development time of up to six months in length (23 units, 32·8%; mean time = four months) or was on-going (23 units, 32·8%).

TABLE 13.10
Typical Development Time Used by Cook–Chill Units

Development time	*No. of units*	%
Up to 6 months	23	32·8
6–12 months	9	12·9
12 months +	4	5·7
On-going	23	32·8
None	11	15·8
Total	70	100·0

Furthermore, there was a direct correlation between the extent of development work and consumer satisfaction, a finding which in itself must convince any doubters as to the value of research and development work.

Clearly, from the discussion so far, product research and development is, essentially, a lengthy and involved process. However, only if this stage of the introduction process is carried out effectively can end-product quality be guaranteed. In short, what all this implies is that innovating organisations transferring production to cook–chill catering systems, need to invest time and resources in effective research and product development programmes.

FURTHER READING

Anon. (1982) Chill for hot food over long distances. *Catering and Hotel Management*, November, 6–8.

Anon. (1983) Cook–chill allows new look menu at Middlesex Polytechnic. *Fast Food*, April, 45–6.

Anon. (1986) Casino cook chill jackpot. *Caterer and Hotelkeeper*, February 6, 31–3.

DHSS (1980) *Guidelines on Precooked, Chilled Foods*. London, HMSO.

Drake, J. (1985) Meals on Wheels Survey, Hampshire County Council Social Services Dept., Winchester.

Droog, P. (1983) Cook–Chill Hospital Systems 'Cuttings from the UK Press'. Regethermic (UK) Ltd.

Electricity Council (1982) *Planning For Cook–Chill*, 3rd Edn., Electricity Council Project Planning Unit.

Fuller, J. (1966) *The Chef's Manual of Kitchen Management*, London, B. T. Batsford Ltd.

Gilbert, R. J., Miller, K. L. & Roberts, D. (1989) *Listeria monocytogenes* and chilled foods. *Lancet*, Feb 18, 383–4.

Kingcott, E. (1988) Update of the DHSS Guidelines for Pre-cooked, Chilled Foods. From notes on a meeting organised by the Electricity Council, October.

Kinton, R. & Cesarani, V. (1978) *The Theory of Catering*, 4th Edn, London, Edward Arnold (Publishers) Ltd.

Lewis, A. & Light, N. (1988) A survey of cook–chill catering in the UK and the need for the extension of product shelf-life. *Food Science & Technology Today,* **2,** 214–17.

Lillicrap, D. R. (1971) *Food and Beverage Service,* London, Edward Arnold (Publishers) Ltd.

Lund, B. M., Knox, M. R. & Cole, M. B. (1989) Destruction of *Listeria monocytogenes* during microwave cooking. *Lancet*, Jan. 28, 218.

Walker, A. E. (1988) The transfer of technology in the catering industry: a study of UK cook–chill catering operations. PhD Thesis, CNAA/Dorset Institute of Higher Education, Poole, Dorset, UK.

Walker, A. E. (1989) *Success or Failure in Cook–Chill Catering—a Study in Technology Transfer*, Bradford, Horton Publishing Ltd.

West Midlands Regional Health Authority, *Chilled Meals Service Handbook*.

Young, H. (1986) Factors affecting the shelf-life and acceptability of cooked, chilled foods. PhD Thesis, Dorset Institute of Higher Education, Poole, Dorset, UK.

CHAPTER 14

People and the Introduction of Cook–Chill Catering Systems

INTRODUCTION

The portrayal of the catering industry as being, historically, a labour-intensive, craft-based industry with relatively little need for new technology, has become something of a cliché. References to it in the popular press often conjure images of an industry locked in a time warp, exploiting its labour-force and utilising antiquated equipment. However, whilst many catering establishments remain labour-intensive and craft-based, the notion that the industry has little use for new technology is becoming an increasingly out-dated view. Despite the fact that until relatively recently the uptake of new technology in the industry has been fairly slow, technological change has had a substantial impact on many catering organisations. For example, micro-computers have found many applications in front office operations in hotels, as point of sale systems in restaurants, for stock control and office systems in all sectors and for temperature control in cook–chill catering units. Technological systems such as cook–chill catering and *sous-vide* have been introduced in all sectors of the industry.

Although the presence of cook–chill systems in the catering industry has been widely acknowledged, the number of units in the UK only represents a small percentage of the industry as a whole. In other countries the uptake of technological systems in catering has not been so slow, for example, the introduction of cook–chill technology and *sous-vide* in France has seen a much faster and more sustained growth than in the UK albeit if only in specific sectors. In France, cook–chill has been a major influence in the health sector, although there are examples of large industrial units producing in excess of 30 000 meals per day.

As in any other industry, technological change is often motivated by cost considerations (see Chapter 12). The overriding factor in the UK catering industry's apparent reluctance to invest in capital-intensive catering such as food manufacturing style catering systems must be the availability of an exploitable pool of cheap labour.

However, the escalating cost of labour has been a major catalyst for technological change in the catering industry. The increasing financial pressures of the 1970s and 1980s has seen a need for caterers to become more competitive, the attraction of economies of scale, increased productivity and a reduced wages bill through the introduction of new technology are often attractive strategies for decision-makers in catering establishments to adopt. It is perhaps the potential to reduce a growing wages bill by reducing their labour-force that lures caterers towards cook–chill catering systems.

The subject of this chapter, therefore, is to investigate the effects that introducing a cook–chill system has on the different groups of people involved in the change process.

WHO IS AFFECTED BY CHANGE?

Until the introduction of computerised control systems and more technologically-based production systems such as cook–freeze and cook–chill in the UK, the use of new technology was rarely made at the expense of significant amounts of labour. Consequently, any introduction of new technology gave the work-force little cause for concern. Indeed some minor changes were seen to improve job quality, freeing workers from tedious tasks in order to concentrate their efforts on more interesting work. For example, the introduction of an automatic potato peeler meant that an employee could be freed from a relatively boring task and put to better use elsewhere in the kitchen.

More recently, however, the introduction of technology has seen more purposeful labour-shedding than had hitherto been experienced by the catering industry. The most obvious examples of this were in catering establishments which had invested in technological systems such as cook–freeze and cook–chill catering. The simple fact that large scale technological change is aimed at improving efficiency, by displacing labour and increasing productivity, is evidence enough that such change will have a major effect on the work-force. Moreover, as an industry, catering is one of the top four employers in the UK and any significant changes within it are, therefore, likely to affect a large number of employees.

However, the work-force are not the only people who may be affected by a major technological change in the catering industry. In service sector industries such as catering, there is a high degree of customer contact. Interactions with catering staff impact on a consumer's judgement of the total meal experience. It is important, therefore, that any changes which take place should not mar a customer's evaluation of the service, including not only food quality but also personal relationships between customer and employee.

It can be said therefore, that a tripartite relationship exists between managers, operatives and customers, all of whom may be affected by a change in the system.

WHO IS AFFECTED BY THE INTRODUCTION OF COOK–CHILL?

There are a number of parties involved in the technological change process. Within an organisation the three way relationship between management, operatives and consumers is perhaps the most important one, although other relationships exist which we will consider briefly in this chapter. Management is predominantly concerned with business objectives aimed at achieving increased profitability through increased efficiency and productivity. It is this function which is a primary motivator for technological change.

The involvement of the work-force as operators of the new technological system often requires changes in hours worked, job content and skill requirements. As a result, the impact of change upon them tends to be concerned with the social and psychological effects rather than the financial and business implications of such a change, which are the dominant effectors for managers early in the process of change.

The relationship between managers and operatives is critical during the process of technological change in any industry. However, as we have seen above, in service industries such as catering, interaction with customers is also an essential ingredient in the process.

The effect on customers brought about by the introduction of a cook–chill catering system will vary depending on the type of system involved. For example, in cook–chill vending situations, the very fact that customers have to interact with a machine rather than a member of the service staff may strongly influence the success of the system. Also, in some hospital catering situations, the consumers (i.e. patients) are unusually removed from the caterer, in that members of the nursing staff have overall responsibility for serving food to patients. The social consequences of such a change need, therefore, to be taken into account and overcome by management.

Thus, there are important implications for the human factor in a change to cook–chill catering. Rejection by any one of these different groups of people could slow or even halt progress towards the success of that system. Indeed, some commentators have argued that caterers may be wrong to automatically seek labour reductions in an industry where the customer/staff contact is a vital part of the customer's appreciation of the whole service (see later).

Figure 14.1 indicates that technological change also involves interaction between members of external organisations (e.g. equipment manufacturers and food suppliers). Obviously, there are many other relationships not shown in the diagram or discussed but which are also important including relationships with consultants, other units, financial overlords, environmental health officers, academics and others. However, it is worth a brief overview of the relationships between the organisation introducing cook–chill and equipment manufacturers and food suppliers respectively.

Equipment suppliers play a prominent role in the technological change process in the catering industry. However, it has been suggested that there is insufficient liaison between caterers and equipment suppliers with regard to the industry's equipment requirements. Consequently, in the early days of cook–chill catering's development and even, to some extent today, equipment manufacturers have tended to supply products which they supposed the caterer wanted, rather than supplying to the caterers needs. Moreover, equipment manufacturer's interests probably take little regard for the wider implications of major technological change within the catering industry. Equally, as results from the UK Cook–Chill Survey have indicated, caterers often fail to take into account the opinions of the user and are often influenced predominantly by cost considerations rather than the capabilities of the equipment coupled with the needs of the operator and consumer. It is up to the catering manager therefore to influence the equipment manufacturer to tailor products to the needs of the industry and for this reason their relationship is an important one.

Likewise, the relationship between managers and food suppliers may affect the success of a cook–chill catering system. Cook–chill catering relies on standardised recipes and methods for obtaining consistent quality in end-products (see Chapter 13). In turn, chefs require

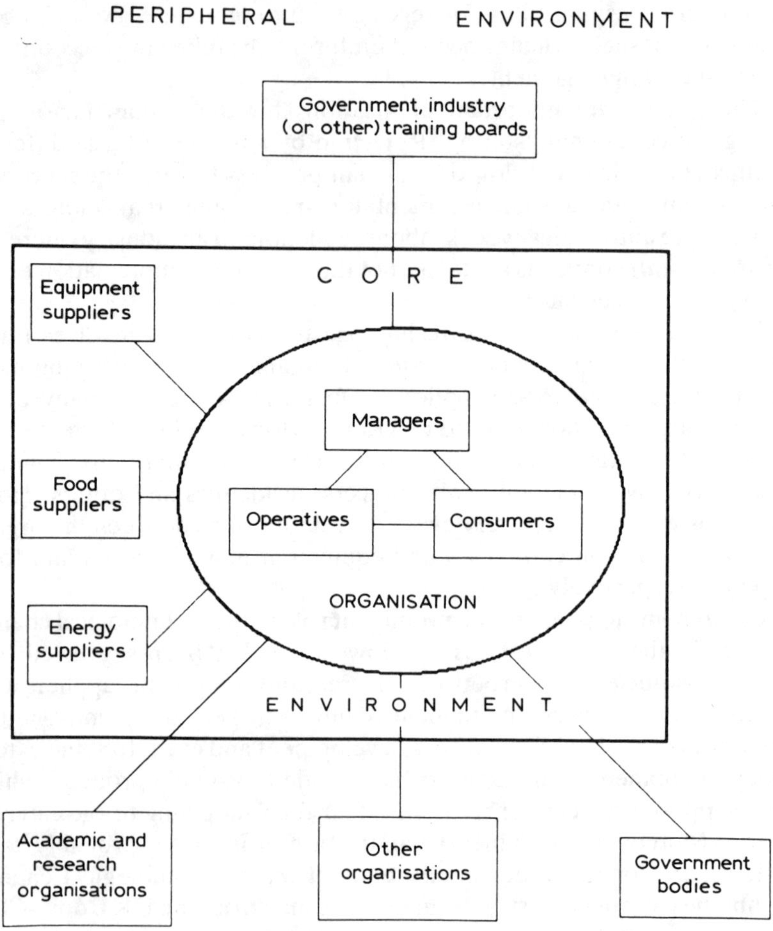

Fig. 14.1. Diagram illustrating the interactions involved internally in a cook–chill catering unit, within the core environment (those involved in servicing the unit) and externally to the unit.

standardised raw materials to a guaranteed and consistent specification. Thus, relationships with suppliers are also important in order to receive the necessary raw ingredients.

In short, the introduction of a cook–chill catering system is likely to affect a large number of people both within and outside the organisation involved. Within the organisation, the introduction of a cook–chill system has implications for management, operatives and consumers alike. The following sections therefore consider both the reaction to and implications for each of these main groups likely to be affected by the introduction of cook–chill catering.

MANAGEMENT INVOLVEMENT AND RESPONSE TO CHANGE

One of the main factors influencing managerial reactions to the often traumatic change which the introduction of cook–chill catering represents, is the amount of control and influence they have in the decision-making process. In other words, the extent to which they were a key decision-maker may influence their level of enthusiasm for a cook–chill catering project. For example, a manager who instigates the idea for a cook–chill catering system is more likely to show dedication to making a success of the project than a catering manager who has been ordered to introduce it by superiors. Consequently, their reactions to change are likely to mirror their involvement and support of the innovatory move.

It is widely accepted that individual managers (in any industry) react to change in different ways and can be categorised as such. For example, they can support it, comply with it, resist it, or ignore it. Those associated with support are often dubbed innovative or proactive; those who comply, reactive; those who resist, laggards; and those who ignore it, traditionalists. In addition, there are others who may be unaware of the potential for change.

The implications of the change to cook–chill catering are far-reaching, not only because of its essentially technological base, but also because it necessitates a dramatic change in the working practices of both the management and work-force. Thus, it can be said that there are basically three reasons for a manager to introduce a cook–chill catering system. First, because the manager is requested to do so from his superiors, second because he/she sees it as a vital move for the organisation and third, because they are reacting to a crisis. It is often easier to have something tangible to react to (such as the threat of competitive tendering forcing catering services out to catering contractors) than to create activity in the absence of stimulation.

These arguments suggest, therefore, that cook–chill catering managers who have introduced new systems are not necessarily a group of highly proactive forward thinking managers, but also include a number of laggards forced to adopt cook–chill catering as a result of other extenuating circumstances, particularly a need to rationalise their systems through sometimes harsh cost-cutting exercises or as a response to crisis demands.

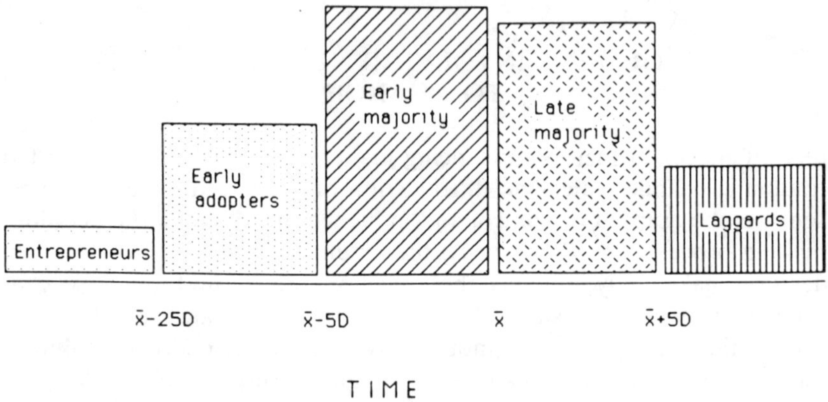

Fig. 14.2. Graphic illustration of the forms of innovation taken in technology transfer. Entrepreneurs only form a statistically small minority whilst the early and late majorities account for the main bulk of innovators.

The model of innovative categories shown in Fig. 14.2 has been shown in other studies of technological innovation to represent managerial reactions to change. Managers in each category are said to exhibit different behavioural characteristics as shown in Table 14.1. A few of the very early adopters of cook–chill catering systems (i.e. before 1979) interviewed in the UK Cook–Chill Survey could be classed as entrepreneurs. They were certainly venturesome and eager to attempt new ideas. Managers in the early adopter category were identified by the extent to which they were utilised as 'model systems' by other managers who fell into the later adopting categories. The early majority category accounted for perhaps the largest proportion of the managers in the UK Cook–Chill Survey, with the remainder of the catering managers falling into the late majority sector. This latter group were characterised by a strong pressure to change through economic necessity. In this category, there were a number of managers from the unsuccessful group identified in the UK Cook–Chill Survey who admitted to being somewhat reluctant cook–chill users.

TABLE 14.1
Characteristics of Different Members of Innovative Categories

Category	*Characteristics*
Entrepreneurs	Venturesome, eager to adopt new ideas
Early adopters	Respected, opinionated leaders, 'the individual to check with', role models
Early majority	Deliberate, highly interactive, 'be the first by which to try, nor the last to lay the old aside', willing followers
Late majority	Sceptical, cautious, act only when safe
Laggards[a]	Traditionalists, loners, suspicious of change

[a] May or may not be the fault of the individual—may have limited resources (therefore, lag rational) or may more accurately reflect the reality of a problem system.
Adapted from Rogers (1971).

Obviously, not all the managers interviewed in the survey fitted neatly into this categorisation system. Some managers showed characteristics which bridged several of these categories, showing later adopter characteristics in their initial moves towards cook–chill, but later showing entrepreneurial, or early adopter characteristics by implementing more sophisticated techniques such as computer controls, HACCP (see Chapter 9) and *sous-vide* (Chapter 8).

IMPLICATIONS FOR MANAGEMENT

In addition to the potential for cost savings previously discussed, the introduction of a cook–chill system offers the opportunity to the manager for greater control not only of the food production system but also, by redefinition of labour tasks, the work-force within it. Managers in the UK Cook–Chill Survey were in agreement that, to be successful, a cook–chill catering system depended on two main factors—management quality and, perhaps more importantly, full commitment from all those involved with and affected by the system. It is important to nurture this commitment throughout the organisation—an exercise which stems from management—through a good communication system which itself depends on management style. Indeed, it is worth noting that management style may have to change as a result of the introduction of a cook–chill catering system.

Burns and Stalker (1968) describe two polar extremes of management style, mechanistic and organic. Both styles may be used by managers to 'exploit the human resources of a concern in the most efficient manner feasible in the circumstances of the concern'.

TABLE 14.2
Summary of the Differences between Mechanistic and Organic Management Systems

Mechanistic	*Organic*
Specialised differentiation of functional tasks	Specialist knowledge and experience, common tasks of concern
Individual tasks isolated from concern as a whole	Nature of tasks seen as part of the total concern
Each level of authority has to reconcile performances to superiors, who in turn are responsible for ensuring that each is relevant to own special part	Adjustment and continual redefinition of individual tasks through interaction with others
Precise definition of role	Responsibilities shared
Rights, obligations and methods responsibility of a functional position	Spread of commitments
Hierarchic structure of control, authority and communication	Network structure of control, authority and communication
Final decisions always at the top of hierarchy reinforcing structure	Knowledge-holders not necessarily head of concern, therefore *ad hoc* centre of control
Interaction: vertical—commands	Interaction: lateral—consultation
Operations governed by orders, instructions and decisions by superiors	Communication consists of information and advice rather than instructions and decisions
Insistence on loyalty and obedience	Commitment is more highly valued than loyalty and obedience
Greater importance attached to internal rather than general knowledge experience and skill	Importance and prestige attached to affiliations and expertise valid to the nature of the firm

Table 14.2 summarises the differences between the two extreme management systems. Some of the more salient points relevant to catering organisations incorporated within these definitions relate to the hierarchical structures within the catering establishment. It has been argued elsewhere that the introduction of cook–chill catering systems reduces the number of hierarchical levels within a kitchen, making the communication links between management and operatives 'closer'. In other words, the system becomes more akin to an organic rather than mechanistic organisational structure.

Many members of organisations who are at the same hierarchical level contribute their own area of expertise into the system. Thus, in organic style establishments a more lateral network of communication is required (as opposed to the more vertical channels in mechanistic ones). It is important, therefore, to remember that whilst organic systems are not hierarchic in the sense that mechanistic systems are, they are essentially stratified and care must be taken for lateral methods of communication to take place.

Thus, although in the UK Cook–Chill Survey, some successful cook–chill systems appeared to approximate to organic rather than mechanistic management systems in that their hierarchical structure was 'flatter' (i.e. the top and bottom levels of an organisation were 'closer' to one another), organic style management does not necessarily mean that communication will be more effective. Information still has to travel laterally as well as vertically involving a number of different link-pin members of the organisation and, thus, at any point communication channels have the potential to break-down.

WORK-FORCE REACTIONS TO CHANGE

Resistance by the work-force is a common reaction to any form of organisational change. There is often a real fear that technological change will mean machine domination, a reduction in skills and job losses. This resistance is perfectly rational because, almost inevitably, change in the work-place has the potential to affect the labour-force substantially. Many new technologies affect employment structures and occupational skills. For example, in other industries the introduction of computer numerically controlled machine tools has meant that jobs previously done by skilled craftsmen are now carried out by machines with a relatively unskilled operative overseeing production. Similarly, the introduction of cook–

chill catering systems often shifts emphasis from a chef's individual skill and creativity in producing menu items to standardised production allowing for little flexibility for a chef to impose his or her own identity on the end-product. Furthermore, it has been suggested by some researchers that the introduction of new technology will increase the separation between the more highly skilled and the less skilled sections of the workforce. This may result because many workers with traditionally manual skills are vulnerable to technological change, whilst those with skills more allied to the new technology will find further opportunities for acquiring more skills or technician status. For example, in cook–chill catering, chefs in a CPU may develop additional skills in temperature and microbiological control in addition to their more traditional craft-based skills, whereas those at satellite units often lose the chance to exhibit their traditional skills and, in effect, may become machine operators rather than skilled craftsmen.

THE NEED FOR LABOUR REDUCTION

Technological change tends to shift concentration from a labour-intensive to a capital-intensive base, often requiring or leading to a reduction in labour. However, not all catering organisations achieve a labour reduction through the introduction of cook–chill systems, nor do all organisations attempt to do so. In some cases, particularly where cook–chill catering is introduced for supplementary feeding systems such as banqueting and on-site full-meal vending facilities, its introduction may require an increase in staff. For example, in the UK Cook–Chill Survey there were some units which experienced an increase in numbers employed because the introduction of cook–chill catering allowed them to accommodate 'extra' production for banqueting and peripheral full-meal vending facilities. Likewise, in establishments where the introduction of a cook–chill system enables the opportunity, through increased productivity, to cater to a wider market, additional staff are often required to 'dispose' of the extra meals produced. This included provision for staffing new outlets, transportation and outside catering. However, an increase in the numbers of full-time equivalents employed is unusual. In the UK Cook–Chill Survey, for example, over 80% of operations experienced a *reduction* in full-time equivalent employees employed (see Table 14.3). Percentage employment change actually ranged from 185 to −86·5% with the median value being −16·5%. The methods used to achieve this

TABLE 14.3
Percentage Employment Change in Establishments in the UK Cook–Chill Survey

Employment change	*No. of units*	*%*	*Cumulative %*
Over 60% reduction	6	7·7	7·7
40–60% reduction	14	17·9	25·6
20–40% reduction	15	19·3	44·9
0–20% reduction	28	35·9	80·8
0–20% increase	4	5·1	85·9
20–40% increase	2	2·6	88·5
40–60% increase	2	2·5	91·0
60–80% increase	7	9·0	100·0
Total	78	100.0	

reduction varied. Some units concentrated on reduction of actual numbers employed whilst others reduced individual employees hours worked. Furthermore, 70% of the managers interviewed said that their main method for recouping expenditure was through the reduction of labour (see Table 14.4).

TABLE 14.4
Methods Used to Recoup Capital Expenditure by Cook–Chill Units in the UK Cook–Chill Survey

Method[a]	*No. of units*	*%*
Reduction of staff numbers	55	68·75
Cheaper purchasing procedures	5	6·25
Increased revenue	17	21·25
Other	14	17·5
Total	80	100·0

[a] Some units used more than one method.

In the UK Cook–Chill Survey, there was little difference in the percentage employment change in units in the successful and unsuccessful groups whose employment levels had decreased as a result of the introduction of cook–chill (median percentage employment change was −40·5 and −42·5% for each group respectively). However, this measurement takes no account of the size of each operation. It is reasonable to assume that larger cook–chill operations are more likely to experience a greater employment loss as a result of economies of scale. Thus, change in employment levels were calculated as a percentage of employment change per unit of capacity. The results of this exercise were revealing. The successful group occupied a much narrower range (minimum value −4·0; maximum value −0·19; median value −1·2). In contrast the unsuccessful group covered a much wider range of values (minimum value −45·5; maximum value −0·78; median value −4·35). This suggests that employment reduction per unit of capacity impacts on success. More specifically, too great an employment loss per unit of capacity was associated with cook–chill units in the unsuccessful group. The units in the successful group limited their employment reduction whereas many operations in the unsuccessful group made too great an employment loss per unit of capacity which, in turn, led to an unsuccessful transfer of cook–chill technology.

In short, it appears that for the majority of cases, some reduction in the work-force seems inevitable, but realistic employment reduction is more effective tailored to the size of reduction than too ambitious reduction programmes.

NUMBERS OF EMPLOYEES AFFECTED BY THE INTRODUCTION OF COOK–CHILL

In the UK Cook–Chill Survey 6026 full-time (equivalent) staff were employed prior to the introduction of cook–chill catering in 75 of the 80 organisations visited. After implementation of the new system, 4828 people were employed in the same organisations. Thus, approximately 20% of the total work-force lost their jobs as a direct result of the introduction of cook–chill catering. Bearing in mind the fact that the majority of units in the UK Cook–Chill Survey introduced it to produce only a small proportion of their total catering output, the implications for staffing levels in cook–chill establishments utilising the system for the majority of their output may be more far-reaching.

In addition, the number of workers remaining in employment in units in the UK Cook–Chill Survey were mainly part-time. There were 967 of these working in central production kitchens, the rest were involved in part-time work at satellite units and transportation. Thus, for the vast majority of employees, the introduction of cook–chill catering had quite a substantial effect. These findings support those of a later study which showed that employees were reduced by some 20% to cater for the different manpower requirements of their new system at a large London hotel.

HOW IS LABOUR REDUCTION ACHIEVED?

It has been suggested repeatedly in this book that the need to introduce new technology (i.e. cook–chill catering systems) is often motivated by the need to save money by reducing labour costs. This can be done in a number of ways. By reducing numbers employed *per se*, reducing the numbers of skilled personnel required, reducing the number of full-time workers employed and placing more emphasis on part-time workers or reducing the amount of 'expensive' work hours at weekends, unsocial daily hours and overtime. Many cook–chill catering units employ all or a combination of these methods to achieve a reduction in their labour-force.

In the UK Cook–Chill Survey there were three main methods employed to reduce labour (see Table 14.5). Most organisations attempted to allay any employee fears of job loss by using natural wastage (i.e. retirement, early retirement, and other voluntary departures) as their main method of labour reduction. However, there was a need in some units to resort to redundancies whilst others were able to relocate staff either physically to other parts of the organisation or 'timewise' by reducing their number of hours worked.

There were some important differences between the successful and unsuccessful groups in the UK Cook–Chill Survey with regard to the methods used to reduce staff. The successful group made significantly

TABLE 14.5
Methods Employed by Managers in the UK Cook–Chill Survey to Accommodate Changes in the Number of Staff Employed

Method	*No. of units*	%
Redundancy	28	35·9
Natural wastage	36	46·2
Relocation of staff	15	19·2
Recruit new staff	15	19·2
Total	78[a]	100·0

[a] Some units used more than one method.

more use of natural wastage methods (15 units, 83·3% of units reducing labour as a result of cook–chill) than the unsuccessful group (6 units, 46·2%). Whereas, the unsuccessful group made greater use of redundancy to reduce employment (9 units, 69·2%) than the successful group (8 units, 44%). Furthermore, redundancy was used by the majority of units in the unsuccessful group as the *only* method of employment reduction (5 units, 55·6%) or their main method of employment reduction (3 units, 33·3%). In contrast few of the successful group (2 units, 25%) utilised redundancy as their only means of staff reduction. Indeed, the majority of units in the successful group who used redundancy did so in rare cases and in addition to their main methods (usually natural wastage).

There were a number of reasons why the methods used to reduce staff numbers affected success. Perhaps the overriding reasons, however, were that redundancy methods were often an expensive alternative to natural wastage methods, in terms of both monetary costs and possible industrial disputes. Furthermore, the use of natural wastage methods presented less of a threat to the work-force and trade unions than the use of redundancy methods. Whilst it could be argued that natural wastage methods add costs for an organisation until those employees leave, it is a more acceptable method to employees themselves. In addition, it has been suggested elsewhere that employees not under a threat of enforced job loss tend to be more productive than those wary of the implications of technological change.

IMPLICATIONS FOR THE WORK-FORCE

The introduction of cook–chill technology often results in the elimination of weekend work, unsocial hours and overtime. The peaks and troughs so familiar in catering work are, in cook–chill catering situations, often replaced by an even production throughout an eight hour shift. This has the potential to affect the amount of pay employees receive. Furthermore, the introduction of cook–chill catering was seen to greatly affect job content particularly in large operations.

The implications for the work-force in terms of pay in an industry with no protectionist body for workers in a cook–chill catering situation are, in theory, potentially serious. For example, consider a catering establishment where, prior to the introduction of cook–chill employees were expected to work weekends, split shifts and plenty of overtime (at higher hourly rates for working extra and unsocial hours). Their take home pay would be increased by a significant amount and could make up the difference between a poor wage and an acceptable one. After the introduction of cook–chill catering, rationalisation of the production process only requires working a straight eight-hour shift, with no opportunity for weekend work, unsocial hours and overtime, i.e. employee's pay becomes limited to their basic rate only and they receive a reduced net income. This, indeed, has shown to be the case in two recent surveys in the UK.

In the UK Cook–Chill Survey, 56 managers said that the hourly pay rates for operatives had stayed the same after the introduction of cook–chill catering, eleven had increased and only six had decreased their hourly pay scales (see Table 14.6). As we have noted earlier, cook–chill catering also allows for a rationalisation of the working day often leading to employees working fewer hours than previously. Thus, with hourly pay rates staying the same or even decreasing, some employees experience a net decrease in income. This finding was supported by recent articles in the trade press which showed that some cook–chill catering outlets were experiencing difficulty in recruiting and retaining staff because their earning potential is much reduced. Furthermore, all but one of the managers interviewed mentioned that there were also a number of extra benefits available to employees (see Table 14.7). The most common of these was, inevitably, free meals on shift, but pension schemes, productivity bonuses and day release courses were also widely used additional benefits.

In the successful group there were more units (5 units, 21·7%) whose basic hourly wage rates had increased following the introduction of cook–chill than in the unsuccessful group (2 units, 9·1%). However, the overwhelming majority in both the successful group and the unsuccessful group (16 units, 69·6% and 18 units, 81·8% respectively) experienced no change

TABLE 14.6
Change in Hourly Pay Rates after the Introduction of Cook–Chill in Units in the UK Cook–Chill Survey

Hourly pay rate	*No. of units*	*%*
Increased	11	15·1
Decreased	6	8·2
Stayed the same	56	76·7
Total	74	100·0

TABLE 14.7
Typical Employee Benefits in Units in the UK Cook–Chill Survey

Benefit	*No. of units*	%
Pension	36	45·0
Productivity bonus	28	35·0
Free meals on duty	54	67·5
Low priced meals	4	5·0
Educational qualifications	27	33·8
Health scheme	8	10·0
Profit sharing	7	8·8
Loans	4	5·0
Free accommodation	3	3·8
Free transport	7	8·8
None	1	1·3
Other	27	33·8
Total	80	100·0

in wage rates. A recent survey of pay in National Health Service cook–chill catering units in the UK revealed that the introduction of new technology had a negative effect on take home pay and as a result there was a higher incidence of labour shortages in cook–chill catering units than in conventional, cook–serve, hospital units. In the UK Cook–Chill Survey some units countered this effect by introducing productivity bonuses to make up for any loss in net pay.

CONDITIONS OF EMPLOYMENT IN COOK–CHILL CATERING UNITS

In theory, the introduction of cook–chill catering should be attractive to employees in that it reduces pressure on them at peak times during the day. One of the advantages of this is that production staff have more time to spend on the visual aspects of the product, training, product development and so on. In addition, some employees should be attracted by the removal of split shifts and work involving unsocial hours. Indeed, a well organised production kitchen can work effectively operating eight hour shifts, for a five day week. On the other hand, some employees might find such a system of catering boring, repetitive and may even miss the crisis

TABLE 14.8
Employment Stability in Units in the UK Cook–Chill Survey

Length of employment	*No. of employees*[a]	%
More than 5 years	1438	46·4
2–5 years	750	24·2
1–2 years	374	12·1
Up to 12 months	537	17·3
Total	3099	100·0

[a] In 70 organisations only.

element of conventional catering systems. In the former, labour turnover would be expected to be low whilst in the latter, labour turnover would be higher.

There is little evidence which suggests that either effect predominates. It appears from results of the UK Cook–Chill Survey that employee turnover in cook–chill catering units is (as far as the catering industry as a whole is concerned) relatively low (median turnover = 34%). Furthermore, the employment pattern in these units tended towards longer-service employees indicating a more stable work-force (see Table 14.8). The figure may become even more stable given that a proportion of those leaving were part of a natural wastage programme of labour reduction planned for in the initiation of cook–chill catering (see earlier). This is reinforced by the figures shown in Table 14.9, which show that 19 of the 70 organisations represented in Table 14.8 were expecting to reduce their labour turnover even further in the future.

TABLE 14.9
Anticipated Future Changes in Employment Numbers

Future employment levels	*No. of units*	%
More staff employed	18	22·5
Less staff employed	14	17·5
No change in staff numbers	46	57·5
Not sure	2	2·5
Total	80	100·0

COOK–CHILL AND EMPLOYEE SATISFACTION

There were differences between units in the UK Cook–Chill Survey which suggested that the reactions of the labour-force are influenced by activities prior to the introduction of cook–chill catering. Employee satisfaction varied between organisations to a fairly high degree, with operatives in the successful group of cook–chill units being more satisfied with their work than employees in the unsuccessful group. Indeed, a number of organisations in the unsuccessful group experienced industrial relations problems which were difficult to overcome. There was a general consensus of opinion amongst the managers of these units that much of the problem was a result of misunderstandings on the part of the work-force. In particular, the threat of pending job and financial insecurity, uncertainty and reduction in skill requirements were often cited as the main grievances of employees worried at the introduction of cook–chill catering.

COMMUNICATION AND CONSULTATION OF EMPLOYEES

From what we have discussed so far, it seems inevitable that the introduction of cook–chill catering will lead to some job losses (see earlier), but many of the other concerns of employees can be dispelled with adequate communication and training.

Some researchers have commented that communication between managers and employees during a technological change is only good in that it provides a realistic picture of the environment. From the results of the UK Cook–Chill Survey alone it appears that if this picture of the environment is not given, undesirable consequences result for all the parties involved. Too often the introduction of cook–chill catering was seen as a purely technical process of introducing a more systematic method of producing food. The psychological contract between the organisation and its work-force was often ignored. Although it was unnecessary for employees to be informed of every single detail involved with the change of system, it was clear from the information given by managers in the survey that employees who were kept well informed of the change process were more likely to accept the new system with the minimum of disruption. Further, these operatives manifested a greater degree of job satisfaction than employees who had not been well informed.

Table 14.10 shows some striking differences between the successful group and the unsuccessful groups in their policy of informing employees during the decision-making phase. Of those cook–chill units falling into the successful group, 19 (90·5%) kept their chefs continually informed of progress with regard to the new system. A similar pattern emerged for other levels of staff, with 17 (80·9%) of the successful group keeping their staff informed continually. This included ensuring that employees were involved from the outset, not only with the methods to be used but also the problems as they saw them. It was also felt important for employees to observe successful operations at work.

TABLE 14.10
Continuity of Involvement in the Decision-Making Process for Different Levels of Staff in the Successful and Unsuccessful Groups in the UK Cook–Chill Survey

Staff level	*Successful group continuous consultation*		*Unsuccessful group continuous consultation*	
	No. of units	*%*	*No. of units*	*%*
Chefs	19	90·5	11	45·8
Cooks	17	80·9	10	41·8
Kitchen assistants	17	80·9	10	41·8
Kitchen porters	17	80·9	10	41·8
Total	21	100·0	24	100·0

In contrast, the unsuccessful group placed less importance on informing their staff. Table 14.10 shows that less than half of the respondents continually informed staff. Nine operations (37·5%) did not involve chefs at all prior to start-up and ten operations (41·7%) did not inform the rest of their employees (cooks, kitchen assistants and porters). In four of the units (18·2%) in the unsuccessful group, information was given to staff, but only on a one-off basis. In other words, the decision to introduce cook–chill catering was mentioned briefly with the full implications of its introduction left unsaid. In some cases it appeared that employees were only informed as an afterthought or as a necessity in order to issue redundancy notices.

The position with regard to the dissemination of information was echoed by the employees themselves. Operatives in the successful group corroborated the evidence given independently by their managers and agreed that communication had been effective. Most operatives stated that successful communication resulted when the project had been well thought out and a clear picture could be given to employees regarding what the introduction of a cook–chill catering system actually meant for the employees. Managers who used this strategy were also able to benefit from constructive comments from operatives on the more practical aspects of introducing the system.

OPERATIVE TRAINING IN COOK–CHILL UNITS

One of the consequences for the work-force of introducing technology is a change of skill requirements. On the one hand, jobs themselves may become more acceptable because cook–chill technology can remove the need to become involved in tedious jobs. However, on the other hand, it can suppress the opportunity for flair and imagination in creating recipes. In addition, some operatives may experience increased feelings of boredom particularly if they become predominantly 'machine operators'. However, the capability to work as part of the system is achieved largely through education and, for this reason, training programmes become an indispensable part of the successful introduction of a cook–chill catering system.

The UK Cook–Chill Survey showed that many of the problems experienced in the early stages of the introduction process were operative-related (see Chapters 11 and 15). The majority of these problems were related to inexperience and lack of knowledge. An important factor in preventing these types of problems was the use of effective training before the installation of the new system. Information on the extent of training carried our before and after the installation was collected from managers and operatives of cook–chill units interviewed in the survey. Not surprisingly, most managers claimed to have involved their staff in a training programme. However, information from operatives themselves revealed some notable differences in the type of training programme given and lead to the assumption that some training methods are more effective than others as far as cook–chill catering systems are concerned.

For the purposes of the survey, employees were divided into three categories: chefs; cooks; and kitchen assistants. Typical members of the chef's group included head chefs, production managers and head cooks, whilst the cooks group consisted of cooks, sous chefs, second and third chefs, kitchen supervisors and production supervisors. Finally, the kitchen assistants group included general assistants, production assistants, porters and apprentices.

For both the successful and unsuccessful groups, 35 employees out of a total of 68 interviewed received pre-installation training of some sort. In addition, 64 out of 68 employees interviewed in both groups were given post-installation training. Table 14.11 shows the type of training received by employees in the successful and unsuccessful groups. There was a marked difference in the type and extent of training provided both pre- and post-installation of the new system between the successful and unsuccessful groups. The differences were most noticeable in the training received before the introduction of cook–chill catering. Although the use of manufacturers' training packages was similar in both groups, the successful group showed a tendency to extend training to include other types. For example, a substantial proportion of chefs (9 chefs, 60%) who received pre-installation training also gained experience from visits to other cook–chill catering units either to work with or obsèrve another cook–chill system in operation. In contrast, of those chefs receiving pre-instal-

TABLE 14.11
Type of Training Received by Chefs in the Successful and Unsuccessful Groups in the UK Cook–Chill Survey

Type of training	*Successful group*		*Unsuccessful group*	
	No. of units	*%*	*No. of units*	*%*
Pre-cook–chill				
Visits elsewhere	9	56·3	2	15·4
Manufacturer's training	10	62·5	4	30·8
Induction/formal	3	18·8	0	0·0
Literature	9	56·3	1	7·7
Films, etc.	5	31·3	1	7·7
Past experience	3	18·8	2	15·4
College course	2	12·5	1	7·7
Pre-cook–chill total	15	93·8	6	46·1
Post-cook–chill				
On-the-job	11	68·8	2	15·4
Observation only	5	31·3	3	23·1
No training	0	0·0	2	15·4
Total interviewed	16	100·0	13	100·0

lation training in the unsuccessful group, only two chefs (33·3%) had the opportunity to train and gain experience from other cook–chill units.

In addition 60% of chefs in the successful group who received pre-installation training received key literature (including the DHSS guidelines (1980) and handouts produced by their own units). In contrast, only 16·6% of the chefs in the unsuccessful group who had any pre-installation training received this type of literature. This constitutes one of the most noticeable differences between the two groups in terms of training.

In contrast to the chefs, the cooks and kitchen assistants received a minimal amount of training prior to the new system being installed. However, the successful group again gave a higher percentage of their cooks and kitchen assistants pre-installation training (36·8% of cooks and 30% of kitchen assistants interviewed) compared with the unsuccessful group (33·3 and 11·1% for cooks and kitchen assistants respectively). The majority of training took place after the system had been installed (see Table 14.11). This was apparent particularly with regard to cooks and kitchen assistants in the successful group and the cooks in the unsuccessful group.

The majority of post-installation training was classed as either 'on-the-job' or learning through 'looking-on'. The major difference between these two was that the former involved a high degree of employee participation, whereas the latter involved only observation. In the successful group there was a greater tendency to utilise on-the-job training rather than looking-on. Furthermore, a higher percentage of employees in each occupational class in the successful group (chefs 68·8%; cooks 52·6%; and kitchen assistants 80%) received on-the-job training than those in the unsuccessful group (chefs 15·4%; cooks 33·3%).

However, the unsuccessful group relied more on observation as a means of post-installation training for each level of staff (chefs 23·1%; cooks 44·1%; and kitchen assistants 100%). Even though the unsuccessful group depended on this type of training, its use was lower than in the successful groups where 31·3% of chefs and 52·6% of cooks learnt by observation. Only one employee in the whole of the successful group claimed to have had no formal training compared with three (13%) in the unsuccessful group.

In summary, these findings suggest that the majority of pre-installation training in successful operations concentrated on chefs. The main methods used were equipment manufacturer's training packages, visits to other units and cook–chill specific literature. Amongst the unsuccessful units there was little evidence of pre-installation training and what there was concentrated solely on manufacturer's training schemes. With regard to post-installation training a greater emphasis was placed on on-the-job training rather than observation for all levels of staff in the successful group, with little difference in emphasis for the unsuccessful group. These data strongly suggest, therefore, that it is the type of training rather than training *per se* which was crucial to eventual success. It is also clear from these results that training must be tailored to the requirements of particular organisations.

OVERALL EFFECTS ON THE LABOUR-FORCE

The introduction of technological systems can result in new or altered jobs for existing employees. This raises the question of whether employees acquire new skills (reskilling) or are deskilled as a consequence of the introduction of technology. Evidence on cook–chill catering systems suggests that some employees experience an increase in skills and others a decrease. For example, chefs at the central production unit (see Chapter 4) are likely to increase their skill level by becoming involved in additional 'technician' status activities such as temperature control of food products. Whereas, workers at satellite units are more likely to experience a reduction in skills, particularly traditional craft skills.

Qualitative results from the UK Cook–Chill Survey show that there was a range of opinion on the effects of introducing cook–chill catering on skill level. The majority of catering managers interviewed in the successful group (13 units, 51·1%) felt that there was an overall increase in skill level after the introduction of the new system (see Table 14.12). Some aspects of these include, new knowledge, production and temperature control, recipe development and more time to apply catering skills.

In contrast, a narrow majority of catering managers from the unsuccessful group thought that overall skill levels had remained unchanged (10 units, 45·5%). However, a substantial proportion (9 units, 40·9%) said they thought overall skill levels had increased (see Table 14.12). The higher incidence of respondents in the successful group who stated an increase in skill requirement was probably linked to a greater use of temperature and microbiological control by employees in the successful group. The successful group were also more likely to have employed or retained operatives with a greater skill level at the regeneration unit than the un-

TABLE 14.12
Change in Skill Requirements as a Result of the Introduction of Cook–Chill

Change in skill level	*Successful group*		*Unsuccessful group*	
	No. of units	*%*	*No. of units*	*%*
Higher	13	59·1	9	40·9
Lower	5	22·7	3	13·6
Same	4	18·2	10	45·5
Total	22	100·0	22	100·0

successful group. A common response from managers of the unsuccessful group was that there was a skills increase at the production unit and a skills decrease at the regeneration end. For example, some managers stated that they thought there was an increased need for skills at the central production unit whilst at the regeneration end less skills were required.

Some workers in cook–chill units investigated by the UK Cook–Chill Survey expressed the view that the introduction of cook–chill catering had reduced their previously relatively exciting jobs (referring to the activity immediately prior to meal times) into boring, repetitive tasks with little chance of variety and use of imagination. However, the most prominent feeling was one of isolation from customers at the point of service. More often than not, service was divorced from production not only physically but also temporally, i.e. customers were receiving food which had been produced as much as five days previously.

IMPLICATIONS FOR CUSTOMERS

The tripartite relationship between managers, operative and customers has already been discussed. Any worker dissatisfaction with new technology will tend to affect their behaviour and thus has the potential to impinge on their relationships with the customer.

This situation may be even more exaggerated where the cook–chill operative is not the deliverer of the food. For example, many hospitals in the UK utilise nursing staff to give patients their meals. It has been noted on a number of occasions that less than willing nursing staff have the ability to influence a patients acceptability of the product. Some examples were given by managers in the UK Cook–Chill Survey who had overheard nursing staff making negative comments regarding products directly to the patient concerned. Obviously, this has implications for training a wider network of employees than those directly employed within the catering departments in hospitals, as these personnel provide the link with the consumer.

A second example is that of a catering outlet where before the introduction of cook–chill catering the food is served by personnel, whereas after its introduction, vending machines are used by customers to receive chilled full-meals which they subsequently regenerate with a microwave. This type of service is used in a number of situations including industrial sites and for some forms of service in hotels. The effect of this change on consumers is enormous as they not only have to cope with a new technology, but they also lose out on the social rapport often associated with cafeteria and canteen staff.

Another important consideration is whether customers should be informed of any decision to change to cook–chill. Opinion was divided into two distinct groups by managers in the 80 units visited in the UK Cook–Chill Survey regarding the relative merits of informing consumers. One group felt that the customer 'has a right to know' what is happening to the food being served to them. Whilst the second had the attitude that 'what they don't know doesn't hurt them'. However, amongst most managers there was a consistent effort to alert customers to the use of cook–chill catering. To a large extent this was decided by sector influence. Caterers with more captive markets (see earlier) were more likely to inform customers than those with more casual clientele.

The level of satisfaction with the food received is probably one of the most informative measures as to the success of a catering operation. If the food does not reach the expectations of the customer, it is likely to be rejected and return custom becomes unlikely. In general, the comments of consumers in the successful group of cook–chill units in the UK Cook–Chill Survey were favourable:

> Excellent choice and tasty (industrial unit) . . . A lovely choice everyday (industrial unit) . . . Quick and reasonably priced . . . A fantastic change, really excellent meals brought in with this new system . . . Good choice of menu and food well cooked. Seldom any cause for complaint . . . We would like the canteen to stay open later!.

However, there were some problems in the unsuccessful group:

> The food is unimaginative and there is a shortage of staff . . . Standards are poor . . . food is usually cold . . . standard of food drops at weekends . . . It does not compare to the food we were used to before [cook–chill] . . . Inconsistent quality.

In short the successful group enjoyed a relatively high level of consumer satisfaction compared with the unsuccessful group.

SUMMARY

It is clear that in any technological change process there is a conflict between the introduction of new technology and the interests of the people concerned whether they are managers, operatives or consumers. The personnel involved in such a change have the potential to affect the eventual success of the operation. Thus, it is essential that they are given careful consideration before any change is instigated. For example, good communication, consultation and negotiation networks should be set up, which provides for an open exchange of views and advice. New contractual agreements often need to be decided and above all appropriate training schedules need to be followed. If this element of the transfer of technology is forgotten or mis-handled, then the impact of the technology could be negated or even turned against the business.

FURTHER READING

Burns, T. & Stalker, G. M. (1968) *The Management of Innovation*, London, Tavistock Publications Ltd.

Byrne, D. (1985) *Wage Councils: The Evidence*. Low Pay Review, 22, Low Pay Unit.

Byrne, D. (1986) *Waiting for Change? Working in the Hotel and Catering Industry*. Low Pay Pamphlet No. 42, Low Pay Unit, London, the Hotel Catering Workers Union, Surrey.

Department of Employment (1988) *Labour Market Quarterly Report*, October.

Duncan, C. (1981) *Low Pay—Its Causes and the Post War Trade Union Response*, Chichester, Research Studies Press, John Wiley and Son.

Greater London Council (1985) *The London Industrial Strategy (The Food Industry)*, GLC, London, 67–94.

Hotel and Catering Training Board (HCTB) (1988) *New Employment Forecasts, Hotel and Catering Industry 1988–1993*. Research Unit, HCTB/MMD, London.

Littler, D. & Pearson, A. (1972) Uncertainty and Technological Innovation. *Management Decisions,* **10,** 111–16.

Mitchell, P. (1988) The structure of labour markets in the hotel and catering industry: what do employment law cases indicate? *Service Industries Journal*, **8**(4).

Pine, R. (1987) *Management of Technological Change in the Catering Industry*, Aldershot, Gower Publishing.

Ryan, D. (1979) Technological change in the tertiary sector. *Work and People,* **5,** 13–18.

Woodward, J. (1968) Resistance to change. *Management Int. Rev.*, **8** (4–5), 137–43.

Worland, D. & Wilson, K. (1988) Employment and labour costs in the hospitality industry: evidence from Victoria, Australia. *Int. J. Hospitality Management,* **7** (4), 363–77.

CHAPTER 15

Benefits, Concerns and the Future of Cook–Chill

INTRODUCTION

The introduction of a cook–chill catering system can provide substantial benefits to the caterer but, as we have seen in previous chapters, its introduction can sometimes give cause for concern. There are many reasons why the process of implementing a new system can be troublesome and this is common to change, not only in catering, but in any industry. Research in other industries implementing change has highlighted instances where this has been the case. For example, in manufacturing the introduction of computer numerically controlled machine tools to carry out work previously done by skilled craftsmen caused problems with the work-force in terms of acceptability, changes in skill level and loss of job satisfaction. In addition, the capabilities of the technology itself were found to be lacking. In particular, the automatic computer controls failed to take into account that the boring tool suffered substantial wear after continued use. This need had been overlooked during the decision-making phase and hence caused problems after start-up. Similarly, the introduction of computerised office systems in the service sector brought problems for secretarial staff through boredom, repetitiveness and loss of skills. Most people when faced with change have a tendency to resist that change for fear of the consequences.

This chapter seeks to summarise the main advantages which have been discussed throughout this book as well as some of the more common problems experienced by caterers who promote change in their organisation through the introduction of a cook–chill catering system. In addition, it attempts to distinguish between those problems experienced by users which were solved relatively easily and those issues which gave real cause for concern. This will lead us towards the formation of a 'blueprint' for future caterers involved with the introduction of a cook–chill catering system, based on the experiences of users. Finally, we will assess the future of cook–chill systems in the catering industry.

PRACTICAL BENEFITS OF A COOK–CHILL CATERING SYSTEM

As we discussed in Chapter 1, there are many advantages associated with the introduction of a cook–chill catering system. For example, in functional terms, the introduction of cook–chill catering allows for easier control of production through the elimination of periods of high activity, just prior to meal times, allowing more efficient usage of equipment, staff and staff skills. In addition, food production can be incorporated within eight hour shifts, five days a week, thus reducing costs on labour for overtime and weekend work. A concentration of activities and skills in a central production unit means that higher productivity can be achieved. Moreover, a smaller floor area required at the satellite units allows food to be regenerated and served in units which would otherwise be unable to support a catering service through lack of space. The ability to regenerate menu items at satellite units allows for improved end-product quality by the elimination of warm-holding.

From a technical viewpoint, the introduction of a systemised process of catering provides a clear route for documenting all the activities involved in the process and thereby introducing a systematic quality assurance programme such as HACCP (see Chapter 9). It also allows the use of a more extensive range of menu items than, for example, its sister system cook–freeze catering. In addition, energy savings are possible through efficient and even usage of equipment throughout the day coupled with a reduction in the need for a full range of equipment at satellite units.

There are also important social benefits which result from the introduction of a cook–chill catering system. For example, the introduction of a time buffer reduces the stress of high pressure work close to meal times. More importantly, staff absenteeism in the CPU will not cause too many problems for the production schedule. In contrast, a similar situation in a conventional catering system would cause substantial problems for the kitchen team.

There are also benefits at management level after the introduction of a cook–chill catering system. For example, the method provides an infrastructure for greater control over the whole production process because of its more systematic base. This enables a more rationalised approach to the catering function, offers greater potential for control of the whole process and supports hygiene control mechanisms such as HACCP.

Obviously, not all caterers introducing a cook–chill catering system will experience all the advantages that the system potentially can offer. In some cases, any improvements are dependent on the catering system in use before the introduction of the new method. For example, in catering establishments where the previous system was some form of hot transportation, the detrimental effects of prolonged storage at high temperatures on food products can be improved substantially by the introduction of a cook–chill catering system. In such cases, there is an elimination of the need for hot transportation and hence a better quality end-product results as the deleterious effects of warm-holding are no longer experienced. On the other hand, if a cook–chill catering system replaces a previously well-thought-of conventional service, then it may be seen by staff and consumers as a retrograde step.

Advantages may be determined by the type of cook–chill catering system introduced. For example, the introduction of a cook–chill vending system can dispense meals to customers where previously catering facilities were not available. Typically, this occurs in staff feeding situations where conventional systems prove costly and impractical.

Clearly, there are some wide-ranging advantages possible through the introduction of cook–chill catering and many others besides those mentioned here have been experienced. However, alongside the advantages there have been a number of problems voiced by users and critics of cook–chill catering systems. The most commonly cited ones are described in the next section.

PERCEIVED PROBLEMS LINKED WITH THE COOK–CHILL CATERING PROCESS

As we have seen in previous chapters, the introduction of a cook–chill catering system is not always problem-free. The results from the UK Cook–Chill Survey have shown that a range of problems can and do arise during the technology transfer process. It is inevitable in any radical change that problems will occur. Many of these will be relatively trivial (i.e. 'teething problems') but a few may be more fundamental. As we shall see later, it is possible to lay down broad guidelines for successful implementation of new technologies but perhaps one of the most important keys to a unit becoming successful is that management must be able to anticipate potential problems and, if problems do occur, these must be quickly and adequately overcome.

Some difficulties or inadequacies which cause problems are relatively easy to rectify, although much depends on the circumstances within the individual organisations concerned. For example, a need to more accurately control temperatures appears to have a simple solution in terms of merely purchasing the relevant control systems. However, it can become a problem if the organisation does not have the resources available to fund such a purchase, especially in situations where the decision to introduce cook–chill catering has been based purely on achieving substantial cost savings.

It has only been fairly recently that the problems associated with cook–chill catering systems have been brought to light. Undoubtedly, this is a result of the surge of interest in cook–chill catering systems in the 1980s which has seen the application of the method in very different catering situations across all sectors of the industry. In addition, the early advocates of cook–chill catering set up their systems introduction on the basis of very little background research. In practice, to succeed in implementing a problem-free cook–chill catering function requires dedicated attention to detail and the aim to establish a highly systemised, strictly controlled and monitored catering operation backed up by sound research and development.

Evidence has shown that any problems experienced with the transfer of cook–chill catering technology in real situations can be roughly divided into three basic categories. Those which are primarily 'functional', those which are 'technical' and those which are 'social' problems. Whilst it is important to consider all of the problems experienced, the areas of most concern have been linked to inadequate training, insufficient or inappropriate allocation of funds, poor customer perception of food products, confusion of operatives working in dual cook–chill/cook–serve catering systems, the fact that the practice of warm-holding still persists and a lack of understanding of the information concerning the microbiological aspects of the system.

In particular, the UK Cook–Chill Survey showed that difficulties arose in new operations where relevant information was difficult to find or unobtainable. Very often, staff found the information available was difficult to understand. It was also laborious for managers to reproduce information in a form suitable for all members of staff to fully grasp the implications for safety procedures in operating cook–chill catering systems. In particular, information on microbiological aspects of food in cook–chill catering situations was difficult to find and that which was available was either too superficial or, at the other extreme, too technical. Over half the managers interviewed said that there was a critical lack of suitable information available to them with regard to microbiological safety. Others said that greater emphasis should be placed on microbiological aspects of cook–chill systems in training courses, trade magazines and any other forum for disseminating material to catering employees.

What was perhaps more alarming in the findings from this survey was the fact that some managers were unaware of the need to seek out further information relating to the microbiological safety of cook–chill units. However, the majority of managers agreed that a basic knowledge of food microbiology was necessary for all users of cook–chill catering, and thus, the need to gain such knowledge prior to installation of a new system was deemed to be of the utmost importance.

OPERATIVE-PERCEIVED PROBLEMS

A frequent occurrence in many cases of technological change is that the decision-makers are not the actual users or implementers of the technological system in question. Thus, a new system is often planned without input from those who will be involved in the practical implementation and day to day running of the system. This section reports the findings of the UK Cook–Chill Survey with respect to the problems perceived by the operatives as a result of the introduction of cook–chill catering into their respective organisations. Many of the problems voiced by the operatives were acknowledged by their respective managers, but, not unexpectedly, the operatives paid greater attention to detailed practical aspects relating to the problems experienced.

Typically, operative-cited problems were related to production schedules, physical layout of food for chilling, specialist equipment, regeneration techniques and the operation of two different catering systems simultaneously. Overall, there was a greater percentage of employees in the unsuccessful group reporting problems in working with the technological change, than in the successful group. More specifically, the employees from the unsuccessful group had problems which were predominantly related to regeneration techniques whereas few of the employees in the successful group reported such difficulties. Most of the problems had been overcome by employees in the successful group at the time of interview whereas employees in the unsuccessful group were still searching for solutions. This was, however, not necessarily a reflection on the quality of the staff. Rather, it was a result of a lack of resources or poor quality management. In particular, the differences manifested between the two groups with regard to information gathering and employee training during the initiation phase had a major impact on the eventual success of the system and the ability to cope with and overcome subsequent problems. An outstanding feature from interviews with employees was that more operatives in the successful group thought that there were no real problems with cook–chill catering itself (chefs, 46·7%; cooks, 63·2%; kitchen assistants, 60%) in comparison to the unsuccessful group (chefs, 18·2%; cooks, 12·5%; kitchen assistants, 0%). For example, operatives from the successful group felt that cook–chill catering was advantageous in terms of its potential to enable more time for concentrating effort on attractive presentation and in its flexibility to cope with sudden changes in demand through food being prepared for the customer 2–3 days in advance, thus, ensuring that further food products could be immediately dispatched from a buffer stock. However, some of the responses from operatives in the unsuccessful group indicated the potential for abuse of the system. In particular, a lack of specialist equipment and having to 'make do' with equipment not technically competent for the job required, e.g. chillers doubling up as extra refrigerators, resulting in poor stock control and unnecessary queues of food waiting to enter the chillers. The unsuccessful group had severe problems with inferior end-product quality, which need not have been a problem had sufficient resources been invested in research and development work and training both operatives and management.

The unsuccessful group had more practical problems associated with recipes and procedures than purely equipment-related problems. At the time of survey, this group were still experiencing difficulties solving these problems. The resultant atmosphere was not conducive to improving staff morale. This, in turn, was causing an escalation of industrial relations problems manifesting itself in increased labour turnover and lack of commitment to the cook–chill catering method.

Having considered some of the most common problems associated with the transfer of cook–chill catering technology, we will now examine the most important differences in the activities carried out by key decision-makers in those units in the UK Cook–Chill Survey who were classed as being successful and unsuccessful. This will enable us to evolve a 'blueprint' for caterers which should maximise the chances of ensuring the successful introduction of a cook–chill catering system.

TOWARDS A BLUEPRINT FOR SUCCESS IN COOK–CHILL CATERING

In-depth analysis of all stages of the transfer of cook–chill technology in the 80 unit samples of the UK Cook–Chill Survey revealed that there are a number of factors which clearly impact on the eventual success of the system. However, activities carried out prior to the implementation of the system appeared to be the most critical. In other words, it was the extent of work carried out during the decision-making phase which held the key to eventual success (see Chapter 11).

TABLE 15.1
Summary of the Main Differences in the Activities Carried Out by the Successful and Unsuccessful Units in the UK Cook–Chill Survey

Activities	
Successful group	*Unsuccessful group*
Carried out thorough background work	Carried out little background work
Training regarded as of paramount importance	Training regarded as an unnecessary expense
Undertaken careful equipment selection	Equipment selection influenced by glossy sales packages
Research and development orientation	No research and development work
Continual evaluation and improvement policy	Introduction seen as a one-off activity
Regular consultation with employees and trade unions	Consultation with employees and trade unions regarded as unnecessary
Strict adherence to the technical and safety requirements of the system	Unawareness of the technical and safety requirements of the system
Stable management force	High management turnover

A close look at the activities carried out in units which fell into the successful and unsuccessful group respectively reveals a range of surprisingly simplistic differences (see Table 15.1). Indeed, most of these appear to be a matter of commonsense. Unfortunately, it must be remembered that sometimes financial constraints are placed upon the innovator such that commonsense may need to be sacrificed. Thus, the differences between the two groups tended to reflect the amount of time and resources which were allocated to the introduction of the system.

The first, essential point is that the UK Cook Chill Survey showed that if the reasons for introducing cook–chill catering are purely financial, the system is likely to fail. Quite simply, this is because development, equipment purchase and support mechanisms for any cook–chill catering system must all be costed into the system and cannot be short-cut. In the UK Cook–Chill Survey, those organisations which had shown total financial commitment had been rewarded with an excellent system of catering.

Success was associated with a policy of thorough background research including a detailed study of a wide range of information sources such as equipment manufacturers, articles in the trade press and relevant publications (particularly those which detailed case studies of similar organisations), other cook chill users, seminars and the DHSS guidelines. Managers of the most successful units found that other users of cook–chill catering systems similar to their own were an extremely useful source of information. Likewise, the DHSS guidelines were well used and adhered to. This was not the case in the unsuccessful group where other users, if visisted, were organisations dissimilar to their own and very few had made any use of the DHSS guidelines.

In short, the successful group were more thorough and carried out more research on the introduction of a cook–chill system than the unsuccessful group.

There was some evidence of an optimum decision time needed in order to assess the true valuc of introducing a cook–chill system into individual operations. However, this was intrinsically linked to thorough in-depth development work. For example, during the decison-making period, the successful group spent more time researching into the type and amount of equipment necessary for the smooth running of their cook–chill system. This included investigating the capabilities of various pieces of equipment on the market in order to assess whether the equipment could adequately fulfil their needs and whether it was priced reasonably compared with other similar models on the market. In addition, visists were made to other organisations already in possession of identical equipment to decide on the relative merits of the different equipment available.

The successful group invested more time and resources in producing meal items specifically tailored to their individual requirements and were, therefore, much more orientated towards research and development than the unsuccessful group.

The successful group ensured effective communication with all levels of employees throughout the decision-making process and, after implementation, consulted employees on future developments and improvements. This included involving recognised trade unions in their discussions which, in turn, enabled a more acceptable approach to the inevitable staff reductions.

The successful group placed a heavy emphasis on providing effective training for all grades of staff both before and after implementation of the new system. The initial training involved background information on the principles of the system and detailed training on safety aspects in particular. In addition, training courses were given at existing cook–chill units to allow experienced cook–chill operatives to transfer their knowledge and skills to new users. Training after introduction consisted of participative on-the-job training, both at the CPU and at satellite units. This training was on-going and incorporated any developments which took place. This policy of effective communication in successful organisations was also evident with regard to consumers who were involved on the basis of a firm strategy of consumer consultation. In contrast the unsuccessful group again relied on more haphazard techniques of communication.

In short, many activities were involved in the build up to implementation of the system in the 80 units visited in the survey. Analysis showed that there were some activities which had little effect on success, whilst others were seen to impact, quite dramatically, on success. It is these latter factors which we shall use in formulating a 'blueprint' for successful introduction of a new cook–chill catering system.

THE BLUEPRINT

From the results of the UK Cook–Chill Survey and the experiences of other users, it is possible to suggest the following blueprint which, if adhered to, should maximise the chances of introducing a successful cook–chill catering unit.

- Carry out thorough and systematic background work before any decision to use cook–chill catering is made. Consult a range of sources in carrying out this work.
- Select equipment tailored to the requirements of the individual organisation in terms of size, selected menu cycle and storage time.
- Do not be lured by a package from just one equipment manufacturer but ensure a method of comparison on order to select the pieces suited to the requirements of the individual system.
- Alter premises in anticipation of the needs of the new system rather than installing the system and then discovering the need to incorporate changes in the superstructure.
- Ensure that in-depth product research and development is carried out in advance of implementation, if possible, and an ongoing commitment after installation.
- Implement effective communication strategies between managers, employees, trade unions and, if necessary, consumers.
- Ensure the adequate provision of pre- and post-implementation training for all levels of staff at CPU and satellite kitchens.
- Ensure that consumer involvement is on a systematic rather than haphazard basis.
- Ensure that efficient temperature controls and monitoring equipment are purchased and utilised in accordance with their design, the requirements of cook–chill catering guidelines and so that their use can be incorporated in an adequate quality assurance system.
- Ensure efficient and effective microbiological controls are in place as part of a complete quality assurance system.
- Ensure that a sensitive approach to employment reduction is established should this be necessary.

COOK–CHILL CATERING SYSTEMS AND THE FUTURE

The use of cook–chill systems in catering is now well established worldwide. Its use in Europe and the USA is well known but recent developments have seen its introduction across the globe to Hong Kong, Singapore, Australia and New Zealand, with the use of *sous-vide* and other similar systems following suit. It is anticipated that cook–chill catering will continue to increase world-wide, despite the recent controversy surrounding the system in the UK. This is not to ignore the recent concerns regarding the safety of the system. Such concerns are real and of paramount importance. The prevention of contamination of food products with food poisoning and/or food spoilage bacteria should be uppermost in the mind of each cook–chill caterer.

The future will see a greater emphasis on the use of quality assurance systems such as HACCP (Chapter 9) which it is hoped will eventually be incorporated by all cook–chill catering users as a systematic method for controlling hygiene. Some companies have already instigated the use of the technqiue. For example, at least one major airline already uses HACCP to control their cook–chill operation, as do a number of other large cook–chill users.

There will be more large scale CPU developments, particularly in the health sector as regional health authorities join together and produce total systems. In addition there will be an increase in the number of units selling excess produce to outside organisations and hence an increased number of satellite units 'buying-in' cooked, chilled products from other cook–chill catering organisations.

There have been many calls from scientists, caterers and educationalists for the UK government to introduce legislation or some form of licensing system for all cook–chill users. The government have recently published an updated version of the UK cook–chill guidelines and have indicated commitment to ensure that all environmental health officers are properly trained in all aspects and limitations of the cook–chill system. It may be important, in the future, for governments to issue, renew or remove specific cook–chill or systems catering licenses by determining whether an operator is using the system properly and safely.

There will be a move from equipment which has been adapted for use in a cook–chill situation to production of more equipment specifically designed with cook–chill catering and its current limitations in mind. This will include specially designed equipment to cope with the needs of different situations, e.g. to deal with liquids. For example, gaseous nitrogen chillers for the rapid cooling of liquids, computer controlled combination ovens operating within extremely accurate temperature ranges with self-monitoring systems and sophisticated packaging machines some of which will be linking activities of pasteurisation, chilling and packaging. Such machinery can already be seen emerging into the market-place.

The use of vacuum packaging, modified atmosphere packing and cool-tank methods will increase. Over a third of all cook–chill catering users in the UK are already using some form of supplementary packaging and it is anticipated that new and more controlled techniques will become available. For example, in 1989 in the UK, a new, long shelf-life ambient storage temperature technique was introduced. In addition, irradiation and short time heat sterilisation may prove to be useful techniques to supplement cook–chill. In the case of irradiation, however, its use must be viewed with some caution because although it may appear effective in extending product shelf-life, the nature of chemical by-products resulting fom its use are by no means clear and much more research may be necessary before the majority of food producers are convinced of its overall safety.

The main developments with regard to social aspects relate to the education and training of personnel and the general public with regard to cook–chill catering systems. One of the problems that cook–chill caterers have had to face recently in the UK is the doubt expressed by some members of society as to the safety of cook–chill catering. Much of this has come about because of adverse publicity relating to cook–chill in the popular press. However, whilst this may have caused a temporary hiatus, chilled food continues its growth in the market. It seems likely, therefore, that as long as cook–chill caterers ensure their adherence to approved government guidelines, its future looks healthy. There is no doubt, however, that there remains a real need for more training courses for all levels of staff which give a thorough grounding in cook–chill catering, use of equipment, temperature control, stock control, quality assurance and other key issues. As in so many walks of life, education and a thorough understanding and appreciation of the cook–chill catering system must remain the primary key to the successful use of this exciting and important method.

Appendix 1

A list of menu items used in cook–chill catering in the UK (personal communication from A. Lewis). Although representative and reasonably comprehensive, this list is not thought to be exhaustive. Frequencies represent the number of times dishes were mentioned in a survey of about 100 UK cook–chill units in 1988.

MEAT AND FISH PRODUCTS

Dish	*Frequency*
Bacon burger	1
Baked gammon	1
Baked herring	1
Barbequed pork slices	1
Beef	9
Beef burger	7
Beef sausages	1
Black pudding	1
Boiled beef	1
Braised lamb & peaches	1
Braised liver	2
Braised liver & bacon	1
Braised sausages	2
Braised silverside	2
Braised steak garni	1
Breaded plaice fillet	1
Chicken	17
Cod	5
Cod Acorienne	1
Cod fish fingers	5
Cod in breadcrumbs	1
Cold silverside of beef	1
Cooked ham	9
Cooked tongue	5
Cooked tuna	12
Corned beef	12
Deep fried cod in breadcrumbs	2
Diced barbeque pork	1
Duck	1
Farmhouse paté	5
Fillet steak	1
Fish	5
Fried cod in batter	2
Garlic sausage	3
Grilled bacon	2
Grilled gammon	8
Grilled lamb chop	2
Grilled pork chop	3
Grilled sausages	4
Grilled sole	1
Grilled steak	1
Grilled steaklet	2
Grilled trout	1
Herring	1
Herring in oatmeal	1
Kidney	1
Lamb	3
Lamb chops	3
Lamb cutlets	1
Luncheon meat	1
Minced beef	7
Minced lamb	6
Minced pork	1
Pilchard	3
Plaice—country style	1
Plaice—breaded & lemon	2
Poached cod	7
Poached plaice	1

Dish	*Frequency*
Rainbow trout	1
Roast beef	9
Roast brisket of beef	1
Roast chicken	5
Roast ham	1
Roast lamb	4
Roast leg of pork	4
Roast pork	8
Roast topside of beef	1
Roast turkey	4
Salami	1
Sardines	4
Sausages	2
Sauté liver	1
Savoury mince	1
Sirloin steak	2
Smoked haddock	1
Smoked mackerel	9
Smoked trout	1
Stewed steak	1
Tinned salmon	1
Tongue	1
Turkey	5

COMPOSITE DISHES

Dish	*Frequency*
Barbeque king rib	1
Barbeque pork casserole	1
Bean & potato pie	1
Bean & vegetable paella	1
Beef & bean hotpot	1
Beef & cider casserole	1
Beef & mushroom casserole	1
Beef & vegetable casserole	1
Beef casserole	1
Beef curry	1
Beef goulash	1
Beef lasagne	1
Beef stroganoff	1
Beef tajine with rice	1
Black eye bean curry	1
Black eye bean dahl	1
Bolognaise	6
Boston pork & beans	1
Canelloni milanaise	1
Carribean steamed cod	1
Cauliflower cheese	5
Celery & cashew nut rissole	1
Cheese & broccoli flan	1
Cheese & mushroom flan	1
Cheese & onion flan	4
Cheese & potato pie with breadcrumb topping	1
Cheese & rice burger	1
Cheese & spinach ravioli	1
Cheese & tomato pizza	3
Cheese burger with tomato	1
Cheese cutlet with fried egg	1
Cheese flan	1
Cheese flan Mexicaine	1
Cheese onion & mushroom slice	1
Cheese scotch egg	1
Cheshire lamb crumble	1
Chick pea & potato curry	1
Chicken & ham pie	2
Chicken & potato pie	1
Chicken Bretonne	1
Chicken casserole	2
Chicken chausseur	3
Chicken fricassée	1
Chicken in lemon sauce	1
Chicken korma	1
Chicken Maryland	1
Chicken masala	1
Chicken nuggets	1
Chicken pie	1
Chicken Portugaise	1
Chicken risotto	1
Chilli beans	1
Chilli burgers	1
Chilli cassoulet	1
Chilli con carne	3
Chinese chicken	1
Chinese range ribs	1
Chinese spring rolls	1
Corn korma	1
Corned beef pie	1
Cornish pasty	4
Cottage cheese & pineapple	1
Cottage pie	7

Dish	*Frequency*
Country vegetable lasagne	1
Curried beef	1
Curried chicken	2
Curried eggs	1
Curried pork	1
Curried prawns	1
Deep fried cod in breadcrumbs	1
Devilled chicken	1
Devilled kidneys	1
Egg & bacon flan	1
Egg & nut cutlet	1
Egg & tomato flan	1
Faggots	1
Fish cakes	3
Fisherman's pie	1
Gala pie	3
Gammon slice with madeira sauce	1
Goujon of plaice & lemon	1
Grilled forespur & egg	1
Grosvenor pie	1
Ham & apple casserole	1
Ham omelette	1
Haricot bean casserole	1
Haricot bean stew	1
Homemade meatloaf	1
Lamb & lentil casserole	1
Lamb & mushroom casserole	1
Lamb & potato pie	2
Lamb chop casserole	1
Lamb chop reforme	1
Lamb curry	2
Lamb hotpot	2
Lamb madras	1
Lamb ratatouille	1
Lamb rissoles	1
Lamb stew & dumplings	1
Lamb tangerine	1
Lancaster hot pot	3
Lasagne	2
Leek & cauliflower cheese	1
Leek & tomato flan	2
Leek, tomato & courgette pie	1
Lentil & vegetable pasty	1
Lentil cottage pie	1
Lentil curry	1
Lentil moussaka	1
Lentil pie	1
Macaroni & mushroom bake	1
Macaroni in tomato sauce	1
Macaroni, tomato & chick pea	1
Meatballs in tomato sauce	1
Minced beef & mushroom hotpot	1
Minced beef & onion pie	1
Minced beef plait	1
Minced beef potato nests	1
Minced ham & egg fricassée	1
Minced lamb fricassée	2
Mincemeat pasty	1
Mixed bean & corn bake	1
Mixed farmhouse grill	1
Mixed vegetable curry	1
Moussaka	1
Mushroom & asparagus pizza	1
Mushroom & tomato flan	1
Mutton curry	1
Mutton lasagne	1
Nasi goreng	1
Noisettes of lamb perigourdine	1
Norfolk stew	1
Nut loaf	2
Onion bhajee	1
Oriental lemon chicken	1
Panir cutlets	1
Pilau rice & egg	1
Pork & mushroom casserole	1
Pork medallions	1
Pork pie	3
Pot roast of beef	1
Prawn palif	1
Quiche	3
Quiche lorraine	2
Ragout of lamb	1
Ravioli	2
Red lentil dhal	1
Risotto	1
Russian pie	1
Samasos	2
Sausage & vegetable stew	1
Sausage casserole	1
Sausage-in-a-hole	1

Dish	*Frequency*
Sausage rolls	3
Sauté beef & mushrooms	1
Sauté chicken & sweetcorn	1
Savoury flan	1
Savoury stuffed sausage	1
Scampi with lemon	1
Scotch egg	3
Scotch mince pie	2
Scotch mutton pie	1
Scrambled egg	1
Seafood quiche	1
Shepherd's pie	4
Smoked fish florentine	1
Sole suchet	1
Spaghetti in tomato sauce	1
Spaghetti & bolognese	1
Spanish omelette	2
Spanish style pork	
Spinach lasagne	1
Spring lamb stew	1
Steak & kidney pie	4
Steak & vegetable casserole	1
Steak pie	1
Steamed eggs & pork	1
Stilton & celery flan	1
Stuffed peppers	1
Stuffed tomatoes	1
Surfburger	1
Sweet & sour pork	1
Sweet & sour turkey	1
Sweetcorn & mushroom quiche	1
Sweetcorn & celery fritter	1
Toad-in-the-hole	1
Tortillas	1
Tuna burger	1
Tuna fish finger rolls	
Tuna pie	2
Turkey casserole	
Turkey fritters	

VEGETABLES

Dish	*Frequency*
Baby carrots	10
Baked beans	17
Baked jacket potato	25
Baked parsnips	1
Boiled potatoes	10
Boulangere potatoes	1
Broad beans	6
Broccoli	10
Brussels sprouts	13
Butter beans	1
Buttered diced swede	1
Cabbage	15
Carrots	18
Cauliflower	17
Chinese stir fry vegetables	1
Chipped potatoes	23
Courgettes	5
Creamed potatoes	102
Croquette potatoes	6
Dauphinoise potatoes	1
Duchess potatoes	2
Fondant potatoes	1
Forestiere potatoes	1
Green pepper	1
Green sliced beans	20
Grilled tomatoes	16
Jacket potato	12
Kidney beans	1
Lyonnaise potatoes	1
Marquis potatoes	1
Minted potatoes	1
Mixed vegetables	21
Mushrooms	4
Nan bread	1
New potatoes	1
Onions	3
Parsley potatoes	33
Peas	35
Peas country style	1
Pilaf rice	1
Plum tomatoes	2
Potato scones	1
Ratatouille	2
Rice savoury	25
Roast parsnips	1
Roast potatoes	25
Runner beans	3
Sauté potatoes	11

Dish	*Frequency*
Savoury potatoes	2
Savoyarde potatoes	1
Scallop potatoes	1
Spinach	2
Swede	11
Sweetcorn	14
Turnip	2
Vegetable & chick pea curry	1
Vegetable & lentil pie	1
Vegetable & tomato curry	1
Vegetable casserole	1
Vegetable chilli	1
Vegetable curry	1
Vegetable fricassée	1
Vegetable goulash & pasta shells	1
Vegetable hot pot	2
Vegetable lasagne	2
Vegetable moussaka	1
Vegetable pancake & mushroom	1
Vegetable pancake mornay	1
Vegetable pancake roll	1
Vegetable pasty	1
Vegetable pie	1
Vegetable pizza	1
Vegetable quiche	1
Vegetable risotto	1
Vegetable sauté omelette	1
Vichy carrots	2
Viennese steaks	1
Wholegrain rice	6
Yellow split pea curry	1

SWEETS

Dish	*Frequency*
Almond slice	1
Apple & apricot turnover	1
Apple & blackberry crumble	1
Apple & strawberry pie	1
Apple charlotte	1
Apple crumble	2
Apple sultana tart	1
Apple tart	1
Apricot condé	1
Apricot crumble	2
Apricot fool	1
Apricot pie	1
Apricot sponge	1
Apricot tart	1
Baked apple	3
Baked egg custard	2
Baked jam roll	1
Baked jam tart	1
Baked rice pudding	13
Bakewell tart	3
Banana custard	3
Blackcurrant cheesecake	2
Bread & butter pudding	4
Bread pudding	1
Butterscotch whip	1
Cabinet pudding	1
Cheesecake	2
Cherry cheesecake	1
Cherry meringue	1
Cherry pie	1
Chocolate blancmange	2
Chocolate éclair	1
Chocolate mint mousse	1
Chocolate sponge pudding	1
Cinnamon apples	1
Cinnamon pears	1
Coconut & cherry sponge	1
Coconut & jam sponge	1
Coconut sponge	1
Coffee & chocolate gateau	1
Coffee mousse	1
Compote of fruit	1
Cream caramel	1
Creamed rice pudding	1
Danish pastry	1
Date & walnut slice	1
Devonshire slice	1
Dutch apple	
Dutch apple flan	2
Dutch apple tart	1
Egg custard tart	1
Eve's pudding	1
Forest fruit crumble	1
Fruit compote	1
Fruit crumble	2

Dish	*Frequency*
Fruit fool	1
Fruit jelly	2
Fruit pie	3
Fruit semolina	1
Fruit trifle	2
Gateaux	1
Ginger sponge	1
Gooseberry crumble	2
Gooseberry fool	2
Gooseberry pie	1
Ground rice	4
Jam & almond slice	1
Jam shortcake	1
Jam sponge	3
Jellied fruit	2
Lemon cheesecake	1
Lemon curd tart	1
Lemon meringue pie	8
Lemon milk jelly	1
Lemon sago	1
Lemon sponge	2
Lemon tapioca	1
Macaroni pudding	4
Manchester tart	3
Mandarin cheesecake	1
Mandarin flan	1
Mandarin slice	1
Marmalade pudding	1
Milk pudding	2
Mince meat tart	1
Orange & lemon jelly	1
Orange & lemon sponge	1
Orange egg custard	1
Orange gateau	1
Orange meringue flan	1
Peach condé	1
Peach flan	1
Peach milk jelly	1
Peach sponge	1
Pear flan	1
Pear jelly condé	1
Pineapple cheesecake	1
Pineapple condé	2
Pineapple crumble	1
Pineapple meringue pie	1
Pineapple sponge pudding	2
Pineapple upsidedown	1
Plum tart	1
Poached pears	1
Raspberry blancmange	1
Raspberry jelly trifle	1
Raspberry milk jelly	1
Raspberry mousse	2
Rhubarb compote	1
Rhubarb crumble	5
Rhubarb pie	1
Rhubarb sponge	1
Rum & peach pavlova	1
Sago pudding	3
Semolina pudding	9
Sherry trifle	2
Spotted dick	1
Steamed jam roll	1
Steamed jam sponge	1
Steamed sultana roll	1
Steamed sultana suet pudding	1
Stewed apples	3
Stewed pears	1
Stewed prunes	2
Strawberry blancmange	1
Strawberry fruit jelly	1
Strawberry jelly	1
Strawberry jelly trifle	1
Strawberry mousse	1
Strawberry whip	1
Stuffed baked apple	2
Sultana egg custard	1
Summer fruit crumble	1
Swedish apple	1
Sweet mince tart	1
Syrup & lemon sponge	1
Syrup sponge	2
Tapioca milk pudding	7
Treacle tart	1
Vanilla slice	1
Waffles	1

SOUPS AND OTHERS

Asparagus	2
Beef & corn	1

Dish	*Frequency*	*Dish*	*Frequency*
Beef & tomato	1	Kidney bean & corn salad	1
Beef & vegetable broth	1	Kidney bean side salad	2
Beetroot & bean salad	1	Leek	1
Beetroot & pepper salad	1	Lentil	3
Boiled noodles	1	Lentil & carrot broth	1
Californian side salad	1	Minestrone	6
Celery	3	Mixed salad	10
Celery & cucumber salad	1	Mushroom	1
Celery/watercress salad	1	Mushroom ball salad	1
Cheese salad	1	Oxtail	3
Chicken	1	Peanut, peach & yoghurt salad	2
Coleslaw	1	Potato & celery	1
Consommé	19	Potato salad	4
Cottage cheese	5	Prawn crackers	1
Cottage cheese & pineapple	2	Rice salad	2
Cream	1	Sage & onion stuffing	1
Cress salad	1	Salad	3
Cucumber & tomato salad		Scotch broth	3
Egg & cress salad	1	Spaghetti	5
Egg mayonnaise salad	2	Spring vegetable	1
Egg salad	2	Stuffing	1
French onion	4	Tomato	7
Grape & walnut salad	1	Tomato & cucumber salad	1
Green pea	1	Tomato & beef	1
Green salad	7	Vegetable	5
Ham salad	1	Waldorf salad	2
Indian rice salad	2	Yorkshire pudding	5

Appendix 2

EXAMPLES OF MENU ITEMS USED IN UK COOK–CHILL CATERING ESTABLISHMENTS

Information gained from customers in establishments in the UK Cook–Chill Catering Survey

Item	*Frequency*
Meat and fish dishes	
Bacon	7
Braised beef	1
Beef Julienne	1
Beef burger	1
Fried fish	10
Grilled plaice	3
Gammon steak	1
Honey ham	2
Lamb chops	1
Liver	5
Meat balls	2
Pork chop	2
Roast beef	7
Roast lamb	3
Roast chicken	5
Sausage	2
Sausage roll	6
Scampi	1
Smoked mackerel	1
Steak	1
Turkey steak	2
Veal	1
Composite products	
Beef curry	7
Beef stew	3
Beef and mushroom pie	1
Beef bourguignonne	1
Corned beef hotpot	1
Chicken curry	7
Chicken and ham pie	3
Chicken risotto	1
Cornish pasty	2
Chicken chasseur	1
Cheese and potato pie	1
Cauliflower cheese	1
Chicken pie	1
Cottage pie	1
Chilli con carne	1
Duck casserole	1
Lasagne	3
Lamb korma	2
Mushroom & turkey pie	2
Quiche	1
Ravioli	1
Sweet & sour pork	6
Seafood pancake	1
Steak & kidney pie	3
Spaghetti bolognaise	1
Tandoori chicken	1
Toad-in-the-hole	1
Vegetable curry	1
Vegetable dishes	
Baked beans	9
Butter beans	2
Beans	5

Item	*Frequency*	*Item*	*Frequency*
Boiled potatoes	1	Apple crumble	2
Broccoli	2	Apple & blackcurrant pie	1
Brussels sprouts	3	Bakewell tart	2
Carrots	12	Bilberry tart	1
Chipped potatoes	24	Bread & butter pudding	2
Cabbage	3	Cabinet pudding	1
Cauliflower	2	Cheesecake	2
Jacket potatoes	2	Cherry meringue	1
Lyonnaise potatoes	1	Cinnamon sponge	1
Marquis potatoes	1	Chocolate sponge	2
Mashed potatoes	7	Ginger sponge	1
Mixed vegetables	6	Gooseberry crumble	1
Mushy peas	1	Jam sponge	1
New potatoes	3	Jam roly poly	1
Peas	9	Lemon meringue pie	1
Roast parsnips	2	Madeira pudding	1
Roast potatoes	7	Pineapple upside down	1
Rice	11	Rice pudding	2
Sauté potatoes	3	Rhubarb pie	1
Swede	5	Rum baba	1
Sweetcorn	4	Sultana sponge	1
Sweet dishes		Treacle sponge	1
Apple pie	6	Trifle	2

Appendix 3

A compilation of the major references used in the research and preparation of this book. This list is not exhaustive as many references on the microbiology of cooked foods, chilling processes, chill technology and other areas have been excluded in order to restrict the list to those papers most pertinent to the caterer

Adams, I. (1985) Chilled foods: wider aspects and developments. In: *Proc. Chill Foods,* Chipping Campden, UK, Campden Food Preservation Research Association, pp. 99–112.

Ainsworth, P. & Shelten, A. (1981) Quality of cook–freeze and cook–chill food. *Hospitality*, June, 21–5.

Alcock, P. A. (1980) *Food Hygiene Manual*, London, H. K. Lewis & Co, Ltd.

Allen, T. J. (1966) Performance of information channels in the transfer of technology. *Ind. Management Rev.,* **8,** 87–98.

American Society of Heating, Refrigeration and Air Conditioning Engineers (ASHRAE) (1977) *Handbook of Fundamentals*, New York, ASHRAE.

Anon. (1970) Pressure cooking vacuum packs. *Food Processing Industry,* **39,** 460.

Anon. (1976) Wide acceptance of cook–chill except in UK. *Staff & Welfare Caterer*, May, 29–32.

Anon. (1982) Chilled saves cash for Social Services. *Chilled Foods*, Oct., 23.

Anon. (1982) Blast chilled boosts hospital hygiene. *Chilled Foods*, Aug/Sept., 27.

Anon. (1982) Developing new cook–chill operations. *Chilled Foods*, Apr., 36–8.

Anon. (1982) Meals on wheels 'down under'. *Chilled Foods*, Apr., 11.

Anon. (1982) Catering for change. *British Business*, 29th Jan., 206–10.

Anon. (1982) Cook–chill uncovered. *Chilled Foods,* **1** (1), 38, 39, 42.

Anon. (1983) Cook–chill fare to be introduced for first class rail travellers. *Caterer and Hotelkeeper*, 13th Oct., 16.

Anon. (1983) Blast chilling and refrigeration at the Mount Royal Hotel. *Caterer and Hotelkeeper*, 17th March, 71.

Anon. (1984) Commercial cook–chill comes to Ealing schools. *Frozen and Chilled Foods*, June, 2.

Anon. (1984) Cook–chill goes to school in Surrey. *Frozen and Chilled Foods,* Dec., 20.

Anon. (1985) Campden Food Preservation Research Association—Symposium on Chilled Foods, 7–9th May, 1985.

Anon. (1984) Update on the gas flush pack. *Food Manufacture,* **59** (3), 22, 23, 25, 65.

Anon. (1986) Cook–chill controversy. *Caterer and Hotelkeeper*, 21st Aug., 7.

Anon. (1987) Cook–chill approved. *Guardian*, 1st Oct.

Anon. (1987) Cooking in a vacuum. *Hospital Caterer*, Jan., 14.

Anon. (1987) Cook–chill system under fire. *Financial Times*, 1st Apr., special survey.
Anon. (1987) Cook–chill right from the start. *Hospital Caterer*, Apr., 17.
Anon. (1987) Report from the Public Health Laboratory Service Communicable Diseases Surveillance Centre. *Brit. Med. J.*, **295,** 1123–5.
Anon. (1987) Desk-top vacuum packing station within the reach of small caterers. *Catering*, Jan., 75.
Anon. (1987) Computerised cook–chill system. *Catering*, May, 117.
Anon. (1987) *Daily Telegraph*, 4th May, 8th July.
Anon. (1988) Microbiology criteria committee may consider *sous-vide* products. *Food Chemical News*, Sept.
Anon. (1988) Central production unit for Oxford. *Hospital Caterer*, Mar./Apr., 18.
Anon. (1988) Cook–chill: getting it right. *Hospital Caterer*, Jan./Feb., 7–8.
Anon. (1988) Listeria bugs cook–chill report. *Caterer and Hotelkeeper*, Sept., 6.
Anon. (1988) Cold comfort and cook–chill. *Hospital Caterer*, Mar./Apr., 10.
Anon. (1988) The *sous-vide* processing method barred to US restaurants. *Food Engineering*, April, 19–21.
Anon. (1988) Danepak enter ready meal business. *Food Trade Review,* **58,** 371–4.
Anon. (1989) Cook–chill 'doubled' Regis turnover. *Caterer and Hotelkeeper*, London Extra, Jan., 111.
Armstrong, R. (1986) Cook–chill catering. *Environ. Health,* **94** (10), 253–5, 257–8.
Armstrong, R. (1986) Problems and benefits of cook–chill catering systems. Paper presented at *The Future of Cook–Chill Technology in Catering*. Workshop, Dorset Institute of Higher Education, Poole, Dorset, UK, April.
Baker, N. R., Siegman, J. & Rubenstein, A. H. (1967) The effects of perceived needs and means on the generation of ideas for industrial research and development projects. *IEEE Trans. of Eng. Mgt.* **EM-14** (4).
Barnes, E. M., Impey, C. S. & Parry, R. T. (1973) The sampling of chickens, turkeys, ducks and game birds. In: *Sampling—Microbiological Monitoring of Environment* (eds Board, R. G. & Lovelock, D. W.), New York, Academic Press, pp. 96–125.
Barnett, H. J., Conrad, J. W. & Nelson, R. W. (1987) Use of laminated high and low density polyethylene flexible packaging to store trout (*Salmo gairdneri*) in modified atmosphere. *J. Food Prot.,* **50,** 645–51.
Basford, A. H. (1963) Public health aspects of food vending. *Roy. Soc. Health J.,* **83,** 103–10.
Bauman, H. E. (1974) The HACCP concept and microbiological hazard categories. *Food Tech.,* **28** (9), 30–4, 74.
Binstead, R. (1983) Why Battle was chosen as cook–chill launch pad. *Caterer and Hotelkeeper*, Jul., 49–57.
Bjorkman, A. & Delphin, K. A. (1966) Sweden's Nacka hospital food system centralises preparation and distribution. *Cornell Hotel and Restaurant Administration Quarterly,* **7,** 84–91.
Blackler, F. & Brown, C. (1986) Alternative models to guide the design and introduction of the new information technologies into work organisations. *Journal of Occupational Psychology,* **59,** 287–313.

Bobeng, B. J. & David, B. D. (1977) HACCP models for quality control of entrée food service systems. *J. Food. Prot.*, **40** (9), 632–8.
Bognar, A. (1980) Nutritive values of chilled meals. In: *Advances in Catering Technology* (ed. Glew, G.), London, Elsevier Applied Science Publishers, pp. 387–407.
Boltman, D. B. & Shaw, V. A. (1977) A computer simulation program for establishing equipment and storage requirements of a catering system. In: *Catering Equipment Systems and Design* (ed. Glew, G.), London, Applied Science Publishers, pp. 463–72.
Boltman, B. (1978) *Cook–Freeze Catering Systems,* London, Applied Science Publishers.
Boddy, D. & Buchanan, D. (1985) New technology with a human face. *Personnel Management*, Apr., 28–31.
Bone, C. (1979) A catering officers view of cook–chill. *Staff and Welfare Caterer*, Apr., 5–7.
Brecht, P. E. (1980) Use of controlled atmospheres to retard deterioration of produce. *Food Tech.*, **34,** 45–50.
Brown, A. (1984) What lessons are to be learned about school meals in Merton? *Caterer & Hotelkeeper*, 19th Jan., 69–75.
Brownridge, E. (1985) Dieticians enjoy test-run of re-thermalized breakfast. *Foodservice & Hospitality,* **18** (5), 29–30.
Bryan, F. L. (1974) Identification of food hazards in food service establishments. *J. Environ. Health*, **36,** 537–40.
Bryan, F. L. (1978) Factors that contribute to outbreaks of foodborne disease. *J. Food Prot.,* **41,** 816–27.
Bryan, F. L. (1981) HACCP approach to epidemiologic rationale and application in food service operations. *J. Environ. Health,* **44,** 7–14.
Bryan, F. L. (1981) Health hazards of food service establishments. *Food. Tech.,* **35,** 78–87.
Bryan, F. L. (1982) Foodborne disease risk assessment of food service establishments in a community. *J. Food Prot.,* **45** (1), 93–100.
Bryan, F. L. & Lyon, J. B. (1984) Critical control points of hospital food operations. *Food Prot.,* **47** (12), 950–63.
Bunch, W. L., Matthews, M. E. & Marth, E. H. (1976) Hospital chilled food service: acceptability and microbiological characteristics of beef soy loaves when processed according to systems procedures. *J. Food Sci.,* **41,** 1273–6.
Bunch, W. L., Matthews, M. E. & Marth, E. H. (1977) Fate of *Staphylococcus aureus* in beef soy loaves subjected to procedures used in hospital chill foodservice systems. *J. Food Sci.,* **42,** 565–6.
Burns, T. & Stalker, G. M. (1961) *The Management of Innovation*, London, Tavistock Publications Ltd.
Byrne, D. (ed.) (1986) *Waiting for Change? Working in the Hotel and Catering Industry*. Low pay pamphlet n. 42. Low Pay Unit, London, the Hotel Catering Workers Union, Surrey.
Campbell, D. (1987) Bug count slammed. *Caterer and Hotelkeeper*, 30th July, 54.
Carr, T. P. & Marchello, J. A. (1987) Growth of aerobic psychrotrophs and colour changes of precooked beef slices as affected by packaging procedure. *J. Food Prot.,* **50,** 733–6.

Chitty, G. (1984) Catering for 90,000 caterers with the help of cook–chill. *Caterer and Hotelkeeper*, Jan. (19), 179.

Ciobanu, A., Lascu, G., Bercescu, V. & Niculescu, L. (1976) *Cooling Technology in the Food Industry*, Tunbridge Wells, UK, Abacus Press.

Cooper, H. (1987) Experts defend cook–chill. *Hotel & Catering Technology*, July/Aug., 14.

Cowling, A. B. (1985) Social trends—their implications for chilled foods. In: *Proc. Chill Foods*, Chipping Campden, UK, Campden Food Preservation Research Association.

Cremer, M. L. & Chipley, J. R. (1977) Satellite food service system assessment in terms of time and temperature conditions and microbiological and sensory quality of spaghetti and chilli. *J. Food Sci.*, **42,** 225–9.

Cremer, M. L. & Chipley, J. R. (1979) Time and temperature, microbiological and sensory quality of meat loaf in a commissary food service system transporting heated food. *J. Food Sci.*, **44,** 317–21.

Cremer, M. L. & Chipley, J. R. (1980) Time and temperature, microbiological and sensory assessment of roast beef in a hospital food service system. *J. Food Sci.*, **45,** 1472–7.

Croft, J. L. (1987) Menu Hazard Analysis; its application to food hygiene in hospitals. *Caterer and Hotelkeeper*, July/Aug., 15–16.

Cummins, S. & Edwards, D. (1989) Cautious welcome for cook–chill guidelines. *Caterer and Hotelkeeper*, 30th June, 14.

Curtis, J. E. & Huskey, J. E. (1974) HACCP analysis in quality assurance. *Food Prod. Devel.*, **8** (3), 19, 24.

Dahl, C. & Matthews, M. (1978) Cook–chill foodservice systems: Microwave heating precooked portions of beef loaf. *J. Microwave Power,* **13** (1), 87–93.

Dahl, C. & Matthews, M. (1979) Hospital cook–chill foodservice systems. *J. Amer. Dietet. Assoc.*, **75,** 34–7.

Dahl, C. & Matthews, M. (1980) Effect of microwave heating in cook–chill foodservice systems. Some data on temperature, yield, moisture and fat content of typical foods. *J. Amer. Dietet. Assoc.*, **77,** 290–5.

Dahl, C., Matthews, M. & Marth, E. (1978) Cook–chill foodservice systems: microbiological quality of beef loaf at five process stages. *J. Food Prot.*, **41** (10), 788–93.

Dahl, C. A., Matthews, M. E. & Marth, E. H. (1980) Survival of *Staphylococcus aureus* in beef loaves, potatoes and frozen and canned green beans after microwave heating in a simulated cook/chill food service system. *J. Food Prot.*, **43,** 916–21.

Dahl, C., Matthews, M. & Marth, E. (1980) Cook–chill foodservice system with a microwave oven: aerobic plate counts from beef loaf, potatoes and frozen green beans. *Journal of Microwave Power,* **15** (2), 95–105.

Dahl, C., Matthews, M. & Lund, D. (1981) Effect of microwave heating in cook–chill foodservice system. Energy balance for heating meat loaf. *J. Amer. Dietet. Assoc.*, **79,** 296–301.

Dahl, C., Matthews, M. & Marth, E. (1981) Cook–chill food service system with a microwave oven: injured aerobic bacteria during food product flow. *Eur. J. Appl. Micro. Biotech.*, **11,** 125–30.

Dahl-Sawyer, C., Jen, J. & Huang, P. (1982) Cook–chill foodservice systems with conduction, convection and microwave reheat subsystems. Nutrient retention in beef loaf, potatoes and peas. *J. Food Sci.*, **47,** 1089–95.
Daniels, J. A., Krishnahmurthi, R. & Rizvi, S. S. H. (1985) A review of the effects of carbon dioxide on microbial growth and food quality. *J. Food Prot.*, **48** (6), 532–7.
Darvell, D. (1983) Cook–chill banquets for new generation Post House. *Caterer and Hotelkeeper*, Aug. (4), 28.
Davidson, C. M. & Webb, G. (1973) The behaviour of *Salmonella* in vacuum packaged cooked cured meat products. *Can. Inst. Food Sci. Technol. J.*, **6,** 41–4.
Dembo, V. (1979) Technology transfer planning. *Research & Development Management,* **9,** 3.
Demetriadi, J. (1986) Method *sous-vide. Hospitality*, July/Aug., 34.
Department of Health & Social Security (1980) *Guidelines on Precooked, Chilled Foods*, London, HMSO.
Department of Health & Social Security (1989) *Guidelines on Precooked, Frozen and Chilled Foods*, London, HMSO.
Douds, C. F. (1971) The state of the art in the study of technology transfer. *Research & Development Management,* **1,** 3.
Downs, G. W. & Mohr, L. B. (1976) Conceptual issues in the study of innovation. *Admin. Sci. Quarterly,* **21,** 700–14.
Dronfield, E. & Soto, P. (1980) Hardship Hotel. The 'counter information services'. CIS report.
Egan, A. F., Ford, A. L. & Shay, B. J. (1980) A comparison of *Microbacterium thermosphactum* and lactobacilli as spoilage organisms of vacuum packed sliced luncheon meats. *J. Food Sci.*, **45,** 1745–8.
Electricity Council, UK (1980) *Planning for Cook–Chill*, 2nd Edn, Electricity Council Project Planning Unit.
Electricity Council, UK (1982) *Planning for Cook–Chill*, 3rd Edn, Electricity Council Project Planning Unit.
Electricity Council (1983) Cook–chill and cook–freeze. Caterelectric. EC document no. 4421/11/83.
Elisgood, T. (1988) Cook–chill; getting it right. Report of a study day held at Penns Hall on 26th November 1987. *Hospital Caterer*, May, 7–8.
Enfors, S.-O. & Molin, G. (1978) The influence of high concentrations of CO_2 on the germination of bacterial spores. *J. Appl. Bact.*, **45,** 279–85.
Enser, M. (1985) Shelf-life of red meat, fish and poultry. *J. Sci. Food Agric.*, **36,** 122–8.
Etheridge, I., Graff, N., Grant, A. & Rudge, T. (1988) Cook–chill under the microscope. *Environmental Health,* **196,** 21–8.
Ettlie, J. E. (1977) Validation of an unobtrusive measure of technology utilisation. *Psychological Reports,* **40,** 123–8.
Ettlie, J. E. (1983) A note on the relationships between managerial change values, innovative intentions and innovative technology outcomes in food sector firms. *Research & Development Management,* **13,** 4.
Euromonitor (1984) *The Catering Report, Section 7; Vending*, 2nd Edn, Euromonitor Publications Ltd.

Fischer, W. A. (1976) Empirical approaches to understanding technology transfer. *Research & Development Management,* **6**, 151–7.

Fletcher, D. (1989) Deadly bacteria found in chilled pre-cooked food. *Daily Telegraph*, 13th Jan.

Frey, W. (1982) Zur Problematik der Haltbarkeitsfristen by vorverpackten Fleischwaren. *Die Fleischerei,* **5,** 303–4.

Fuller, J. (1966) *The Chefs Manual of Kitchen Management*, London, B. T. Batsford Ltd.

Fusco, L. (1987) Is cook–chill the saviour? *Hospital Caterer*, Nov./Dec., 9–10.

Galbraith, N. S., Barrett, N. J. & Sockett, P. N. (1987) The changing pattern of foodborne disease in England and Wales. *Public Health,* **101,** 319–28.

Garcia, G. W., Genigeorgis, G. & Lindroth, S. (1987) Risk of growth and toxin production by *Clostridium botulinum* nonproteolytic types B, E and F in salmon fillets stored under modified atmospheres at low and abused temperatures. *J. Food Prot.,* **50,** 330–6.

Gilbert, R. J. (1985) Food poisoning problems. In: *Proc. Chill Foods,* Chipping Campden, UK, Campden Food Preservation Research Association.

Gillespie, R. (1984) Major concepts in institutional feeding. *Dairy and Food Sanitation,* **4,** 302–5.

Ginsberg, B. (1963) Packaging for food vending. *Cornell Hotel and Restaurant Administration Quarterly,* **4,** 69.

Glanfield, P. (1980) *Applied Cook–Freezing*, London, Elsevier Applied Science Publishers.

Glew, G. (1979) New systems for catering. *Food Flavourings, Ingredients & Processing,* **4,** 203–5.

Glew, G. (1979) Catering using new systems—a critical appraisal. Cook–chill Catering Conference (European Study Conference Ltd), 27th Feb. 1979.

Glew, G. (1980) The contribution of large scale catering operations to nutrition. *World Rev. Nut. & Dietetics,* **34,** 1–45.

Glew, G. (1981) Rising fuel costs and the chill factor. *Caterer and Hotelkeeper*, April, 89–93.

Glew, G. (1982) New uses for refrigeration in the catering industry. *Int. J. Refrig.,* **5,** 1.

Glew, G. (1984) Future technologies in the catering industry. Paper presented to the Committee on Agricultural Problems, Economic Committee for Europe at the Symposium on Technological and Economic Aspects of Catering, Budapest, Hungary, April, 1983.

Glew, G. (1985) Chilled foods in catering. In: *Proc. Chill Foods*, Chipping Campden, UK, Campden Food Preservation Research Association.

Glew, G., Berg, R. & Yin Yam (1980) Some effects of the presence or absence of oxygen on stored precooked chilled chicken and pork meat. In: *Advances in Catering Technology* (ed. Glew, G.), London, Elsevier Applied Science Publishers, pp. 417–23.

Gold, B. (1980) Adoption of technology decision making in industry. *Int. J. of Management Sci.,* **8,** 505–16.

Grayson, D. (1982) Job evaluation and changing technology. Work Research Unit. Dept. Employment Occasional Paper 23.

Greater London Council (1985) The London industrial strategy. *The Food Industry*, GLC, London, pp. 67–94.
Grose, D. (1988) Public perception of food safety. *J. Roy. Soc. Health,* **108,** 138–41.
Grosser, Von. J. & Nitsch, G. (1982) Zum Erhitzen der Brühwurst. *Die Fleischerei,* **1/82,** 45–8.
Guise, B. (1986) Exchanging gases to extend life. *Food Processing,* Sept., 35–7.
Guise, B. (1986) Food irradiation: the dream becomes a reality. *Food Processing*, April, 27–9.
Halliday, D. (1985) Looking back on five years-benefits and experiences. Paper presented at cook–chill conference, Middlesex, Poly., Apr. 25th.
Hammond, P. H. (1980) The transfer of computing and control technology into manufacturing industry. *IEEE Proceedings, Part A,* **127** (1), 27–37.
Hanna, M. O., Stewart, J. C., Zink, D. L., Carpenter, Z. L. & Vanderzant, C. (1977) Development of *Yersinia enterolitica* on raw and cooked beef and pork at different temperatures. *J. Food Sci.,* **42,** 1180–4.
Harris, N. D. & Lindsay, R. C. (1972) Flavour changes in re-heated chicken. *J. Food Sci.,* **37,** 19–22.
Hayes, P. R. (1985) *Food Microbiology and Hygiene*, London, Elsevier Applied Science Publishers.
Hayward, B., Glew, G., Gilbert, R., Pinegar, J., Armstrong, R. & Majewski, C. (1987) An interim report to the Yorkshire Regional Health Authority. By the regional expert group on the safety of the cook–chill system.
HCTB (1986) *Employee Relations in the Hotel and Catering Industry*, 6th Edn, London, Hills & Lacy Ltd.
HCTB (1988) *New Employment Forecasts. Hotel and Catering Industry 1988–1993,* A Research Unit HCTB/MMD, London.
Hekkens-Klaassens, J. (1985) Getting information to the people who can use it. In: *Advances in Catering Technology* (ed. Glew, G.) Vol. 3, London, Elsevier Applied Science Publishers, pp. 217–30.
Hempel, S. (1983) Cook–chill system allows new look at Middlesex Polytechnic. *Fast Food*, Apr., 45–6.
Herrmann, H. S. (1987) The big chill—fact or fiction? Paper given at cook–chill conference in Aberdeen.
Hintlian, C. B. & Hotchkiss, J. H. (1987) Comparative growth of spoilage and pathogenic organisms on modified atmosphere-packed cooked beef. *J. Food Prot.,* **50,** 218–23.
Hotchkiss, J. H. (1988) Experimental approaches to determining the safety of food packaged in modified atmospheres. *Food Tech.,* **42,** 55–64.
Hunt, C. (1984) Nutrient losses in cook–freeze and cook–chill catering. *Human Nutrition: Applied Nutrition,* **38,** 50–9.
Institute of Food Science and Technology (UK) (1982) *Guidelines on Pre-cooked Chilled Foods*.
Ivory, M. (1987) Feeding thc 2,000. *Caterer and Hotelkeeper,* 11th June, 63–5.
Ivory, M. (1989) Cook–chill challenge. *Caterer and Hotelkeeper,* 2nd Feb., 61–2.
Jacobson, M. & Koehler, H. H. (1970) Development of rancidity during short term storage of cooked poultry meat. *J. Agr. Food Chem.,* **18,** 1069–72.

James, S. J. & Bailey, C. (1985) Principles and problems in cooling foodstuffs. In: *Proc. Chill Foods,* 7–9th May, Stratford-upon-Avon. Chipping Campden, UK, Campden Food Preservation Research Association.

Jamieson, W. (1980) Use of hypobaric conditions for refrigerated storage of meats, fruits, and vegetables. *Food Tech.,* March, 64–71.

Jenner, G. (1987) NHS market meals. *Caterer and Hotelkeeper*, 23rd July, 12.

Jenner, G. (1987) What future for cook–chill? *Caterer and Hotelkeeper*, 30th July, 16.

Jonsson, L. & Danielsson, K. (1981) Vitamin retention in foods handled in food-service systems. *Lebensm Wiss. u. Technol.,* **14,** 94–6.

Kader, A. A. (1980) Prevention of ripening of fruits by use of controlled atmospheres. *Food Tech.,* March, 51–4.

Karlstrom, B. (1982) Sensory quality of peas handled in food-service systems. *Lebensm. Wiss. u. Technol.,* **15,** 326–9.

Kereluk, K., Peterson, A. C. & Gunderson, M. F. (1961) Effect of different temperatures on various bacteria isolated from frozen meat pies. *J. Food Sci.,* **26,** 21–5.

Kimber, A. (1984) Update on gas flush pack. *Food Manufacture*, March, 22–5, 65.

Kinton, R. & Cesarani, V. (1978) *The Theory of Catering*, 4th Edn, London, Edward Arnold Publishers Ltd.

Konopka, C. (1983) Blast chilling and refrigeration at the Mount Royal Hospital. *Caterer and Hotelkeeper*, 17th March, 71.

Konopka, C. (1987) Living in a vacuum. *Caterer and Hotelkeeper*, 15th Oct., 91, 93.

Kossovitas, C., Navab, M., Chang, C. & Livingston, G. (1973) A comparison of chilled-holding versus frozen storage on quality and wholesomeness of prepared foods. *J. Food Sci.,* **38,** 901.

Larmour, R. (1982) Problems faced by managers in the hotel and catering industry. *Int. J. Hospitality Management,* **1,** 45–7.

Lebert, F. & Rosset, R. (1974) Plats cuisines, refrigeres ou congeles: Technologie et hygiene, orientations actuelles. *Revue Generale du Froid*, 11th Dec., 1127–43.

Leith, P. (1986) Cooking *sous-vide* in France. *Caterer and Hotelkeeper*, 15th May, 28.

Lemaire, W. H. (1988) Too many cooks. *Food Engineering*, April, 70.

Leonard-Barton, D. & Klaus, W. (1985) Implementing new technology. *Harvard Business Review*, Nov.–Dec.

Lewis, A. & Light, N. (1988) A survey of cook–chill catering in the UK and the need for the extension of product shelf-life. *Food Science & Technology Today,* **2,** 214–17.

Light, N. D. (ed.) (1986) *The Future of Technology in Cook–Chill Catering*, The Food and Hospitality Research Centre, Dorset Institute, Poole, Dorset, UK.

Light, N. D. (1986) *Sous-vide* cuisine—the way ahead? In: *The Future of Technology in Cook–Chill Catering* (ed. Light, N.), The Food and Hospitality Research Centre, Dorset Institute, Poole, Dorset, UK.

Light, N., Young, H. & Youngs, A. (1987) A qualitative survey of the use of chilled vending machines as a means of food service in the UK. *Food Science & Technology Today,* **1,** 247–51.

Light, N. D., Young, H. & Youngs, A. (1987) Operating temperatures in chilled food vending machines and risk of growth of food poisoning micro-organisms. *Food Science & Technology Today,* **1,** 252–6.

Light, N., Hudson, R., Williams, J., Barrett, J. & Schafheitle, J. (1988) A pilot study on the use of *sous-vide* cooking as a production system for high quality foods in catering. *Int. J. of Hospitality Management,* **7,** 21–7.

Lillard, H., Klosse, A., Hegge, R. & Che, V. (1973) Microbiological comparison of steam (at sub-atmospheric pressure) and immersion-scalded boilers. *J. Food Sci.,* **38,** 90–6.

Lioutas, T. (1988) Challenges of controlled and modified atmosphere packaging: a food company's perspective. *Food Tech.,* Sept., 78–86.

Littler, R. & Pearson, A. (1972) Uncertainty and technical innovation. *Management Decision,* **10,** 111–16.

Lott, G. (1973) Mikrobiologische untersuchungen von fleisch fertiggerichten, die nach dem Nacka—Verfahren hergestellt wurden. *Praventivmedizin,* **18,** 371–4.

Lucke, F.-K. (1981) The relevance to meat products of *Clostridium botulinum.* In: *Psychrotrophic Micro-organisms in Spoilage and Pathogenicity* (eds Roberts, T. A., Hobbs, G., Christian, J. H. B. & Skovgaard, N.), New York, Academic Press, pp. 141–62.

Macdonald, S. (1985) Technology beyond machines. In: *Implementing New Technology* (eds Rhodes, E. & Wield, D.), Oxford, Blackwell Open University Press, pp. 41–9.

Madden, R. H. & Moss, B. (1987) Extension of shelf-life of minced beef by storage in vacuum packages with carbon dioxide. *J. Food Prot.,* **50,** 229–33.

Mansfield, E. (1969) *The Economics of Technological Change*, London, Longmans.

Massey, D. & Meegan, R. (1982) *The Anatomy of Job Loss*, London & New York, Methuen & Co Ltd.

Matthews, M. (1985) Microwave ovens: Effects on food quality and safety. *Commentary,* **85** (8), 919–21.

McDaniel, M. C., Marchello, J. O. & Tinsley, A. M. (1984) Effect of different packaging treatments on microbiological and sensory evaluation of precooked beef roasts. *J. Food Prot.,* **47,** 23–6.

McDermid, K. (1986) Catering for senior citizens. *Caterer and Hotelkeeper*, 10th Apr., 45–8.

McDermid, K. (1987) Contract coup. *Caterer and Hotelkeeper*, 5th Feb., 44–5.

McGuckian, T. (1969) AGS food system—chilled pasteurised food. *Cornell Hotel and Restaurant Administration Quarterly,* **10,** 87–99.

Mead, G. C. (1983) In: *Food Microbiology Advances and Prospects, Soc. Appl. Bact. Symp. series 11* (eds Roberts, T. A. & Skinner, F. A.), New York, Academic Press, pp. 24–38.

Mikklesen, B. (1985) The quality of potatoes in four different meals-on-wheels systems. *J. Foodservice Sys.,* **3,** 251–6.

Millross, J., Speht, A., Holdsworthy, K. & Glew, G. (1973) *The Utilization of the Cook–Freeze Catering System for School Meals.* University of Leeds.

Minor, L. J. (1972) Today's food production systems. *Cornell Hotel and Restaurant Administration Quarterly,* **13,** 43–6.
Mintel (1981) *The Mintel Report on Automatic Vending*, London, Mintel Publications Ltd.
Mobbs, C. C. (1986) Cook–chill production. *Hospital Caterer*, Sept., 20, 22.
Morgan, R. (1983) Perfecting cook–chill for fast and hot meals. *Fast Food*, Mar., 41–2.
Morris, N. (1987) Designing for hygiene? *Design 460*, April.
Morton, J. A. (1967) A Model of the Innovative Process. Proceedings of a conference on technology transfer and innovation. National Science Foundation, pp. 67–5.
Mulder, R. W. A. W. & Gerrits, A. R. (1974) The microbiological shelf-life of vacuum broiled chickens. *Archiv fuer Lebensmittelhygiene,* **25** (5), 108–11.
Munce, B. A. (1981) The role of the food service industry in the microbiological safety of food: a review. *Food Technol. Aus.,* **33** (7), 326–31.
Munce, B. A. (1984) Hazard analysis and critical control points and the food service industry. *Food Technol. Aus.,* **36** (5), 214–17, 222.
Munce, B. A. & Buckle, K. A. (1981) Microbial changes in some prepared frozen foods during 5°C storage. In: *Psychrotrophic Micro-Organisms in Spoilage and Pathogenicity* (eds Roberts, T. A., Hobbs, G., Christian, J. H. B. & Skovgaard, N.), New York, Academic Press.
Naidu, Y. M., Sawyer, C. A. & Thompson, S. (1983) Cook/chill foodservice systems: microbiological quality and end-point temperature of beef loaf, peas and potatoes after reheating by conduction and microwave regeneration. *J. Food Prot.,* **46,** 1036–43.
Napleton, L (1989) Cook–chill foodservice technique has history of successful applications. *Industrial Caterer*, 10–12.
Nelson, R. & Winter, S. (1977) In search of a useful theory of innovation. *Research Policy,* **6,** 36–76.
Nicholanco, S. & Matthews, M. E. (1978) Quality of beef stew in a hospital chilled food service system. *J. Amer. Dietet. Assoc.,* **72,** 31–7.
Nielsen, H. J. S. (1983) Influence of temperature and gas permeability of packaging film on development and composition of microbial flora in vacuum packed bologna type sausage. *J. Food Prot.,* **46,** 693–8.
NOP (1982) *Attitudes to Vending Machines*. A report on a study by National Opinion Poll market research for the Automatic Vending Association of Britain. NOP/8518.
Nystrom, H. (1979) *Creativity and Innovations*, Chichester, Wiley & Sons.
Ohlsson, T. (1977) Pasteurisation of meat, fish and convenience products. In: *Food Quality and Nutrition* (ed. Downey, W. K.), London, Applied Science Publishers, pp. 163–90.
O'Neill, R. (1967) Why technological innovations fail. Proceedings of a conference on technology transfer and innovation. National Science Foundation, pp. 67–5.
Osner, R. & Hudson, P. (1987) *The Application of the HACCP Approach to the Prevention of Hygiene and Spoilage Hazards in Catering*. Occasional Paper Series Vol. 1. The Food and Hospitality Research Centre, Dorset Institute, Poole, Dorset, UK.

Oxford Regional Health Authority (1986) Cook–chill appraised. Report prepared by RM Associates, July 1986.
Pallot, P. (1987) Warmed up food at hospital wastes one million pounds. *Daily Telegraph*, 8th June.
Pallot, P. (1989) Listeria poison risk is bigger than admitted. *Daily Telegraph*, 16th Jan.
Palumbo, S. A. (1986) Is refrigeration enough to restrain foodborne pathogens? *J. Food Prot.*, **49** (12), 1003–9.
Patterson, J. T. & Gibbs, P. A. (1973) Observations on the microbiology of cooked chicken carcasses. *J. App. Bact.*, **36,** 689–97.
Paulus, K., Piekarski, J. & Leistner, L. (1980) Chilled meals: possibilities and limits. In: *Advances in Catering Technology* (ed. Glew, G.), London, Applied Science Publishers, pp. 375–86.
Pavitt, K. (1980) *Technical Innovations and British Economic Performance*, Science Policy Review Unit, Sussex, London, Macmillan Press Ltd.
Pearson, A. M., Gray, J. I., Wolzak, A. M. & Horenstein, N. A. (1983) Safety implications of oxidised lipids in muscle foods. *Food Tech.*, **37,** 121–9.
Peterson, A. C. & Gunnerson, R. E. (1974) Microbiologial critical control points in frozen foods. *Food Tech.*, **28,** 37–44.
Pinaga, F., Carbonell, J. V., Valles, S. & Ampara Feria, M. (1979) Pasteurisation of ready to serve meals: quality changes during refrigeration of beef goulash in air atmosphere. *Revista de Agroquimica Y Technologiade Alimenta,* **19,** 520–32.
Pine, R. (1987) *Management of Technological Change in the Catering Industry*, Aldershot, Gower Publishing.
Pokorny, J., Vozenilenilkova, B., Kovarova, H., Marcin, A. & Dividek, J. (1982) Effects of cold storage on the quality of fried chicken and duck muscle. *Nahrung,* **26,** 689–95.
Pralus, G. (1985) Une histoire d'Armour—La Cuisine *Sous-Vide*—La Cuisine de l'Or 2000, France.
Pring, A. (1986) *Sous-Vide*; Boon or threat? *Caterer and Hotelkeeper*, 24th July, 19.
Raffael, M. (1984) Revolution in the kitchen. *Caterer & Hotelkeeper*, Aug. 16th, 34–5.
Raffael, M. (1985) Cooking in a vacuum—perhaps the best thing since sliced brioche. *Guardian*, 27th Sept.
Ray, G. (1969) The diffusion of new technology. A study of 10 processes in 9 industries. *Nat. Inst. Econ. Rev.*, **48** (May), 40–60.
Ray, G. (1983) The diffusion of mature technologies. *Nat. Inst. Econ. Rev.*, **106,** 56–62.
Rhodes, E. & Wield, D. (1985) Implementing New Technologies, Blackwell Open University Press.
Riemann, H. & Bryan, F. L. (1979) *Food-Borne Infections and Intoxications*, 2nd Edn, London, Academic Press Ltd.
Rinke, W. (1976) Three major systems reviewed and evaluated. *Food Service,* **50,** 73–8.
Roberts, D. (1982) Factors contributing to outbreaks of food poisoning in England and Wales 1970–79. *J. Hyg. Camb.*, **89,** 491–8.

Rogers, E. (1983) *Diffusion of Innovation*, 3rd Edn, New York, The Free Press, Macmillan Publishing Co. Ltd.
Rollin, J. & Matthews, M. (1977) Cook–chill foodservice systems. Temperature histories of a cooked ground beef product during the chilling process. *J. Food Prot.*, **40** (11), 782–4.
Rollin, J., Matthews, M. & Lund, D. (1979) Cook–chill foodservice systems: predicting cooling time of a ground beef entreé chilled in bulk. *J. Amer. Dietet. Assoc.*, **75**, 440–5.
Rothwell, R. (1977) The characteristics of successful innovators and technically progressive firms. *Research & Development Management,* **7** (3), 191–205.
Rothwell, R. & Robertson, A. (1975) The contribution of poor communication to innovative failure. *Aslib Proceedings,* **27** (10), 393–400.
Rothwell, R. & Zegveld, W. (1979) *Technical Change and Employment*, London, Francis Pinter.
Ryan, D. (1979) Technical change in the tertiary sector. *Work and People,* **5,** 13–18.
Sawyer, C. A. & Naidu, Y. M. (1984) Sensory evaluation of cook–chilled products reheated by conduction, convection and microwave radiation. *J. Foodservice Sys.*, **3,** 89–106.
Schafheitle, J. M. (1987) An investigation of the factors affecting the quality of chilled, stored foods prepared by the *sous-vide* vacuum cooking technique. Research Diploma Dissertation. The Food and Hospitality Research Centre, Dorset Institute, Poole, Dorset, UK.
Schafheitle, J. & Light, N. (1986) *An Introductory Handbook to Cuisine en Papillote* Sous-Vide, Department of Catering & Hotel Management, Dorset Institute, Poole, Dorset.
Schafheitle, J. M. & Light, N. (1989) *Sous-vide* preparation and chilled storage of chicken ballotine. *Int. J. Food. Sci. Tech.*, **24,** 199–205.
Schafheitle, J. M. & Light, N. (1989) *Sous-vide* cooking and its application to cook–chill—what does the future hold? *J. Contemporary Hospitality Management,* **1,** 5–10.
Schoeffler, S., Buzzell, R. & Heany, D. (1974) Impact of strategic planning on profit performance. *Harvard Business Review,* Mar./Apr.
Schon, D. (1967) *Technology and Change*, New York Delacorte Press.
Schwebel, W. (1977) Pasteurised cold stored prepared meals with shelf life of up to 3 weeks. *GV-Praxis mit Tiefkuehlpraxis,* **17** (2), 69–70.
Scott, C. (1986) An interim report on costs and manpower requirements. British Telecom, Brighton February.
Sheard, M. (1986) Working party on food poisoning in hospitals, institutions and residential establishments. *Environ. Health,* **93** (10), 263–6.
Sheppard, J. (1987) The big chill. A report on the implications of cook–chill for public services. London Food Commission. Report no. 15.
Searle, G. & Crozier, M. J. (1981) An investigation of low temperature food storage in large hotels in the city of Westminster. City of Westminster, Dept. of the Environment.
Sessions, P. (1987) Sensible *Sous-Vide. Hospital Caterer*, May/June, 14.
Sieper, A. & Likert, R. (1973) Organisational climate for successful innovation. *Public Management*, May.

Silla, H. & Simonsen, B. (1985) Shelf-life of chilled, cooked and cured meat products. *Fleischwirtschaft,* **65,** 66.

Silliker, J. H. & Wolfe, S. K. (1980) Microbiological safety considerations in controlled atmosphere storage of meats. *Food Tech.,* **34,** 59–63.

Simonsen, B., Bryan, F., Christian, J., Roberts, T., Tompkin, R. & Silliker, J. (1987) Prevention and control of food-borne salmonellosis through application of Hazard Analysis Critical Control Point (HACCP). *Int. J. Food Micro.,* **4,** 227–47.

Snyder, P. O. (1986) Microbiological quality assurance in food service operations. *Food Tech.,* **40,** 122–30.

Snyder, P. O. & Matthews, M. E. (1984) Microbiological quality of food service menu items produced and stored by cook/chill, cook/freeze, cook/hot hold and heat/serve. *J. Food Prot.,* **47,** 876–85.

Souder, W. (1974) Autonomy, gratification and research and development outputs. A small-sample field study. *Management Science,* **20** (8).

Spahl, A., Reineccius, G. & Tatini, S. (1981) Storage life of pork chops in CO_2-containing atmospheres. *Appl. Environ. Microb.,* **41,** 816–18.

Stacey, C. (1985) Roux is set to fill the vacuum. *Caterer & Hotelkeeper,* Nov. 7th, 65, 67, 70, 73, 75.

Stacey, C. (1986) Feeding the night shift. *Caterer and Hotelkeeper,* May 15th, 41–3.

Stacey, C. (1986) Remploys cook–chill joint venture. *Caterer and Hotelkeeper,* March 6th, 32.

Stewart, A. (1985) Cook–chill in operation. A look at Barnet school meals. *Catering Times,* June 24th, 25.

Stiles, M. E. & Ng, L.-K. (1979) Fate of pathogens inoculated onto vacuum packaged sliced hams to stimulate contamination during packaging. *J. Food Prot.,* **42,** 464–9.

Stone, H. (1983) Chill or freeze? *Catering & Hotel Management,* **55** (7), 13.

Stormont, G. J. & Harvey, J. F. (1971) An assessment of the Regethermic meals service system at the Royal London Homeopathic Hospital, July.

Tändler, K. (1972) In Folienbehältnissen pasteurisierte Fleischfertiggerichte für die Germeinschaftsverpflegung. *Die Fleischwirtschaft,* **52** (7), 845–50.

Tändler, K. (1972) In Folienbehältnissen pasteurisierte Fleischfertiggerichte fur die Gemeinschaftsverpflegung, *Die Fleischwirtschaft,* **52** (9), 1105–13.

Tändler, K. (1973) In Folienpackungen erhitzte Fleischgerichte. *Die Fleischwirtschaft,* **53** (9), 1241–4.

Tändler, K. (1976) In Folienpackungen erhitzte Fleischgericht—Mitteilung-Kontrollen beim Abfüllen, Evakuieren und Verschließen der Behältnisse. *Die Fleischwirtschaft,* **56** (4), 539–47.

Tändler, K. (1976) In Folienpackungen erhitzte Fleischgerichte—Mitteilung: Kontrolle der Verschlußdichte der Behältnisse. *Die Fleischwirtschaft,* **56** (10), 1473–84.

Taubert, C. A. (1982) Defining sanitation hazards and critical control points in food service operations. *J. Foodservice Sys.,* **2,** 171–5.

Thatcher, F. S. & Clark, D. S. (1968) *Micro-organisms in Food: Their Significance and Methods of Enumeration,* University of Toronto Press, pp. 49–64, 151–8.

Thompson, J. & Johnson, D. (1963) Food temperature preferences of surgical patients. *J. Amer. Dietet. Assoc.*, **43,** 209–11.

Trollope, K. (1987) Cook–chill on hold. *Caterer and Hotelkeeper,* 18th June, 9.

Tuomi, S., Matthews, M. & Marth, E. (1974) Temperature and microbial flora of refrigerated ground beef gravy subjected to holding and heating as might occur in a school foodservice operation. *J. Milk Food Technol.*, **37,** 457–62.

Tuomi, S., Matthews, M. & Marth, E. (1974) Behaviour of *Clostridium perfringens* in precooked chilled ground beef gravy during cooling, holding and reheating. *J. Milk Food Technol.*, **37,** 494–8.

Unklesbay, N. & Unklesbay, K. (1978) Energy expended in alternate foodservice systems for chicken menu items. *J. Amer. Dietet. Assoc.*, **73,** 20–6.

Wadey, C. (1987) What future for cook–chill? *Caterer and Hotelkeeper*, 30th July, 16.

Walker, A. E. (1988) The transfer of technology in the catering industry: a study of UK cook–chill catering operations. PhD Thesis, CNAA/Dorset Institute of Higher Education, Poole, Dorset, UK.

Walker, A. (1989) *Success or Failure in Cook Chill Catering: A Study in Technology Transfer*, Bradford, Horton Publishing.

Walker, A. & Light, N. (1988) Cook–chill catering: How to make it work. *Catering,* **VIII** (3), 45–7.

Walker, A. & Light, N. (1988) Is cook–chill being put to its best use? *Hospitality*, Feb., 15–17.

Watson, B. (1983) Courtaulds Acetate opts for cryogenic cook–chill. *Chilled Foods*, Mar., 41–2.

Werner, R. (1977) Schulverpflegung aus der sicht von Schülern und Eltern Bundesministerium für Ernahrung, Landwirtschaft und Forsten, Bonn. In: *Advances in Catering Technology* (ed. Glew, G.), London, Elsevier Applied Science Publishers, p. 415.

West, P., Hughes, W., Witham, T. & Hull, S. (1986) Report on energy costs for the implementation of cook–chill at Heatherwood Hospital, Ascot, file no. Wl530, Dec.

Whitehall, B. (1987) *Sous-vide* report. *Caterer and Hotelkeeper*, 23rd July, 48–51.

Wenlock, R. (1988) Public perception of food safety. *J. Roy. Soc. Health,* **108** (4), 130.

Wenlock, R. (1988) Healthy foods and unhealthy diets—public misconceptions. *J. Roy. Soc. Health,* **108** (4), 132–3.

Whitaker, M. (1986) The diffusion of information technology across the hotel and catering industry (UK). *Int. J. Hospitality Management,* **5** (2), 95–6.

Whitehall, B. (1983) Warming to cook–chill. *Catering Times,* June, 25–8.

Whitehall, B. (1985) Co-ordination and cook–chill. *Caterer and Hotelkeeper*, May, 76.

Whitehall, B. (1987) Cool way to quality. *Caterer and Hotelkeeper*, 30th July, 33–4.

Whitehall, B. (1987) All systems go, *Sous-Vide* report. *Caterer and Hotelkeeper*, 23rd July, 48–51.

Whitehall, B. (1987) Casino cook–chill jackpot. *Caterer and Hotelkeeper*, 6th Feb., 31–3.

Wolfe, S. K. (1980) Use of CO- and CO_2-enriched atmospheres for meats, fish and produce. *Food Tech.*, March, 55–8.
Wood, A. (1989) Cook–chill put on ice. *Caterer and Hotelkeeper*, 2nd March, 8.
Wood, A. (1988) Cold war over cook–chill. *Caterer and Hotelkeeper*, 8th Sept., 16.
Wood, L. (1984) The chilling story of an M & S chicken. *Financial Times*, 2nd Feb., 42.
Woodward, J. (1968) Resistance to change. *Management Int. Rev.*, **8** (4–5), 137–43.
Young, H. (1986) Factors affecting the shelf-life and acceptability of cooked chilled foods. PhD Thesis, Dorset Institute, Poole, Dorset, UK.
Young, H., Youngs, A. & Light, N. D. (1987) The effects of packaging on the growth of naturally occurring microflora in cooked, chilled foods used in the catering industry. *Food Microbiol.*, **4,** 317–27.
Young, H., MacFie, H. & Light, N. (1989) The effect of packaging and storage on the sensory quality of cooked chicken menu items served from chilled vending machines. *J. Sci. Food Agric.*, **48** (3), 323–38.
Young, L., Reviere, R. & Cole, B. (1988) Fresh red meats: A place to apply modified atmospheres. *Food Tech.*, Sept., 65–9.
Zacharias, R. (1980) Chilled meals: sensory quality. In: *Advances in Catering Technology* (ed. Glew, G.), London, Elsevier Applied Science Publishers, pp. 409–16.
Zagary, D. & Kader, A. (1988) Modified atmosphere packaging of fresh produce. *Food Tech.*, Sept., 70–7.
Zallen, E., Hitchcock, J. & Goertz, G. (1975) Chilled food systems. *J. Amer. Dietet. Assoc.*, **67,** 552–7.

Appendix 4

THE 1989 UK DEPARTMENT OF HEALTH AND SOCIAL SECURITY GUIDELINES ON COOKED, FROZEN AND CHILLED FOODS IN CATERING

The revised guidelines on the preparation and use of cooked, frozen and chilled foods in catering, based on similar guidelines published in 1980, were published by the UK Department of Health and Social Security in July 1989 when this book had gone to press. Because it was, therefore, not possible to include a detailed description of these guidelines (which remain substantially very similar to those originally published in 1980), in the text of this book, this Appendix has been included to summarise the main points of difference between the two sets of guidelines.

In terms of the more general principles of the guidelines, most points remain the same. For example, the maximum shelf-life of five days (including the days of production and service) remains in place. However, a new aspect is a brief review of the nutritional quality of cooked, chilled food with particular reference to the detrimental effects of over-cooking and prolonged delay between re-heating and service. Also, production schedules are now required for each menu item and all relevant equipment must have accurate temperature measurement devices fitted.

Overall, the new guidelines are more specific with regard to many of the processes involved in cook–chill catering. In particular the following points are new:

There are tighter controls on the purchase and use of raw materials (including pre-cooked foods). The premises and the standards of products of food suppliers must be inspected by qualified staff. The preparation of raw materials must be done in areas physically separated from those parts of the production unit dedicated to cooking and post-cooking procedures. Also, different machines and utensils must be used for the manipulation and preparation of raw and cooked foods. The previous recommendation of the use of disposable gloves for food handling is withdrawn although it is stated that these may be used in conjunction with proper hand-washing. Finally, in terms of food preparation, rapid thaw equipment is recommended for some frozen foods prior to cooking.

The prime cooking (pasteurisation) of foods is dealt with more specifically in that it is stated that the food must reach an internal (core) temperature of 70°C for not less than 2 min. Also, automatic controls and sensors are required on cooking equipment.

A more specific description of the chilling process is included. The new guidelines also require that the total rapid chilling capacity of a unit is equal to or greater than the maximum expected peak loads. Specifically-designed chillers for cooked foods must be used and they must be fitted with temperature monitoring devices with alarms and remote recording devices. Immersion chillers are mentioned as is a reference to liquid chillers and especially the cleaning problems associated with them. The effect of lidding cooked food is discussed and it is noted that this could reduce contamination. However, longer chilling times are not allowable for lidded foods (the recommendation remains at 90 min from cooking temperature to 3°C). The cooling of meat joints is dealt with more specifically with recommendations on size.

In terms of the management of cooked, chilled food, labels must state the day of production as well as the expiry date of the chilled food. Bought-in cooked, chilled food also has only a five day shelf-life. There is a more specific description of the use of foods stored for a short time at 5–10°C. Storage at these temperatures must not be used as an alternative form of cook–chill catering when less than a five day shelf-life is required. Also, foods which are stored at such temperatures must be consumed within 12 h. A 10°C environment must be used for cold-plating of food.

Insulated containers are allowable only for short distribution trips to end (satellite) kitchens.

Standard microwave ovens are not recommended as they may not re-heat food evenly. Food must be re-heated close to or at the point of service and it must be re-heated within 30 min of being removed from the chill store. A gap of 15 min between re-heating and service is allowable.

The guidelines fail to deal with more sophisticated methods, especially those which lead to products with, potentially, a longer shelf-life than the standard five days. Thus, cook-tank methods and *sous-vide*/cook–chill are both specifically excluded. Also ignored is the special case of airline catering. It must be inferred from this that airline catering which uses cook–chill methods should be regarded as standard cook–chill catering and follow the guidelines as they are written.

The establishment of quality assurance methods is strongly recommended and the principles of hazard analysis critical control points is specifically mentioned.

Index